T0296440

Infinite-Dimensional Dynamical Systems

This book develops the theory of global attractors for a class of parabolic PDEs that includes reaction–diffusion equations and the Navier–Stokes equations, two examples that are treated in detail. A lengthy chapter on Sobolev spaces provides the framework that allows a rigorous treatment of existence and uniqueness of solutions for both linear time-independent problems (Poisson's equation) and the nonlinear evolution equations that generate the infinite-dimensional dynamical systems of the title. Attention then turns to the global attractor, a finite-dimensional subset of the infinite-dimensional phase space that determines the asymptotic dynamics. In particular, the concluding chapters investigate in what sense the dynamics restricted to the attractor are themselves "finite-dimensional."

The book is intended as a didactic text for first-year graduate students and assumes only a basic knowledge of elementary functional analysis.

James Robinson is a Royal Society University Research Fellow in the Mathematics Institute at the University of Warwick.

Cambridge Texts in Applied Mathematics

Infinite-Dimensional Dynamical Systems

An Introduction to Dissipative Parabolic PDEs and the Theory of Global Attractors

JAMES C. ROBINSON

University of Warwick

CAMBRIDGE
UNIVERSITY PRESS

To Tania

CAMBRIDGE UNIVERSITY PRESS
Cambridge, New York, Melbourne, Madrid, Cape Town, Singapore,
São Paulo, Delhi, Dubai, Tokyo, Mexico City

Cambridge University Press
The Edinburgh Building, Cambridge CB2 8RU, UK

Published in the United States of America by Cambridge University Press, New York

www.cambridge.org
Information on this title: www.cambridge.org/9780521635646

First published 2001

A catalogue record for this publication is available from the British Library

Library of Congress Cataloguing in Publication data
Robinson, James C. (James Cooper), 1969–
Infinite-dimensional dynamical systems : an introduction to dissipative parabolic PDEs
and the theory of global attractors / James C. Robinson.
p. cm. – (Cambridge texts in applied mathematics)
Includes bibliographical references.
ISBN 0-521-63204-8 – ISBN 0-521-63564-0 (pbk.)
1. Attractors (Mathematics)   2. Differential equations, Parabolic.   I. Title.   II. Series.
QA614.813.R63   2001
514′.74 – dc21   00-041413

ISBN 978-0-521-63204-1 Hardback
ISBN 978-0-521-63564-6 Paperback

Contents

Preface

Inspired by the success of dynamical systems theory in treating finite-dimensional systems, a similar approach has been developed for partial differential equations over the past two decades. In particular, this book focuses on the way in which the long-term dynamics of certain models can be studied by means of their global attractor, and it aims to give a didactic introduction to the subject at a level suitable for a first-year UK graduate student.

As such, there are various disparate topics that need to be covered, and I have attempted to provide what I hope is more than just an "introduction" to the theory of global attractors by giving a systematic development of the subject from its most basic foundations. There are, however, many excellent texts that treat some of these topics individually, and I would particularly like to mention Renardy & Rogers (1992), Evans (1998), and Gilbarg & Trudinger (1983), which have all had a large influence on the organisation of material in the earlier chapters, and Temam (1988), which contains, in some form, much of the material in later chapters.* Finally, Doering & Gibbon (1995) cover some of the same material in the context of the Navier–Stokes equations, and this book can be viewed in part as a retelling of their results in a language applicable to many other examples.

When I started writing I was planning to produce the kind of book that I would like to have read when I started my Ph.D.; I hope that, in the end, I have managed to produce the sort of text that I should have read. Although my natural inclination was to assume the minimum analytical background (having in mind my own inadequacies as a graduate student), considerations of space

---

* His book, *Infinite-Dimensional Dynamical Systems in Mechanics and Physics*, has influenced my choice of title, even though both books treat only a subset of such systems. The theory of global attractors relies on some kind of dissipation, automatically excluding Hamiltonian systems, for example. The main focus here is on the Navier–Stokes equations and related models.

have forced me to assume some previous familiarity with the theory of Lebesgue integration, of which just a brief outline is given in Chapter 1. Since knowledge of general measure theory is not necessary, a readable treatment of integration such as that given in Priestley (1997) is more than adequate.

The first part of the book covers the functional analysis needed throughout all that follows. Chapters 1–4, which concern Banach and Hilbert spaces, existence and uniqueness theory for ordinary differential equations, linear maps, spectral theory for compact symmetric operators, dual spaces, and weak convergence, may be revision for many readers. This section concludes with a lengthy chapter that treats Sobolev spaces, a cornerstone of the theory, in detail.

Chapter 6 treats questions of existence, uniqueness, and regularity for Poisson's equation. Although restricted to this case for ease of presentation, the analysis there can easily be adapted to more general elliptic equations.

From here on all the machinery is in place to consider existence and uniqueness for time-dependent equations. Chapter 7 introduces the Galerkin method as a means of proving these properties, using a linear parabolic equation as an example. These ideas are then applied in Chapter 8 to a scalar reaction–diffusion equation and in Chapter 9 to the two-dimensional Navier–Stokes equations with periodic boundary conditions.

Chapters 10–12 introduce the global attractor and show how to prove its existence for the two examples of Chapters 8 and 9. Chapter 13 defines the fractal and Hausdorff measures of dimension and provides a method to estimate these dimensions for global attractors, which is then applied to our two examples.

The next three chapters investigate how the finite dimensionality of the attractor affects the asymptotic dynamics. Chapter 14 concerns the squeezing property, covering "determining modes," approximate inertial manifolds, and exponential attractors. Chapter 15 treats the theory of inertial manifolds, using the geometric "strong squeezing property" as a basis for the analysis. Chapter 16 shows that the attractor can be parametrised using a finite number of parameters, and it gives a proof that there is a finite-dimensional system that reproduces the attractor dynamics.

The final chapter consists of a series of exercises that apply many of the techniques learned throughout the book to an analysis of the Kuramoto–Sivashinsky equation.

It is a peculiarity of the British tradition that the rigorous study of partial differential equations is not a common element in the syllabus of many mathematics departments. Even to students with a strong background in functional analysis the material in Chapters 5–9 is likely to be new. As such, teaching from this book requires some judicious editing of material. When I gave one of the lecture courses from which this book is partly derived, I restricted attention to Sobolev spaces of periodic functions (the majority of Chapter 5 can then be replaced with Appendix A) and similarly avoided the many calculations involved in the regularity theory of the Laplacian in general domains. In this way one can get on to the more interesting questions raised in the second half of this book at an earlier stage than would be possible if treating Sobolev spaces and elliptic regularity in detail.

Many people have been helpful, knowingly and unknowingly, in the creation of this book over the past four years. Thomas Ransford's elegant Cambridge Part III lecture course on distribution theory was revelatory and completely changed my view of pure mathematics, while Colin Sparrow's course on dynamical systems was equally inspirational. Paul Glendinning, my Ph.D. supervisor, was a great encouragement as I then tried to combine my two new enthusiasms.

Some years later I was very fortunate to have the opportunity of lecturing on some of the material here as a Part III course in two consecutive years, and without David Crighton's initial encouragement to turn my lecture notes into a book I would not even have started this long task. Rebecca Hoyle managed to sit through the first of these courses (of which thankfully little trace remains here) and was extremely supportive throughout. A kind invitation from José Langa, Tomás Caraballo, and Enrique Fernández Cara to give a series of lectures at the University of Seville over the Easter period in 1997 was the germ of Chapters 9–16; during my visit I enjoyed unparalleled hospitality, particularly from José.

Ciprian Foias, John Gibbon, Edriss Titi, and Roger Temam have all been continually inspiring and maintained my enthusiasm for the subject as a whole, and Brian Hunt very kindly provided me with a proof I needed for Chapter 13, which is reproduced in Appendix B.

Tania, who has been very patient as my head has filled up with Sobolev spaces, has made me laugh and smile for the past three years; while Daisy, our cat, has sat repeatedly on whichever pile of papers I most needed, this particular talent keeping me cheerful throughout the final weeks of constant revision.

Finally, my utmost gratitude goes to Peter Friz and Robert Mackay, both of whom read through the manuscript very closely and critically, and without whom there would be many more errors and inconsistencies than those which undoubtedly remain. The responsibility for these is, sadly, entirely my own.

Introduction

Since Poincaré initiated the qualitative study of solutions of differential equations at the end of the nineteenth century, the theory of dynamical systems has had an important rôle to play in our understanding of many physical models.

However, until relatively recently the application of such ideas was mainly restricted to finite-dimensional systems, such as those that arise in the study of ordinary differential equations (ODEs) or iterated low-dimensional maps. It is only within the past two decades that similar techniques have been systematically applied to the infinite-dimensional systems that arise from partial differential equations (PDEs).

This book develops the dynamical systems approach to a certain class of PDEs, dissipative parabolic equations, and investigates their asymptotic behaviour by means of an object called the global attractor.

Provided that the function f is sufficiently smooth (see Chapter 2), the solutions of a finite set of coupled ODEs,

$$\dot{x} = f(x), \qquad x \in \mathbb{R}^m, \tag{1}$$

($\dot{x}$ is short for dx/dt) give rise to a dynamical system on the finite-dimensional phase space $\mathbb{R}^m$. If $x(t; x_0)$ is the solution of (1) with $x(0) = x_0$, then we can define a solution operator $T(t) : \mathbb{R}^m \to \mathbb{R}^m$ by

$$T(t)x_0 = x(t; x_0).$$

The dynamical system generated by (1) is specified by the pair

$$(\mathbb{R}^m, \{T(t)\}_{t \in \mathbb{R}}) \tag{2}$$

of the phase space $\mathbb{R}^m$ and the family of solution operators $\{T(t)\}_{t \in \mathbb{R}}$. We sometimes abbreviate (2) to $(\mathbb{R}^m, T(t))$.

1

Although the inclusion of the phase space in the definition might seem some-what pedantic, in the case of PDEs the specification of an appropriate phase space is a large part of the resolution of the problem. Since this phase space will be an infinite-dimensional space of functions, we will need various tools from functional analysis, and these are developed in Chapters 1–5.

We illustrate some ideas with the simple example of the heat equation on a one-dimensional domain,

$$u_t = u_{xx}, \qquad u(0) = u(\pi) = 0. \tag{3}$$

(The notation u_t is shorthand for $\partial u/\partial t$, and u_{xx} is shorthand for $\partial^2 u/\partial x^2$.)

To look at this equation a little more closely, we will expand the solution $u(x, t)$ of (3) in terms of a Fourier series on $[0, \pi]$. The boundary conditions mean that we need only sine terms, so we can write

$$u(x, t) = \sum_{n=1}^{\infty} c_n \sin nx. \tag{4}$$

Substituting this expansion into (3) and comparing coefficients leads to an infinite set of ODEs for the $\{c_n\}$:

$$\frac{dc_n}{dt} = -n^2 c_n. \tag{5}$$

This is one way in which we can think of (3) as giving rise to an "infinite-dimensional" problem.

However, it is not immediately clear what constraints we ought to put on the variables $\{c_n\}$ to specify our phase space completely. One possible choice would be to restrict to solutions for which the energy

$$E = \int_0^{\pi} |u(x)|^2 \, dx$$

is finite. Since the functions $\{\sin nx\}$ are orthogonal, we can calculate E in terms of the $\{c_n\}$:

$$E = \frac{\pi}{2} \sum_{n=1}^{\infty} |c_n|^2.$$

In this way finite-energy solutions correspond to choices of Fourier coefficients that are square summable:

$$\sum_{n=1}^{\infty} |c_n|^2 < \infty. \tag{6}$$

Although this choice of "finite-energy" solutions is a natural one, it does pose a problem. If we are prepared to treat any function $u(x)$ that corresponds to a set of coefficients satisfying (6), we end up with the Lebesgue space of square integrable functions $L^2(0, \pi)$,

$$L^2(0, \pi) = \left\{ u : \int_0^\pi |u(x)|^2 \, dx < \infty \right\}.$$

Now, there are functions in $L^2(0, \pi)$ (or, equivalently, functions whose Fourier coefficients satisfy (6)) that are not continuous, let alone twice differentiable. Although (5) makes sense for the coefficients, the original equation (3) does not make sense for the corresponding function u given by (4).

If we are not going to exclude certain choices of $\{c_n\}$ that satisfy (6) but do not correspond to twice differentiable functions (which would produce a very convoluted definition of our phase space), then we need a way of understanding (3) even if $u(x, t)$ is not twice differentiable. The idea, loosely speaking, is that if

$$u = \sum_{n=1}^{\infty} c_n \sin nx$$

then we can *define* u_{xx} by

$$u_{xx} = \sum_{n=1}^{\infty} -n^2 c_n \sin nx \tag{7}$$

even if the series in (7) does not converge to a function in any classical sense. The distribution derivative and the Sobolev spaces of functions whose derivatives are in L^2, covered in Chapter 5, offer one way of doing this rigorously.

Related is the concept of a weak solution, which essentially allows the classical derivatives we expect in (3) to be replaced with such "generalised derivatives" (they are actually weaker than this!). This idea is introduced in Chapters 6 and 7, which deal with Poisson's equation,

$$-\Delta u = f, \qquad u|_{\partial \Omega} = 0$$

("weak in space") and a linear parabolic equation,

$$\frac{\partial u}{\partial t} + \Delta u = f(x, t), \qquad u|_{\partial \Omega} = 0$$

("weak in space and time").

We can interpret (3) as a dynamical system for the coefficients $\{c_n\}$, but we can also interpret it as a dynamical system on $L^2(0, \pi)$. In this case we treat the spatial and temporal dependence of $u(x, t)$ in fundamentally different ways. If we take a "snapshot" of $u(x, t)$ at a particular time τ, then the result is a function of x alone, $u(x, \tau)$. If this is an element of L^2 for each τ then the evolution of $u(x, t)$ in time traces out a trajectory in L^2. In later chapters we make almost exclusive use of this interpretation, suppressing the x dependence and writing things such as "$u(\tau) \in L^2$" as a convenient shorthand. This gives a second, and more useful, sense in which we can understand (3) as an infinite-dimensional problem.

If we are going to use (3) to define a dynamical system on $L^2(0, \pi)$, then we need to obtain existence and uniqueness of solutions, and in particular we have to be sure that if the initial condition $u(x, 0)$ is an element of $L^2(0, \pi)$ then so is $u(x, t)$. To check this for (3) is simple, since we can use (5) to find the solution $u(x, t)$ explicitly. If

$$u(x, 0) = \sum_{n=1}^{\infty} c_n \sin nx,$$

then (5) gives

$$u(x, t) = \sum_{n=1}^{\infty} c_n e^{-n^2 t} \sin nx.$$

Assuming that $u(x, 0) \in L^2$, so that $\sum_{n=1}^{\infty} |c_n|^2 \, dx < \infty$, we want to check that

$$\sum_{n=1}^{\infty} |c_n|^2 e^{-2n^2 t} < \infty. \tag{8}$$

We now come across another potential problem. It is easy to see that if $t > 0$ then the sum in (8) is finite. In fact, we have

$$\sum_{n=1}^{\infty} |c_n|^2 e^{-2n^2 t} < e^{-2t} \sum_{n=1}^{\infty} |c_n|^2,$$

and so the energy (which is proportional to this expression) decays to zero as $t \to \infty$. [Since

$$\frac{\pi}{2} \sum_{n=1}^{\infty} |c_n(t)|^2 = \int_0^{\pi} |u(x,t)|^2 \, dx \equiv \|u(x,t)\|_{L^2}^2,$$

we could also write this in terms of the function $u(x,t)$ as

$$\|u(x,t)\|_{L^2} \le e^{-t} \|u(x,0)\|_{L^2}, \tag{9}$$

showing that $u \to 0$ in the sense of L^2 convergence.]

However, if we consider the case $t < 0$, then the exponentials in the sum, which were responsible for the dissipation of energy as $t \to \infty$, now cause the coefficients to increase dramatically with n. So dramatically, in fact, that if, for example

$$u(x,0) = \sum_{n=1}^{\infty} \frac{1}{n} \sin nx$$

(this has finite energy $\pi^3/12$), then

$$\sum_{n=1}^{\infty} \frac{1}{n^2} e^{-2n^2 t} \tag{10}$$

does not converge for any $t < 0$. The dissipation, which leads to "nice" behaviour as $t \to \infty$, is directly responsible for the blowup of solutions in $t < 0$.

Because of this we are forced to restrict our discussions to positive times. The best we can hope for is to be able to define a "semidynamical system," replacing the group of solution operators $\{T(t)\}_{t \in \mathbb{R}}$, which we could define in the ODE case, with a semigroup of solution operators $\{S(t)\}_{t \ge 0}$. In this way we are led to consider the semidynamical system

$$\left(L^2(0, \pi), \{S(t)\}_{t \ge 0} \right). \tag{11}$$

In contrast to the case of ODEs, for which there exists a general theory guaranteeing existence and uniqueness for a wide class of problems (this is covered in Chapter 2), no such unified approach is possible for PDEs. Each equation usually needs to be studied in its own right if we are to show that it has unique solutions, which can then be used to define a semidynamical system as in (11). We do this for two examples: a scalar reaction–diffusion equation,

$$\frac{\partial u}{\partial t} - \Delta u = f(u), \qquad u|_{\partial \Omega} = 0,$$

in Chapter 8, and the Navier–Stokes equations with periodic boundary conditions in two dimensions,

$$\frac{\partial u}{\partial t} - \nu \Delta u + (u \cdot \nabla u) + \nabla p = f(x), \qquad \nabla \cdot u = 0,$$

in Chapter 9.

The restriction that we can only define the solution for $t \geq 0$ does not interfere with one of the fundamental insights from the theory of dynamical systems, which motivates much of our approach: the complexity of the problem is significantly reduced if we are prepared to neglect any transient phenomena and concentrate only on describing the limiting behaviour as $t \to \infty$. For example, it follows from (9) that the limiting state of the system is $u(x) \equiv 0$, and a description of the "asymptotic dynamics" becomes as simple as can be.

In Chapter 10 we show that in some situations, which correspond physically to systems in which there is dissipation (friction, viscosity, etc.), the asymptotic dynamics all take place on a compact subset $\mathcal{A}$ of the original phase space. We call this set the "global attractor," and we verify that the scalar reaction–diffusion equation from Chapter 8 and the two-dimensional Navier–Stokes equations from Chapter 9 are both "dissipative" and so have a global attractor (Chapters 11 and 12). We also discuss the situation for the three-dimensional (3D) Navier–Stokes equations in some detail.

The remainder of the book concentrates on properties of this set $\mathcal{A}$ and how we can use its existence to deduce important consequences for the behaviour of the underlying PDE. For example, under certain mild conditions, it turns out that on $\mathcal{A}$ we can define $S(t)$ sensibly for all $t \in \mathbb{R}$, and so the dynamics restricted to the attractor define a standard dynamical system,

$$(\mathcal{A}, \{S(t)\}_{t \in \mathbb{R}}).$$

One of the most important properties of many of these attracting sets is that they are *finite-dimensional* subsets of the original, infinite-dimensional phase space. We prove this property for our two examples in Chapter 13, and it is the implications of this surprising result that occupy Chapters 14–16, where we investigate in what sense we can conclude that the original infinite-dimensional system is "in effect" finite-dimensional.

The final chapter consists of a long series of exercises that apply many of the techniques covered in the book to analyse the one-dimensional Kuramoto–Sivashinsky equation,

$$u_t + u_{xxxx} + u_{xx} + uu_x = 0, \qquad u(x, t) = u(x + L, t).$$

A Note on the Exercises

Each chapter ends with a series of exercises, which should be considered an integral part of the book. Indeed, several results proved in the exercises are used in the main body of the text. Some of the exercises in Chapters 14–17 are considerably more involved than those in the preceding chapters, since the material here is much more recent and is all still a focus of current research.

A full set of solutions to the exercises is available on the World Wide Web, at the following address:

$$\text{http}: //\text{www.cup.org/titles/0521635640.html}$$

I would welcome any comments, suggestions, or errata; these can be e-mailed to me at

$$\text{jcr@maths.warwick.ac.uk}$$

Errata will be posted periodically at the above web address.

Part I
Functional Analysis

1

Banach and Hilbert Spaces

The purpose of this chapter is to review some results from basic functional analysis, emphasising those aspects that will be useful in what follows. We introduce various examples of Banach spaces, in particular the spaces C^r of continuous functions and, after a brief outline of the construction of the Lebesgue integral, the L^p spaces of Lebesgue integrable functions. The final section treats some basic results concerning Hilbert spaces, of which L^2 is our most important example.

The material here is somewhat dry, but these are essential foundations. If the ideas are familiar, it is still worth paying attention to the Young and Hölder inequalities (Lemmas 1.8 and 1.9) and the technique of mollification (introduced in Section 1.3.1) by which we will prove various density results (Proposition 1.6, Theorem 1.13, and Corollary 1.14).

1.1 Banach Spaces and Some General Topology

A *norm* on a vector space X is a function $\|\cdot\| : X \to [0, \infty)$ satisfying

(i) $\|x\| = 0$ if and only if (iff) $x = 0$,
(ii) $\|\lambda x\| = |\lambda| \|x\|$ for all $x \in X$, $\lambda \in \mathbb{R}$, and
(iii) $\|x + y\| \le \|x\| + \|y\|$ for all $x, y \in X$ (the "triangle inequality").

A normed space is *complete* if every Cauchy sequence converges, and a *Banach space* is a complete normed space. Although this means that strictly a Banach space is a pair $(X, \|\cdot\|_X)$ of a space and its norm, we will usually use the more convenient notation X alone and specify the norm separately. Most common Banach spaces have a standard norm associated with them, and this is almost always the norm we will use, denoting it by $\|\cdot\|_X$. If there is any ambiguity we will specify the definition of the norm explicitly.

11

A subset Y of a Banach space X is *dense* if its closure in X, written $\overline{Y}$, is all of X. Equivalently, Y is dense in X if every element of X can be approximated arbitrarily closely by an element of Y, so that for any $\epsilon > 0$ there exists a $y \in Y$ such that $\|x - y\| < \epsilon$. In particular it follows that if $x \in X$ we can find a sequence $y_n \in Y$ such that

$$y_n \to x \quad \text{in} \quad X \quad \text{as} \quad n \to \infty,$$

(i.e. such that $\|y_n - x\|_X \to 0$ as $n \to \infty$). Showing that spaces of smooth functions are dense in spaces of Lebesgue integrable functions is one of the major topics of this chapter (Theorem 1.13 and Corollary 1.14).

If a Banach space X has a dense subset that is also countable then we say that X is *separable*. Occasionally the existence of such a countable dense subset is used to simplify an argument, so we will be careful in this chapter to point out any spaces that are separable. It follows easily that the finite product of separable spaces is also separable and a linear subspace of a separable space is separable (see Exercise 1.1).

Finally, the topological property of *compactness* will often be vital in our applications. Recall that a subset E of a Banach space X is *compact* if every open cover of E contains a finite subcover (see Exercise 1.2 for one application of this definition). An equivalent characterisation, which we will find much more useful, is that E is compact iff every sequence in E contains a convergent subsequence. That is, if we know that $\{x_n\} \in E$ then there exists a subsequence $x_{n_j} \to x^*$, with $x^* \in E$. Often we will be trying to solve a problem P that we cannot treat directly. Instead we will consider a sequence of easier problems P_n, which approximate P (in some sense), and for which we can find a solution x_n. If we can also show that the solutions x_n lie in a compact set, then we can hope that the limit of some convergent subsequence x_{n_j} will be a solution of our original problem P. This is the fundamental idea behind the approach we will use to prove the existence of solutions of our model PDEs in Chapters 7–9.

We will now cover various examples of Banach spaces, all of which will be needed in what follows. We start with the simplest.

1.2 The Euclidean Space $\mathbb{R}^m$

We write $x \in \mathbb{R}^m$ as $x = (x_1, \ldots, x_m)$. Then properties (i)–(iii) are easily checked for the standard Euclidean norm

$$|x| = \left(\sum_{j=1}^{m} x_j^2 \right)^{1/2} \tag{1.1}$$

and also for the alternative norms

$$\|x\|_{l^\infty} = \max_{j=1,\dots,m} |x_j| \qquad \text{and} \qquad \|x\|_{l^1} = \sum_{j=1}^m |x_j|.$$

(The reason for the subscripts on the norms is that they are part of a whole family of possible norms indexed by p,

$$\|x\|_{l^p} = \left(\sum_{j=1}^m |x_j|^p\right)^{1/p}. \tag{1.2}$$

We will come across these again in a different context later (Definition 1.19).) The completeness of $\mathbb{R}^m$ is part of its construction, and it is clearly separable (the set of all m-tuples of rational numbers, $\mathbb{Q}^m$, is a countable dense subset).

Two norms $\|\cdot\|_1$ and $\|\cdot\|_2$ on a Banach space X are said to be *equivalent* if there exist positive constants a and b such that

$$a\|x\|_2 \le \|x\|_1 \le b\|x\|_2, \qquad x \in X. \tag{1.3}$$

We now show that although there are many possible norms on $\mathbb{R}^m$, they are all equivalent. This follows from the Heine–Borel theorem, which asserts that a subset of $\mathbb{R}^m$ is compact iff it is closed and bounded.

Theorem 1.1. *All norms on* $\mathbb{R}^m$ *are equivalent.*

Proof. Let $|x|$ be the standard norm (1.1) on $\mathbb{R}^m$, and let $\|x\|$ be another norm. We first show that $x \mapsto \|x\|$ is a continuous map from $\mathbb{R}^m$ (with the standard norm) into $\mathbb{R}$. If we denote the standard basis of $\mathbb{R}^m$ by $\{e_i\}$ (i.e., $e_j = (0, \dots, 1, \dots, 0)$, with the 1 in the jth place) then

$$\left|\|x\| - \|y\|\right| \le \|x - y\| = \left\|\sum_{i=1}^m (x_i - y_i)e_i\right\|$$

$$\le \sum_{i=1}^m |x_i - y_i|\|e_i\|$$

$$\le \left(\sum_{i=1}^m \|e_i\|^2\right)^{1/2} \left(\sum_{i=1}^m |x_i - y_i|^2\right)^{1/2}$$

$$= C|x - y|,$$

using the Cauchy–Schwarz inequality

$$\left| \sum_{i=1}^{m} a_i b_i \right| \leq \left(\sum_{i=1}^{m} a_i^2 \right)^{1/2} \left(\sum_{i=1}^{m} b_i^2 \right)^{1/2}$$

(which is just $|a \cdot b| \leq |a||b|$) to split the sum.

So, certainly $x \mapsto \|x\|$ is also a continuous map from the closed set

$$S = \{x \in \mathbb{R}^m : |x| = 1\}$$

into $\mathbb{R}$. Since S is closed and bounded it is compact, and so $\|x\|$ is bounded on S, and attains its bounds: there exist $b \geq a \geq 0$ such that

$$a \leq \|x\| \leq b \qquad \text{for all } x \text{ with } |x| = 1.$$

Since the bound is attained, and $\|x\| = 0$ iff $x = 0$, it must be the case that $a > 0$.

Now for a general $y \in \mathbb{R}^m$, set $x = y/|y|$, so that $x \in S$. Then

$$a \leq \left\| \frac{y}{|y|} \right\| = \frac{\|y\|}{|y|} \leq b,$$

which gives (1.3) and shows that $|\cdot|$ and $\|\cdot\|$ are equivalent. It follows immediately that any two norms on $\mathbb{R}^m$ are equivalent. ∎

If $(X, \|\cdot\|_X)$ and $(Y, \|\cdot\|_Y)$ are normed spaces, it follows from this result that we have a wide choice of norms to use on the product space $X \times Y$. Generally, we will use the simplest choice,

$$\|(x, y)\|_{X \times Y} = \|x\|_X + \|y\|_Y,$$

the exception being the occasional times when we consider $\mathbb{R}^{m+n}$ as the product $\mathbb{R}^m \times \mathbb{R}^n$, and we nonetheless use the standard $\mathbb{R}^{m+n}$ norm.

In contrast to Theorem 1.1, we will now see that there are many different possible norms on infinite-dimensional spaces and that these need not be equivalent.

1.3 The Spaces C^r and $C^{r,\gamma}$ of Continuous Functions

Throughout this book we will denote an open subset of $\mathbb{R}^m$ by the Greek letter Ω, using $\overline{\Omega}$ for its closure and $\partial\Omega$ for its boundary.

The space $C^0(\Omega)$ consists of all continuous functions on Ω, and $C^0(\overline{\Omega})$ consists of all continuous functions on $\overline{\Omega}$. Note that functions in $C^0(\Omega)$ need not be bounded, whereas if Ω is bounded those in $C^0(\overline{\Omega})$ are both bounded and uniformly continuous. The space of bounded continuous functions on Ω is written $C_b^0(\Omega)$. Unless Ω is unbounded (e.g., $\Omega = \mathbb{R}^m$) this is a somewhat artificial space, since we can identify $C_b^0(\Omega)$ with $C^0(\overline{\Omega})$ when Ω is bounded.

The standard norm on $C^0(\overline{\Omega})$ (or $C_b^0(\Omega)$ when Ω is unbounded) is the supremum ("sup") norm,

$$\|u\|_\infty = \sup_{x\in\Omega} |u(x)|. \tag{1.4}$$

(The reason for the ∞ subscript will be seen in Section 1.4.3 below.) It is straightforward to check that this norm satisfies properties (i)–(iii) above. If no other norm is specified, the space "$C^0(\overline{\Omega})$" will denote the space equipped with the sup norm of (1.4). This space is complete – the uniform limit of continuous functions is itself continuous – and so is a Banach space. It is also separable, as the following simple argument shows.

Proposition 1.2. *If Ω is bounded then $C^0(\overline{\Omega})$ is separable.*

Proof. This follows from the Weierstrass approximation theorem (see, for example, Theorem 7.32 in Rudin (1976)): the set of all polynomials is dense in $C^0(\overline{\Omega})$, and the set of all polynomials with rational coefficients (a countable set) is dense in the set of all polynomials. ∎

Other spaces of continuous functions involve higher orders of differentiability. For functions on general subsets of $\mathbb{R}^m$, the use of multi-index notation is an elegant way to express mixed partial derivatives. We will use the notation D_j to denote $\partial/\partial x_j$, so that, for example,

$$|\nabla u|^2 = \sum_{j=1}^{m} |D_j u|^2. \tag{1.5}$$

A multi-index α is a vector consisting of m nonnegative integers $(\alpha_1, \ldots, \alpha_m)$; we write $|\alpha|$ for the sum of the entries,

$$|\alpha| = \alpha_1 + \cdots + \alpha_m.$$

For any vector $k = (k_1, \ldots, k_m)$, we define k^α by

$$k^\alpha = k_1^{\alpha_1} \ldots k_m^{\alpha_m}.$$

We also set

$$D^\alpha = D_1^{\alpha_1} \dots D_m^{\alpha_m}$$

(as if D were the "vector" $(D_1, \dots, D_m)$) so that we have

$$D^\alpha f = \frac{\partial^{|\alpha|} f}{\partial x_1^{\alpha_1} \dots \partial x_m^{\alpha_m}}.$$

If we define

$$\alpha! = \alpha_1! \dots \alpha_m!$$

and

$$\binom{\alpha}{\beta} = \frac{\alpha!}{(\alpha - \beta)!\beta!} = \binom{\alpha_1}{\beta_1} \dots \binom{\alpha_m}{\beta_m}$$

then we can write, for example, Leibniz' theorem as

$$D^\alpha(fg) = \sum_{\beta \le \alpha} \binom{\alpha}{\beta} D^\beta f \, D^{\alpha - \beta} g, \tag{1.6}$$

where we say that $\beta \le \alpha$ if $\beta_i \le \alpha_i$ for all $1 \le i \le m$.

Making use of these multi-indices we can now define the spaces $C^r(\Omega)$, which consist of functions f, all of whose derivatives up to and including order r are continuous,

$$C^r(\Omega) = \{f : D^\alpha f \in C^0(\Omega) \text{ for all } |\alpha| \le r\},$$

and the space of "smooth" functions $C^\infty(\Omega)$ that are infinitely differentiable on Ω,

$$C^\infty(\Omega) = \bigcap_{r=0}^{\infty} C^r(\Omega).$$

Since Ω is open, it is hard to apply compactness arguments to functions in $C^r(\Omega)$. To get around this, we introduce the notion of the support of a function,

$$\text{supp } f = \overline{\{x : f(x) \ne 0\}},$$

where $\overline{X}$ denotes the closure of X in $\mathbb{R}^m$. This definition gives the smallest closed set supp f such that $f \equiv 0$ on $\mathbb{R}^m \setminus \text{supp } f$. We now define a family of spaces consisting of continuous functions with compact support in Ω. However, these spaces are not complete (see Exercise 1.3).

Definition 1.3. *The space $C_c^r(\Omega)$ consists of all functions in $C^r(\Omega)$ whose support is a compact subset of Ω. [The most useful members of this family are $C_c^0(\Omega)$ and $C_c^\infty(\Omega)$.]*

Since compact subsets of Ω will be used frequently, we use the special notation

$$K \subset\subset \Omega$$

to mean that K is a compact subset of Ω. Note that if $K \subset\subset \Omega$ then K is bounded away from $\partial\Omega$; i.e., there exists an $\epsilon > 0$ such that

$$B(x, \epsilon) \subset \Omega \qquad \text{for all} \qquad x \in K,$$

where $B(x, \epsilon)$ denotes the open ball centred at x and of radius ϵ,

$$B(x, \epsilon) = \{y : |y - x| < \epsilon\}.$$

(To denote a closed ball we simply add the closure overbar to B (rather than over the whole expression $B(x, \epsilon)$)

$$\overline{B}(x, \epsilon) = \{y : |y - x| \le \epsilon\}.)$$

Note that any function $f \in C_c^r(\Omega)$ is in fact in $C^r(K)$, for some $K \subset\subset \Omega$. Spaces of continuous functions on closed sets become Banach spaces when endowed with an appropriate norm.

Definition 1.4. *For Ω bounded, $C^r(\overline{\Omega})$ consists of all r-times continuously differentiable functions, with norm*

$$\|f\|_{C^r} = \sum_{|\alpha| \le r} \sup_{x \in \Omega} |D^\alpha f(x)|.$$

The space $C^\infty(\overline{\Omega})$ is defined by

$$C^\infty(\overline{\Omega}) = \bigcap_{r=1}^{\infty} C^r(\overline{\Omega}).$$

Similar arguments to those above (Proposition 1.2) show that $C^r(\overline{\Omega})$ is a separable Banach space for each $r < \infty$. There is no norm that makes $C^\infty(\overline{\Omega})$ into a Banach space (see Exercise 1.4 for one possible remedy).

With a little further notation, we can also introduce a family of separable Banach spaces $C^{r,\gamma}$ whose definition includes some information about the modulus of continuity of the rth-order derivatives (you are asked to prove in Exercise 1.5 that they are indeed Banach spaces). We say that a function $g : X \to Y$ is Hölder continuous with exponent $0 < \gamma \le 1$ if there exists a constant C such that

$$\|g(x) - g(y)\|_Y \le C \|x - y\|_X^\gamma, \qquad x, y \in X.$$

If $\gamma = 1$ we say that g is a Lipschitz function, with Lipschitz constant C.

Definition 1.5. *$C^{r,\gamma}(\overline{\Omega})$ consists of all functions in $C^r(\overline{\Omega})$ whose rth-order derivatives are Hölder continuous with exponent γ. That is, there exists a constant C (which may depend on f) such that*

$$|D^\alpha f(x) - D^\alpha f(y)| \le C |x - y|^\gamma \qquad whenever \qquad |\alpha| = r.$$

The norm is

$$\|f\|_{C^{k,\gamma}} = \|f\|_{C^k} + \sup_{x,y \in \overline{\Omega},\ |\alpha|=r} \frac{|D^\alpha f(x) - D^\alpha f(y)|}{|x - y|^\gamma}.$$

For example, the space $C^{0,1}(\overline{\Omega})$ consists of Lipschitz continuous functions on $\overline{\Omega}$. If Ω is convex then any $C^1(\overline{\Omega})$ function is Lipschitz (see Exercise 1.6), but a Lipschitz function need not be C^1 (although a Lipschitz function is differentiable almost everywhere; see Theorem 6 in Section 5.8 of Evans (1998)). In this way, the spaces $C^{r,\gamma}, 0 < \gamma \le 1$ can be viewed as intermediate between C^r and C^{r+1}.

1.3.1 Mollification and Approximation by Smooth Functions

We will now use the technique of *mollification* to show that functions in $C_c^0(\Omega)$ can be approximated arbitrarily closely by functions in $C_c^\infty(\Omega)$: given any $f \in C_c^0(\Omega)$ and any $\epsilon > 0$, there exists a function $\phi \in C_c^\infty(\Omega)$ with

$$\|f - \phi\|_\infty < \epsilon.$$

[Note that this does not show (and cannot) that $C_c^\infty(\Omega)$ is dense in $C_c^0(\Omega)$, since $C_c^0(\Omega)$ is not a Banach space.]

Mollification provides an explicit way of finding a smooth function that approximates a given less regular function. We define $\rho(x)$ by

$$\rho(x) = \begin{cases} c\exp\left(\frac{1}{|x|^2-1}\right), & |x| \le 1, \\ 0, & |x| \ge 1, \end{cases}$$

where c is chosen so that

$$\int_{\mathbb{R}^m} \rho(x)\,dx = 1. \tag{1.7}$$

Note that $\rho \in C_c^\infty(\mathbb{R}^m)$. The mollification of u, u_h, is

$$u_h(x) = h^{-m} \int_\Omega \rho\left(\frac{x-y}{h}\right) u(y)\,dy. \tag{1.8}$$

[If we were to write

$$\rho^h(z) = h^{-m}\rho(z/h)$$

then the expression for the mollification u_h is exactly the convolution of ρ^h with u, $u_h = \rho^h * u$. Many authors (e.g., Evans (1998)) choose to adopt this notation.]

Proposition 1.6 (Mollification of C_c^0 Functions). *Let $u \in C_c^0(\Omega)$. Then $u_h \in C_c^\infty(\Omega)$ if $h < \text{dist}(\text{supp } u, \partial\Omega)$, and $u_h \to u$ uniformly on Ω as $h \to 0$.*

Proof. For $h < \text{dist}(\text{supp } u, \partial\Omega)$, (1.8) is a function in $C_c^\infty(\Omega)$, since (i) we can differentiate under the integral sign with respect to x and (ii) the support of u_h lies within an h-neighbourhood of the support of u, and so within a compact subset of Ω.

Now, the functions u_h tend uniformly to u as $h \to 0$ since by (1.7) we can write

$$u_h(x) - u(x) = h^{-m} \int_\Omega \rho\left(\frac{x-y}{h}\right)[u(y)-u(x)]\,dy,$$

which gives, using (1.7) once again,

$$|u_h(x) - u(x)| \le \sup_{|y-x|\le h} |u(y)-u(x)|.$$

Since the support of u is compact, u is uniformly continuous on Ω, and so given $\epsilon > 0$, h can be chosen sufficiently small to guarantee that $\|u_h - u\|_\infty \leq \epsilon$. ∎

Mollification preserves other properties of smooth functions, for example their modulus of continuity (see Exercise 1.7).

1.4 The L^p Spaces of Lebesgue Integrable Functions

We will quickly review the fundamentals of the theory of Lebesgue integration, in particular giving some results that will be useful in studying spaces of integrable functions.

1.4.1 Lebesgue Integration

We follow the presentation in Priestley (1997) and start the construction of the Lebesgue integral by defining the integral of simple functions for which there can be no argument as to the correct definition. First we define the volume, or measure, of a simple block in Ω,

$$I = \prod_{k=1}^{m} [a_k, b_k], \qquad (1.9)$$

by

$$\mu(I) = |I| = \prod_{k=1}^{m} (b_k - a_k).$$

We will say that a set A has "measure zero" if, given any $\epsilon > 0$, one can write

$$A \subset \bigcup_j I_j,$$

with the I_j m-dimensional cuboids as in (1.9), such that $\sum_j |I_j| < \epsilon$.

The class $S(\Omega)$ of *simple functions* on Ω consists of all those functions $s(x)$ that are nonnegative and piecewise constant on a finite number of blocks,

$$s(x) = \sum_{j=1}^{n} c_j \chi[I_j](x), \qquad (1.10)$$

where $c_j > 0$, each I_j is a cuboid in Ω as in (1.9), and $\chi[A]$ denotes the characteristic function of the set A,

$$\chi[A](x) = \begin{cases} 1, & x \in A, \\ 0, & x \notin A. \end{cases}$$

We define the integral of $s(x)$ by

$$\int_\Omega s(x)\,dx = \sum_{j=1}^{n} c_j \mu(I_j). \tag{1.11}$$

It is tedious but fairly elementary to check that this integral is well defined on $S(\Omega)$, so that if $s(x)$ is given by two possible expressions (1.10) then the integrals in (1.11) agree.

Now, if $s_n(x)$ is a monotonically increasing sequence of functions in $S(\Omega)$ ($s_{n+1}(x) \geq s_n(x)$ for each $x \in \Omega$), then it follows from the Definition (1.11) that the sequence

$$\int_\Omega s_n(x)\,dx \tag{1.12}$$

is also monotonically increasing. Provided that the integrals in (1.12) are uniformly bounded in n, then

$$\lim_{n \to \infty} \int_\Omega s_n(x)\,dx$$

exists.

One can show that each monotonic sequence $s_n(x)$ for which the integrals in (1.12) are uniformly bounded tends pointwise to a function $f(x)$ *almost everywhere* (a.e.), that is, except on a set of measure zero. We denote the set of all functions that can be arrived at in this way by $L_{\text{inc}}(\Omega)$, and for such functions we can define

$$\int_\Omega f(x)\,dx = \lim_{n \to \infty} \int_\Omega s_n(x)\,dx.$$

Again, we have to check that this definition does not depend on exactly which sequence $\{s_n\}$ we have chosen.

Finally, we define the space of measurable functions on Ω, written $L^1(\Omega)$, to be all functions of the form $f(x) = f_1(x) - f_2(x)$ with f_1 and f_2 in $L_{\text{inc}}(\Omega)$, and we set

$$\int_\Omega f(x)\,dx = \int_\Omega f_1(x)\,dx - \int_\Omega f_2(x)\,dx.$$

We now state some fundamental convergence results from the theory of Lebesgue integration. Note that "a.e." is also used for "almost every"; and "for a.e. $x \in \Omega$" means the same as "almost everywhere in Ω."

Theorem 1.7. *If Ω is an open subset of $\mathbb{R}^m$ then:*

(i) *(**Monotone Convergence Theorem**) If $\{f_n\}$ is an increasing sequence of measurable functions with*

$$0 \le f_1(x) \le f_2(x) \le \cdots \qquad for\ a.e. \qquad x \in \Omega$$

then

$$\lim_{n \to \infty} \int_\Omega f_n(x)\,dx = \int_\Omega \left(\lim_{n \to \infty} f_n(x) \right) dx.$$

(ii) *(**Fatou's Lemma**) If $\{f_n\}$ is a sequence of nonnegative measurable functions, then*

$$\int_\Omega \left(\liminf_{n \to \infty} f_n(x) \right) dx \le \liminf_{n \to \infty} \int_\Omega f_n(x)\,dx.$$

(iii) *(**Dominated Convergence Theorem**) Let $\{f_n\}$ be a sequence of measurable functions converging pointwise a.e. to a limit on Ω. If there is a function $g \in L^1(\Omega)$ such that $|f_n(x)| \le g(x)$ for every n and a.e. $x \in \Omega$, then*

$$\lim_{n \to \infty} \int_\Omega f_n(x)\,dx = \int_\Omega \left(\lim_{n \to \infty} f_n(x) \right) dx.$$

1.4.2 The Lebesgue Spaces $L^p(\Omega)$ with $1 \le p < \infty$

We now study the family $L^p(\Omega)$ of spaces of Lebesgue integrable functions, for $1 \le p < \infty$. We will find that $L^2(\Omega)$ has a particularly important part to play in what follows (cf. Introduction). The L^p norm is given by the integral (to be understood in the Lebesgue sense)

$$\|f\|_{L^p} = \left(\int_\Omega |f(x)|^p\,dx \right)^{1/p}, \tag{1.13}$$

and $L^p(\Omega)$ consists of all functions with finite L^p norm.

We want to show that (1.13) does indeed define a norm. However, note that axiom (i) for a norm requires that

$$\|f\| = 0 \qquad \text{iff} \qquad f = 0.$$

In other words, two functions are equal iff the norm of their difference is zero. Now, the Lebesgue theory of integration attaches a zero value to any integral

$$\int_\Omega f(x)\,dx$$

for which f is zero *almost everywhere*, that is, except on a set of measure zero. Thus we have to view each element of L^p not as a function, but as an equivalence class of functions that agree almost everywhere. In practise this distinction causes few problems, but we must always bear in mind that "$u = 0$ in $L^p(\Omega)$" means only that $u = 0$ *almost everywhere* in Ω.

Axiom (ii) (that $\|\lambda f\| = |\lambda|\,\|f\|$) is straightforward, but the triangle inequality (iii) requires some work. We will need to use Hölder's inequality, which is based on the simpler inequality given in the next lemma.

Lemma 1.8 (Young's Inequality). *If $a, b \geq 0$, $p, q > 1$ with $p^{-1} + q^{-1} = 1$, then*

$$ab \leq \frac{a^p}{p} + \frac{b^q}{q}. \tag{1.14}$$

Sometimes we will use "Young's inequality with ϵ,"

$$ab \leq \epsilon a^p + \epsilon^{-q/p} b^q, \tag{1.15}$$

which follows from (1.14) applied to $ab = [(\epsilon)^{1/p} a][(\epsilon)^{-1/p} b]$ since $p, q > 1$.

Proof. Consider the function

$$f(t) = \frac{t^p}{p} + \frac{1}{q} - t.$$

Then for $t \geq 0$ this has the minimum value $f(t) = 0$ when $t = 1$. Setting $t = ab^{-q/p}$ it follows that

$$\frac{a^p b^{-q}}{p} + \frac{1}{q} - ab^{-q/p} \geq 0,$$

which simplifies to give

$$\frac{a^p}{p} + \frac{b^q}{q} \geq ab^{-q/p} b^q = ab. \qquad \blacksquare$$

When p and q are as in the above lemma we say that they are "conjugate indices" (sometimes it will be convenient to write "(p, q) are conjugate indices").

Lemma 1.9 (Hölder's Inequality). *Let (p, q) be conjugate indices with $1 < p < \infty$, and suppose that $f \in L^p(\Omega)$ and $g \in L^q(\Omega)$. Then $fg \in L^1(\Omega)$, with*

$$\|fg\|_{L^1} \leq \|f\|_{L^p} \|g\|_{L^q}.$$

A more general version of this inequality is given in Exercise 1.8.

Proof. Consider

$$\int_\Omega \frac{|f(x)|}{\|f\|_{L^p}} \frac{|g(x)|}{\|g\|_{L^q}} \, dx,$$

so that Young's inequality (Lemma 1.8) gives

$$\int_\Omega \frac{|f(x)|}{\|f\|_{L^p}} \frac{|g(x)|}{\|g\|_{L^q}} \, dx \leq \int_\Omega \frac{1}{p} \frac{|f(x)|^p}{\|f\|_{L^p}^p} + \frac{1}{q} \frac{|g(x)|^q}{\|g\|_{L^q}^q} \, dx = \frac{1}{p} + \frac{1}{q} = 1,$$

as required. ∎

We note that one immediate consequence of this result is that, when Ω has finite volume,

$$|\Omega| = \int_\Omega 1 \, dx < \infty,$$

then $L^r(\Omega) \subset L^s(\Omega)$ for $r > s$, since, taking $(p, q) = (r/(r-s), r/s)$, we have

$$\int_\Omega |f(x)|^s \, ds = \int_\Omega 1 \times |f(x)|^s \, dx$$

$$\leq \left(\int_\Omega 1 \right)^{(r-s)/r} \left(\int_\Omega |f(x)|^r \, dx \right)^{s/r}$$

$$= |\Omega|^{(r-s)/r} \|f\|_{L^r}^s,$$

so that

$$\|f\|_{L^s} \leq |\Omega|^{(r-s)/rs} \|f\|_{L^r}. \tag{1.16}$$

A similar argument can be used to prove the L^p interpolation inequality,

$$\|u\|_{L^p} \leq \|u\|_{L^q}^{q(r-p)/p(r-q)} \|u\|_{L^r}^{r(p-q)/p(r-q)}$$

when $q < p < r$ (see Exercise 1.9).

Using Hölder's inequality, we can now establish the triangle inequality for the L^p spaces, which is important enough that it has its own name.

Lemma 1.10 (Minkowski's Inequality). *If $f, g \in L^p(\Omega)$, $1 \le p < \infty$, then $f + g \in L^p(\Omega)$, with*

$$\|f + g\|_{L^p} \le \|f\|_{L^p} + \|g\|_{L^p}.$$

Proof. First note that

$$|f(x) + g(x)|^p \le (|f(x)| + |g(x)|)^p \le 2^p(|f(x)|^p + |g(x)|^p),$$

so $f + g \in L^p(\Omega)$. Also, if (p, q) are conjugate then $(p - 1)q = p$, and so $|f + g|^{p-1} \in L^q$. It follows that we can integrate the inequality

$$|f(x) + g(x)|^p \le |f(x) + g(x)|^{p-1}|f(x)| + |f(x) + g(x)|^{p-1}|g(x)|$$

and use Hölder's inequality on each of the two terms on the right-hand side to obtain

$$\int_\Omega |f(x) + g(x)|^p \le \left(\int_\Omega |f(x) + g(x)|^p \right)^{1/q} \left(\|f\|_{L^p} + \|g\|_{L^p} \right),$$

which yields, on dividing (since $1 - q^{-1} = p^{-1}$),

$$\|f + g\|_{L^p} \le \|f\|_{L^p} + \|g\|_{L^p}. \qquad \blacksquare$$

We have shown that expression (1.13) is a proper norm on L^p. We now show that the L^p spaces are complete and so are in fact Banach spaces.

Theorem 1.11. *If $\{u_n\}$ is a Cauchy sequence in $L^p(\Omega)$, $1 \le p < \infty$, then there exists a $u \in L^p(\Omega)$ such that $u_n \to u$ in $L^p(\Omega)$.*

Proof. Since $\{u_n\}$ is Cauchy we can find a subsequence u_{n_j} such that

$$\left\| u_{n_{j+1}} - u_{n_j} \right\|_{L^p} < 2^{-j}.$$

It follows that

$$\left\| \sum_{j=1}^k |u_{n_{j+1}} - u_{n_j}| \right\|_{L^p} \le \sum_{j=1}^k 2^{-j} < 1$$

for each k, and so if we define

$$v(x) = \lim_{k \to \infty} \sum_{j=1}^{k} \left| u_{n_{j+1}}(x) - u_{n_j}(x) \right|,$$

allowing $v(x) = \infty$ if need be, we can use Fatou's lemma from Theorem 1.7 to deduce that

$$\|v\|_{L^p} \leq 1.$$

It follows that $v(x) < \infty$ almost everywhere and that the series

$$u_{n_1}(x) + \sum_{j=1}^{\infty} \left(u_{n_{j+1}}(x) - u_{n_j}(x) \right)$$

converges absolutely for almost every x. Since the partial sums of this series are just $u_{n_{k+1}}(x)$, we can define a function $u(x)$ for almost every $x \in \Omega$ by

$$u(x) = \lim_{j \to \infty} u_{n_j}(x),$$

and however we wish at the other points of Ω. Finally, we have to show that $u \in L^p(\Omega)$ and that $\|u_n - u\|_{L^p} \to 0$.

Choosing an N such that

$$\|u_n - u_m\|_{L^p} \leq \epsilon/2 \qquad \text{for all} \qquad n, m \geq N,$$

we use Fatou's lemma (Theorem 1.7) again to deduce that

$$\|u - u_{n_j}\|_{L^p} \leq \liminf_{k \to \infty} \|u_{n_k} - u_{n_j}\|_{L^p} \leq \epsilon/2,$$

provided that $n_j \geq N$. It follows that $u - u_{n_j} \in L^p$, and since

$$u = (u - u_{n_j}) + u_{n_j} \qquad \text{with} \qquad u_{n_j} \in L^p,$$

we have $u \in L^p(\Omega)$. For $n > N$, choosing j such that $n_j \geq N$ too, we have that

$$\|u_n - u\|_{L^p} \leq \|u_n - u_{n_j}\|_{L^p} + \|u_{n_j} - u\|_{L^p}$$

$$\leq \epsilon/2 + \epsilon/2 = \epsilon,$$

and so u_n converges to u in $L^p(\Omega)$ as required. ∎

Note that in the course of the proof we have obtained the following very useful result.

Corollary 1.12. *If $u_j \to u$ in $L^p(\Omega)$ then there is a subsequence that converges pointwise to u almost everywhere in Ω.*

We now use the definition of the Lebesgue integral sketched above to show that continuous functions are dense in the Lebesgue spaces.

Theorem 1.13. *$C_c^0(\Omega)$ is dense in $L^p(\Omega)$ for $1 \leq p < \infty$.*

Proof. Take $u \in L^p(\Omega)$ and $\epsilon > 0$. Then there exists a $K \subset\subset \Omega$ such that $u|_K$ satisfies

$$\|u - u|_K\|_{L^p} \leq \epsilon/2.$$

Since $L^p(K) \subset L^1(K)$ (see (1.16)), we can use the definition of L^1 to write $u|_K = u_1 - u_2$, where both $u_i \in L_{\text{inc}}(K)$. If we approximate u_1 and u_2 within $\epsilon/4$ in the $L^p(K)$ norm then Minkowski's inequality shows that we can approximate $u|_K$ within $\epsilon/2$; so assume that $u \in L_{\text{inc}}(K)$.

By the definition of L_{inc} there exists a sequence of simple functions $s_n(x)$ converging pointwise and monotonically to u. Since $s_n(x) \leq u(x)$ it follows that $s_n \in L^p$, and then, since $(u(x) - s_n(x))^p \leq u(x)^p$, the dominated convergence theorem shows that $s_n \to u$ in L^p.

Since (see Exercise 1.10) a blockwise constant function can be approximated arbitrarily closely in the L^p norm by a function in $C_c^0(\Omega)$, the result follows. ∎

One could therefore define $L^p(\Omega)$ ($1 \leq p < \infty$) as the completion of $C_c^0(\Omega)$ with respect to the L^p norm. This means that for any $f \in L^p(\Omega)$ there is a sequence $\phi_n \in C_c^0(\Omega)$ with $\phi_n \to f$ in $L^p(\Omega)$. This yields the completeness of L^p essentially by definition, and it gives a way to define the Lebesgue integral in terms of limits of integrals of continuous functions rather than just step functions.

Using Proposition 1.6 we see immediately that

Corollary 1.14. *$C_c^\infty(\Omega)$ is dense in $L^p(\Omega)$ provided that $1 \leq p < \infty$.*

It also follows easily from Proposition 1.2 that

Proposition 1.15. *$L^p(\Omega)$ is separable provided that $1 \leq p < \infty$.*

Proof. Let

$$\Omega_n = \{x \in \Omega : \operatorname{dist}(x, \partial\Omega) > 1/n \text{ and } |x| < n\}.$$

Then $\overline{\Omega}_n$ is a compact subset of Ω and $\Omega = \cup_{j=1}^{\infty} \overline{\Omega}_j$. Let P_n be the countable dense subset of $C^0(\overline{\Omega}_n)$ ensured by Proposition 1.2. If we take $\mathcal{P} = \cup_j P_j$ then this is a countable subset of $L^p(\Omega)$. Since $C_c^0(\Omega)$ is dense in $L^p(\Omega)$, for a given $f \in L^p(\Omega)$ and $\epsilon > 0$ choose $\phi \in C_c^0(\Omega)$ such that

$$\|f - \phi\|_{L^p} \le \epsilon/2.$$

Since $\phi \in C^0(\overline{\Omega}_n)$ for some n, there is a $\pi \in P_n \subset \mathcal{P}$ such that

$$\|\phi - \pi\|_{\infty} \le \epsilon/2|\Omega_n|^{-1/p}.$$

It follows immediately that $\|\phi - \pi\|_{L^p} \le \epsilon/2$, and so $\|f - \pi\|_{L^p} \le \epsilon$. ∎

1.4.3 The Lebesgue Space $L^{\infty}(\Omega)$

One "Lebesgue" space that does not arise naturally from the integration theory is $L^{\infty}(\Omega)$, the space of "essentially bounded" functions. We say that $f \in L^{\infty}(\Omega)$ if its essential supremum, given by

$$\|f\|_{\infty} = \operatorname{ess\,sup}_{\Omega} |f(x)|, \qquad (1.17)$$

is finite. The right-hand side of (1.17) denotes the smallest value that bounds f almost everywhere:

$$\operatorname{ess\,sup}_{\Omega} |f(x)| = \inf \left\{ \sup_{x \in S} |f(x)| : S \subset \overline{\Omega}, \text{ with } \Omega \setminus S \text{ of measure zero} \right\}.$$

In particular $|f(x)| \le \|f\|_{\infty}$ almost everywhere, and it follows that if $f \in C_b^0(\Omega)$ then the essential supremum of f is the same as its supremum (which explains the notation in (1.4)).

Partly because of its more convoluted definition, $L^{\infty}(\Omega)$ does not share all the nice properties of the other L^p spaces. Although Hölder's inequality can be extended to cover the case $p = 1$, $q = \infty$ (these indices are clearly still conjugate, and the proof is very simple; see Exercise 1.11), $L^{\infty}(\Omega)$ is not separable. Nor is $C_c^0(\Omega)$ a dense subset, since the completion of $C_c^0(\Omega)$ in the L^{∞} norm (which is the same as its completion in the sup norm) is a subspace of $C^0(\overline{\Omega})$ (see Exercise 1.3).

Although it is not based on an integral, it is possible to relate the L^{∞} norm to the L^p norms, $1 \le p < \infty$.

Proposition 1.16. *Let Ω have finite volume. If $u \in L^\infty(\Omega)$ then we have*

$$\|u\|_\infty = \lim_{p \to \infty} \|u\|_{L^p}, \tag{1.18}$$

and if $u \in L^p(\Omega)$ for every $1 \le p < \infty$, with

$$\|u\|_{L^p} \le K, \tag{1.19}$$

then $u \in L^\infty(\Omega)$, with $\|u\|_\infty \le K$.

Proof. First we use (1.16), which shows that

$$\|u\|_{L^p} \le |\Omega|^{1/p} \|u\|_\infty,$$

and hence

$$\limsup_{p \to \infty} \|u\|_{L^p} \le \|u\|_\infty. \tag{1.20}$$

Also, we know that for any $\epsilon > 0$ there exists a subset A of Ω, which has positive measure, such that

$$|u(x)| \ge \|u\|_\infty - \epsilon \qquad \text{for all} \qquad x \in X.$$

Therefore

$$\int_\Omega |u(x)|^p \, dx \ge \int_A |u(x)|^p \, dx \ge |A| \big(\|u\|_\infty - \epsilon \big)^p.$$

It follows that

$$\|u\|_{L^p} \ge |A|^{1/p} \big(\|u\|_\infty - \epsilon \big),$$

from which we have

$$\liminf_{p \to \infty} \|u\|_{L^p} \ge \|u\|_\infty. \tag{1.21}$$

Combining (1.20) and (1.21) gives (1.18).

Now, suppose that we have $\|u\|_{L^p} \le K$ for all $1 \le p < \infty$, but do not have $\|u\|_\infty \le K$. Then we can find a constant $K_1 > K$ and a positive measure subset of Ω, A, such that

$$|u(x)| \ge K_1 \qquad \text{for all} \qquad x \in A.$$

Following the argument that gave us (1.21) above, we have

$$\liminf_{p \to \infty} \|u\|_{L^p} \geq K_1 > K,$$

which contradicts (1.19). Thus $\|u\|_\infty \leq K$ as claimed, and so $u \in L^\infty(\Omega)$. ∎

Finally, we show that $L^\infty(\Omega)$ is a Banach space.

Theorem 1.17. $L^\infty(\Omega)$ *is complete.*

Proof. If $\{u_n\}$ is a Cauchy sequence in $L^\infty(\Omega)$ then there exists a measure zero subset A of Ω such that

$$|u_n(x)| \leq \|u_n\|_\infty \qquad \text{and} \qquad |u_n(x) - u_m(x)| \leq \|u_n - u_m\|_\infty$$

for $x \notin A$ and every $n, m = 1, 2, \ldots$. It follows that $\{u_n\}$ converges uniformly on $\Omega \setminus A$ to a bounded function u, and setting $u(x) = 0$ for $x \notin A$ we obtain a function $u \in L^\infty(\Omega)$ such that $u_n \to u$ in $L^\infty(\Omega)$. ∎

1.4.4 The Spaces $L^p_{\text{loc}}(\Omega)$ of Locally Integrable Functions

Finally, for $1 \leq p \leq \infty$ we say that a function is locally in L^p on Ω, or in $L^p_{\text{loc}}(\Omega)$, if* it is in $L^p(K)$ for every compact subset K of Ω:

$$L^p_{\text{loc}}(\Omega) = \{f : f \in L^p(K) \text{ for every } K \subset\subset \Omega\}.$$

Note that it follows from (1.16) that $L^r_{\text{loc}}(\Omega) \subset L^s_{\text{loc}}(\Omega)$ if $r > s$, for any Ω, bounded or unbounded.

We finish this section on the Lebesgue spaces by showing that if $u \in L^p_{\text{loc}}(\Omega)$ then the mollified function u_h introduced in (1.8) converges to u in $L^p_{\text{loc}}(\Omega)$. We say that u_n converges to u in $L^p_{\text{loc}}(\Omega)$ provided that

$$u_n \to u \qquad \text{in} \qquad L^p(\Omega')$$

for any Ω' with $\Omega' \subset\subset \Omega$.

Proposition 1.18 (Mollification of L^p_{loc} Functions). *If $u \in L^p_{\text{loc}}(\Omega)$, with $1 \leq p < \infty$, then $u_h \in L^p_{\text{loc}}(\Omega)$, and $u_h \to u$ in $L^p_{\text{loc}}(\Omega)$. Furthermore,*

---

* We usually consider the Lebesgue spaces $L^p(\mathcal{O})$, where $\mathcal{O}$ is open, since the value of a function on the boundary does not affect its integral. Nonetheless, it is occasionally notationally convenient to consider Lebesgue spaces on closed sets, such as $L^p(K)$ above. To be entirely consistent one could take this to be shorthand for "$L^p(\mathcal{O})$," where $\overline{\mathcal{O}} = K$, but this is unnecessarily complicated.

$u_h \in C^\infty(\Omega)$, *and if u has compact support in Ω and $h <$ dist(supp u, $\partial\Omega$) then $u_h \in C_c^\infty(\Omega)$.*

Clearly we cannot expect this result to hold for L^∞: since the L^∞ norm corresponds to the supremum norm for continuous functions, it would show that L^∞ consists of continuous functions.

Proof. We first show that if $\Omega' \subset\subset \Omega$ then

$$\|u_h\|_{L^p(\Omega')} \le \|u\|_{L^p(\Omega'')} \tag{1.22}$$

for some Ω'' with $\Omega' \subset \Omega'' \subset\subset \Omega$. We take $h < \frac{1}{2}\text{dist}(\Omega', \partial\Omega)$, and then change variables in (1.8) to $z = (x - y)/h$ so that

$$u_h(x) = \int_{|z|\le 1} \rho(z)u(x - hz)\,dz.$$

Therefore, using (1.7) we get

$$
\begin{aligned}
|u_h(x)|^p &= \left(\int_{|z|\le 1} \rho(z)u(x - hz)\,dz \right)^p \\
&\le \left(\int_{|z|\le 1} \rho(z)^{q(1-1/p)}\,dz \right)^{p/q} \left(\int_{|z|\le 1} \rho(z)|u(x - hz)|^p\,dz \right) \\
&= \left(\int_{|z|\le 1} \rho(z)\,dz \right)^{p/q} \left(\int_{|z|\le 1} \rho(z)|u(x - hz)|^p\,dz \right) \\
&= \int_{|z|\le 1} \rho(z)|u(x - hz)|^p\,dz.
\end{aligned}
$$

Now, if we take $\Omega'' = \{x : \text{dist}(x, \Omega') < h\}$ then

$$
\begin{aligned}
\int_{\Omega'} |u_h(x)|^p\,dx &\le \int_{\Omega'} \int_{|z|\le 1} \rho(z)|u(x - hz)|^p\,dz\,dx \\
&= \int_{|z|\le 1} \rho(z) \int_{\Omega'} |u(x - hz)|^p\,dx\,dz \\
&\le \int_{\Omega''} |u(x)|^p\,dx,
\end{aligned}
$$

which gives (1.22).

We now complete the proof using Theorem 1.13 and Proposition 1.6. For a given $\epsilon > 0$, choose $\phi \in C_c^0(\Omega)$ with

$$\|u - \phi\|_{L^p(\Omega'')} \le \epsilon/3.$$

For h small enough, Proposition 1.6 guarantees that

$$\|\phi - \phi_h\|_{L^p(\Omega'')} \le \epsilon/3,$$

and so if we use (1.22) on $u - \phi$ we get

$$\|u - u_h\|_{L^p(\Omega')} \le \|u - \phi\|_{L^p(\Omega')} + \|\phi - \phi_h\|_{L^p(\Omega')} + \|u_h - \phi_h\|_{L^p(\Omega')}$$

$$\le 2\epsilon/3 + \|u - \phi\|_{L^p(\Omega'')}$$

$$\le \epsilon,$$

as required.

The final properties of u_h follow as in Proposition 1.6. ∎

1.4.5 The l^p Sequence Spaces, $1 \le p \le \infty$

Another family of spaces, with properties similar to the L^p spaces, is the l^p spaces of sequences. We will use these in Chapter 4 as examples when we discuss dual spaces.

Definition 1.19. *For $1 \le p < \infty$, l^p is the space of all infinite sequences $\{x_n\}_{n=1}^{\infty}$ such that the l^p norm (cf. (1.2))*

$$\|x\|_{l^p} = \left(\sum_{j=1}^{\infty} |x_j|^p \right)^{1/p}$$

is finite. l^{∞} consists of all bounded sequences, with norm

$$\|x\|_{l^{\infty}} = \sup_{j \in \mathbb{Z}^+} |x_j|.$$

(This is the reason for the notation for the norms on $\mathbb{R}^m$ in Section 1.2.)

One can check as above (Lemma 1.10) that $\|\cdot\|_{l^p}$ really is a norm, and it is relatively straightforward to show that l^p is complete.

Theorem 1.20. *l^p is complete.*

Proof. We give the proof for the case $1 \le p < \infty$; the proof for $p = \infty$ is simpler and is given as an exercise (Exercise 1.12). Let $x^{(n)}$ be a Cauchy sequence in l^p. Then given $\epsilon > 0$ there exists an N such that for all $n, m \ge N$,

$$\left\| x^{(n)} - x^{(m)} \right\|_{l^p} \le \epsilon,$$

i.e.

$$\sum_{i=1}^{\infty} \left| x_i^{(n)} - x_i^{(m)} \right|^p \le \epsilon^p. \tag{1.23}$$

In particular, for each fixed i the sequence $x_i^{(n)}$ is a Cauchy sequence and hence converges to a limit x_i. We clearly need to show that $x^{(n)} \to x = \{x_i\}$ with $x \in l^p$.

Certainly, from (1.23),

$$\sum_{i=1}^{k} \left| x_i^{(n)} - x_i^{(m)} \right|^p \le \epsilon^p,$$

and letting $m \to \infty$ shows that

$$\sum_{i=1}^{k} \left| x_i^{(n)} - x_i \right|^p \le \epsilon^p \tag{1.24}$$

for all $n \ge N$. Then the triangle inequality in l^p shows that

$$\left(\sum_{i=1}^{k} |x_i|^p \right)^{1/p} \le \epsilon + \left\| x^{(n)} \right\|_{l^p}. \tag{1.25}$$

Letting $k \to \infty$ in (1.25) shows that $x \in l^p$, and letting $k \to \infty$ in (1.24) shows that

$$\left\| x^{(n)} - x \right\|_{l^p} \le \epsilon$$

for all $n \ge N$. So $x^{(n)} \to x$, with $x \in l^p$. ∎

1.5 Hilbert Spaces

The space $\mathbb{R}^m$ not only has a notion of length, but it also has a notion of the angle between two elements, derived from the dot product $x \cdot y = \sum_{j=1}^{m} x_j y_j$. This is the most familiar example of an inner product.

An inner product (over $\mathbb{R}$) on a general vector space X is a map $(\cdot, \cdot) : X \times X \to \mathbb{R}$, such that

(i) $(\lambda x + \mu y, z) = \lambda(x, z) + \mu(y, z)$,
(ii) $(y, x) = (x, y)$ for all $x, y \in X$, and
(iii) $(x, x) \ge 0$ for all $x \in X$, with equality iff $x = 0$.

Note that there is a natural norm associated with the inner product,

$$\|x\| = (x, x)^{1/2}. \tag{1.26}$$

The Cauchy–Schwarz inequality,

$$|(x, y)| \leq \|x\|\|y\|,$$

follows almost immediately from the axioms and provides the triangle inequality for the norm defined by (1.26) (See Exercise 1.13).

A complete inner product space is called a *Hilbert space*. By defining a norm via (1.26), any Hilbert space is also a Banach space. This is not necessarily true the other way round, since the norm in a Hilbert space has to satisfy the parallelogram law

$$\|u + v\|^2 + \|u - v\|^2 = 2\|u\|^2 + 2\|v\|^2. \tag{1.27}$$

(For a proof of this equality see Exercise 1.14. In fact, this property characterises those Banach space norms that *can* be derived from an inner product.)

The two most important examples of infinite-dimensional Hilbert spaces are the sequence space l^2, whose inner product is a direct generalisation of that on $\mathbb{R}^m$,

$$(x, y) = \sum_{n=1}^{\infty} x_n y_n,$$

and the Lebesgue space $L^2(\Omega)$, for which the inner product is given by the integral

$$(f, g) = \int_\Omega f(x)g(x)\,dx.$$

Note that, for $p = q = 2$, Hölder's inequality (Lemma 1.9) becomes the Cauchy–Schwarz inequality

$$|(f, g)| \leq \|f\|_{L^2}\|g\|_{L^2}.$$

We will write $|f|$ for $\|f\|_{L^2}$ in what follows, being specific only when care is needed to distinguish this norm from the modulus. The context should usually make it clear which is meant, and generally we will write $|f|$ for the L^2 norm of the function f and $|f(x)|$ for the modulus of $f(x)$.

1.5.1 The Orthogonal Projection onto a Linear Subspace

If M is a subset of H, then the *orthogonal complement* of H, $M^\perp$, is given by

$$M^\perp = \{u \in H : (u, v) = 0 \text{ for all } v \in M\}.$$

When M is a closed linear subspace of H we can decompose any vector uniquely into an element of M plus an element of its orthogonal complement.

Proposition 1.21. *If M is a closed linear subspace of H, then every $x \in H$ has a unique decomposition as*

$$x = u + v, \qquad u \in M, \ v \in M^\perp.$$

Proof. We will choose u to be a point (or perhaps the point) in M such that

$$\|x - u\| = \delta \equiv \inf_{y \in M} \|x - y\|.$$

The first stage is to show that such a point exists.

There exists a sequence $u_n \in M$ such that

$$\|x - u_n\|^2 \leq \delta^2 + \frac{1}{n}. \tag{1.28}$$

We will show that $\{u_n\}$ is a Cauchy sequence. To this end, set $u = u_m - x$ and $v = u_n - x$ in the parallelogram law (1.27) to obtain

$$\|u_m - u_n\|^2 = 2\|u_n - x\|^2 + 2\|u_m - x\|^2 - 4\left\|x - \tfrac{1}{2}(u_n + u_m)\right\|^2.$$

Since M is a linear subspace, $\frac{1}{2}(u_n + u_m) \in M$, and so

$$\left\|x - \tfrac{1}{2}(u_n + u_m)\right\| \geq \delta.$$

It thus follows using (1.28) that

$$\|u_m - u_n\|^2 \leq \frac{2}{m} + \frac{2}{n},$$

and so $\{u_n\} \subset M$ is a Cauchy sequence, tending to some $u \in M$ since M is closed.

Now consider $v = x - u$; the claim is that $v \in M^\perp$, i.e. that

$$(v, y) = 0 \qquad \text{for all} \qquad y \in M.$$

Consider $\|x - (u - ty)\| = \|v + ty\|$; then

$$\Delta(t) = \|v + ty\|^2 = \|v\|^2 + 2t(v, y) + \|y\|^2.$$

We know from the construction of u that $\|v + ty\|$ is minimal when $t = 0$, and so $\Delta'(0) = 2(v, y) = 0$. Thus $v \in M^\perp$ is claimed.

Finally, the uniqueness follows easily: If $x = u_1 + v_1 = u_2 + v_2$, then $u_1 - u_2 = v_2 - v_1$, and so

$$|v_1 - v_2|^2 = (v_1 - v_2, v_1 - v_2) = (v_1 - v_2, u_2 - u_1) = 0,$$

since $u_1 - u_2 \in M$ and $v_2 - v_1 \in M^\perp$. ■

One can use this proposition to define the *orthogonal projection of x onto M, P_M,* by

$$P_M x = u.$$

Clearly $P_M^2 = P_M$, and it follows from the definition of u that

$$\|x\|^2 = \|u\|^2 + \|x - u\|^2,$$

thus ensuring that

$$\|P_M x\| \leq \|x\|:$$

the projection can only decrease the norm.

1.5.2 Bases in Hilbert Spaces

In $\mathbb{R}^m$ we can write any vector in terms of a finite set of basis elements, for example the coordinate basis

$$\{(1, 0, \ldots, 0), (0, 1, \ldots, 0), \ldots, (0, \ldots, 0, 1)\}.$$

We now want to investigate when we can find a countable basis for an infinite-dimensional Hilbert space, in terms of which to expand any point in H. A set of vectors is *orthogonal* (in H) if

$$(e_i, e_j) = 0, \qquad i \neq j,$$

and *orthonormal* (in H) if

$$(e_i, e_j) = \delta_{ij} = \begin{cases} 1, & i = j, \\ 0, & i \neq j. \end{cases}$$

Definition 1.22. *An orthonormal set $\{e_j\}$ is a basis for H if*

$$x = \sum_{j=1}^{\infty} (x, e_j) e_j \tag{1.29}$$

for every $x \in H$.

We will now give a characterisation of a basis. Note that the equality in (1.30) below is extremely useful,

Proposition 1.23. *Let $\{e_j\}$ be an orthonormal set. Then it is a basis for H iff*

$$\|x\|^2 = \sum_{j=1}^{\infty} (x, e_j)^2 \qquad \text{for all} \qquad x \in H. \tag{1.30}$$

Proof. If $\{e_j\}$ form a basis then for any finite n

$$\left\| \sum_{j=1}^{n} (x, e_j) e_j \right\|^2 = \sum_{j=1}^{n} (x, e_j)^2,$$

and (1.30) follows by taking limits, since the sum (1.29) converges in H.

Now suppose that (1.30) holds for every $x \in H$. We will set

$$Y = \left\{ x : x = \sum_{j=1}^{\infty} (x, e_j) e_j \right\}$$

and show that in fact $Y = H$, by showing that Y is closed and dense. To show that Y is closed, take a Cauchy sequence $\{y_n\}$ in Y. Since $Y \simeq l^2$ it is complete (see Theorem 1.20), so y^*, the limit of the y_n, lies in Y, and Y is closed.

Now suppose that Y is not dense. Then $H \setminus Y$ must contain a nonzero vector, and by Proposition 1.21 this implies that $Y^\perp$ must contain a nonzero vector. However, if $x \in Y^\perp$ then it is orthogonal to each e_j, so it follows from (1.30) that $\|x\| = 0$, and therefore $x = 0$, that is, $Y^\perp = \{0\}$, a contradiction.

Since Y is closed and dense in H, $Y = H$. ∎

We discussed separable Banach spaces briefly earlier in this chapter. Separable Hilbert spaces are particularly well behaved, as the following result shows. Although not all Hilbert spaces are separable (see Exercise 1.15), all those that we will come across in the remainder of this book are.

Proposition 1.24. *A Hilbert space is separable iff it has a countable basis.*

Proof. If a Hilbert space has a countable basis then we can construct a countable dense set by taking finite combinations of the basis elements with rational coefficients, and so it is separable.

If H is separable, let $\{x_n\}$ be a countable dense subset. We will use the Gram–Schmidt process to turn x_n into a countable orthonormal basis. First omit all those elements of $\{x_n\}$ that can be written as a linear combination of the preceding ones, and set

$$e_1 = \frac{x_1}{\|x_1\|}.$$

Then we can define our basis inductively by setting

$$y_n = x_n - \sum_{i=1}^{n-1} (x_n, e_i) e_i,$$

and then

$$e_n = \frac{y_n}{\|y_n\|}.$$

The linear span of the $\{e_n\}$ is the same as that of the x_n and so is dense in H. The argument of the previous proposition shows that the $\{e_n\}$ form a basis. ∎

By using these basis elements we can construct an isomorphism between any separable Hilbert space and l^2, so that in some sense l^2 is "the only" separable infinite-dimensional Hilbert space.

1.5.3 Noncompactness of the Unit Ball

Since l^2, and so any separable Hilbert space, shares many properties in common with $\mathbb{R}^m$, it is tempting to think of it as "$\mathbb{R}^\infty$." However, there is one very important caveat: unlike $\mathbb{R}^m$, closed bounded sets are not compact.

Proposition 1.25. *The unit ball in an infinite-dimensional Hilbert space is not compact.*

The same result is true, with a harder proof, for general infinite-dimensional Banach spaces.

Proof. For any sequence $\{w_n\}_{n=1}^{\infty}$ of orthonormal elements we have

$$\|w_n - w_m\|^2 = 2$$

if $n \neq m$. It follows that no subsequence of $\{w_n\}$ is Cauchy and thus the unit ball cannot be compact. ∎

It is extremely important to be vigilant about compactness when working in infinite-dimensional spaces. In much of what follows we will be developing techniques to cope with the fact that the unit ball is not compact, and this is what is largely responsible for the idea of *weak convergence* to be introduced in Chapter 4.

Exercises

1.1 Show that the product of a finite number of separable Banach spaces is also separable and that a linear subspace of a separable Banach space is separable. [Hint: Use the norm $\|\cdot\|_X + \|\cdot\|_Y$ on $X \times Y$.]

1.2 Show that if X is compact, then given $\epsilon > 0$, there exists a finite set of points $\{x_i\} \in X$ such that any point $x \in X$ can be approximated to within ϵ by one of the x_i: for each $x \in X$ there is an x_i such that

$$|x - x_i| \leq \epsilon.$$

(This is called a "finite ϵ-net.")

1.3 If Ω is bounded, what is the completion of $C_c^0(\Omega)$ in the supremum norm? Deduce that $C_c^0(\Omega)$ is not a Banach space with this norm. Treat similarly the case $\Omega = \mathbb{R}^m$.

1.4 There is no norm that makes $C^\infty(\overline{\Omega})$ into a Banach space. However, there are various subspaces of $C^\infty(\overline{\Omega})$ that are Banach spaces. For example, for any sequence $\mathbf{c} = \{c_n\}_{n=1}^{\infty}$ define the norm

$$\|f\|_{\mathbf{c}} = \sum_{n=1}^{\infty} c_n \|f\|_{C^n(\overline{\Omega})}.$$

Show that the subspace of $C^\infty(\overline{\Omega})$ consisting of all those f with $\|f\|_{\mathbf{c}}$ finite is a Banach space. [This idea is not just artificial. If $c_n = \tau^n/n!$, for some $\tau > 0$, then functions f with $\|f\|_{\mathbf{c}}$ finite are real analytic. See, for example, Theorem 2.12 in Renardy & Rogers (1992).]

1.5 Show that $C^{r,\gamma}(\overline{\Omega})$ is a Banach space (see Definition 1.5).

1.6   Show that if Ω is convex and bounded then any $C^1(\overline{\Omega})$ function is Lipschitz.

1.7   Show that if the function u in Proposition 1.6 is an element of $C^{0,\gamma}(\Omega)$ then so is u_h.

1.8   Prove the generalised Hölder inequality: if $p_1, \ldots, p_n$ are such that

$$\sum_{j=1}^{n} \frac{1}{p_j} = 1$$

and $f_j \in L^{p_j}(\Omega)$ then their product $f_1 \ldots f_n \in L^1(\Omega)$, with

$$\int_\Omega |f_1(x) \ldots f_n(x)| \, dx \leq \|f_1\|_{L^{p_1}} \ldots \|f_n\|_{L^{p_n}}. \qquad (1.31)$$

1.9   Use Hölder's inequality to obtain the L^p interpolation inequality,

$$\|u\|_{L^p} \leq \|u\|_{L^q}^{q(r-p)/p(r-q)} \|u\|_{L^r}^{r(p-q)/p(r-q)}$$

when $q < p < r$ and $u \in L^r(\Omega)$.

1.10 Show that given $s \in S(\Omega)$, $1 \leq p < \infty$, and $\epsilon > 0$, there exists an $f \in C_c^0(\Omega)$ such that

$$\|f - s\|_{L^p} < \epsilon.$$

1.11 Prove Hölder's inequality in the case $f \in L^1(\Omega)$, $g \in L^\infty(\Omega)$,

$$\left| \int_\Omega fg \, dx \right| \leq \|f\|_{L^1} \|g\|_\infty.$$

1.12 Show that l^∞ is complete.

1.13 By considering $\|x + \lambda y\|^2$ show that the Cauchy–Schwarz inequality holds. Deduce the triangle inequality.

1.14 Show that the parallelogram law

$$\|u + v\|^2 + \|u - v\|^2 = 2\|u\|^2 + 2\|v\|^2$$

holds in a Hilbert space.

1.15 The space $l^2(\Gamma)$ consists of real-valued functions on Γ with

$$\sum_{\gamma \in \Gamma} |f(\gamma)|^2 < \infty,$$

and with the inner product

$$(f, g)_{l^2(\Gamma)} = \sum_{\gamma \in \Gamma} f(q)g(q)$$

it is a Hilbert space. Show that $l^2(\Gamma)$ is not separable if Γ is uncountable.

Notes

The results in this chapter are all standard, although as presented they are taken from a variety of sources. As well as treatments in dedicated functional analysis texts, most graduate-level PDE books have some (more or less abbreviated) treatment of these topics.

The basic topology briefly reviewed in Section 1.1 can be found in Bollobás (1990), Kreyszig (1978), Rudin (1976), or Sutherland (1975), among many others. We have not mentioned metric spaces at all in this chapter; these are treated in detail in Sutherland's book but are weaker than normed spaces and will not be needed in what follows. A metric on X is a function $d_X : X \times X \to \mathbb{R}^+$ satisfying

(i) $d_X(x, x) = 0$ for all $x \in X$,

(ii) $d_X(x, y) = d_X(y, x)$ for all $x, y \in X$,

(iii) $d_X(x, z) \leq d_X(x, y) + d_X(y, z)$ for all $x, y, z \in X$.

Clearly a norm gives rise to a metric via $d(x, y) = \|x - y\|$.

The treatment of mollification follows that in Gilbarg & Trudinger (1983); this is an advanced monograph, which nonetheless contains some very clear expositions of such background material.

Priestley's book (1997) offers a very readable introduction to the Lebesgue integral, while the last chapter of Rudin (1976) is more concise and so more suited for anybody requiring some speedy revision of the subject. The treatment in Renardy & Rogers' lovely book (1992) does not require knowledge of the Lebesgue integral, instead defining $L^p(\Omega)$ as the completion of $C^0(\overline{\Omega})$ in the L^p norm (see Theorem 1.13). As discussed in the Introduction, it seemed best to assume some familiarity with the Lebesgue theory in this book.

Adams (1975) covers all the properties of C^r and L^p spaces given here, and many others, in two very nicely presented introductory chapters. The appendices in Evans (1998) are a useful reference, providing some detailed proofs. Those in Taylor (1996) are also useful, although the proofs are in some cases extremely concise.

Unsurprisingly, Hilbert spaces form the main subject matter of Young's nice textbook *Hilbert Spaces* (1988).

2

Ordinary Differential Equations

In this chapter we will show that the solutions of ordinary differential equations (ODEs) can be used to define dynamical systems. The first major aim of this book is to follow through a similar programme for partial differential equations. We will assume some familiarity with the basic theory of dynamical systems, such as given, for example, by Glendinning's book (1994).

We will consider an ordinary differential equation (ODE) defined on the whole* of $\mathbb{R}^m$,

$$dx/dt = f(x), \qquad x \in \mathbb{R}^m, \tag{2.1}$$

and consider the solutions with specified initial condition $x(0) = x_0$. We want to show that corresponding to each x_0 there is a unique solution $x(t; x_0)$, which is C^1 in t (so that (2.1) makes sense) and varies continuously with respect to x_0.

More formally, we want to show the existence and uniqueness of solutions of (2.1) and their continuous dependence on initial conditions. Unlike the theory for partial differential equations (PDEs), which usually has to be tailored in some way to the particular equation under consideration, these problems can be solved in a straightforward way for a large, general class of ODEs. However, these methods can be viewed as prototypes for the arguments that we will use later on for PDEs.

Once this is done, we can use the solutions of (2.1) to construct a dynamical system on $\mathbb{R}^m$. It is convenient to express the solution in terms of a "solution operator" $T(t)$. This is defined, for any $t \in \mathbb{R}$, by

$$T(t)x_0 = x(t; x_0), \tag{2.2}$$

---

* Ordinary differential equations (ODEs) on finite-dimensional smooth manifolds also give rise to finite-dimensional systems, but considering equations defined on the whole of $\mathbb{R}^m$ avoids the machinery of differential geometry while illustrating the methods we will apply later.

where $x(t; x_0)$ is the solution of (2.1) with $x(0) = x_0$. It follows from the existence and uniqueness of solutions that $T(t)$ is well defined and satisfies

$$T(0) = 0 \quad \text{and} \quad T(t)T(s) = T(t+s). \tag{2.3}$$

Furthermore, since the solutions $x(t; x_0)$ depend continuously on both t and x_0,

$$T(t)x_0 \text{ depends continuously on } t \text{ and } x_0.$$

Formally, the pair consisting of the *phase space* $\mathbb{R}^m$ and the *flow* $T(t)$, $(\mathbb{R}^m, T(\cdot))$, is our dynamical system.

2.1 Existence and Uniqueness – A Fixed-Point Method

First, we want to investigate the existence and uniqueness of solutions of (2.1). Rather than analyse (2.1) directly, it is in fact more convenient to work with an alternative form of the equation. The following lemma shows that solving (2.1) and solving the integral equation

$$x(t) = x_0 + \int_0^t f(x(s))\, ds$$

are equivalent problems.

Lemma 2.1. *Assume that* $f : \mathbb{R}^m \to \mathbb{R}^m$ *is continuous. Then solutions of the differential equation*

$$dx/dt = f(x), \qquad x(0) = x_0 \tag{2.4}$$

are equivalent to continuous solutions of the integral equation

$$x(t) = x_0 + \int_0^t f(x(s))\, ds. \tag{2.5}$$

Proof. That solutions of (2.4) give rise to solutions of (2.5) follows by integration from 0 to t. If $x(t)$ is continuous, $f(x(t))$ is continuous, and then it follows from the fundamental theorem of calculus that

$$F(t) = \int_0^t f(x(s))\, ds$$

is C^1 and has derivative $\dot{F}(t) = f(x(t))$. Clearly, (2.5) satisfies the initial condition in (2.4). ∎

2.1.1 The Contraction Mapping Theorem

The advantage of expressing the problem like this is that it is in a convenient form in which to apply the contraction mapping theorem. This result guarantees (under certain conditions, of course) that a map h has a fixed point, that is, there exists an x^* such that $h(x^*) = x^*$.

Theorem 2.2 (Contraction Mapping Theorem). *Let X be a closed subset of a Banach space $(Y, \|\cdot\|)$, and $h : X \to X$ a function satisfying*

$$\|h(x) - h(y)\| \le k\|x - y\|, \qquad \text{for all} \qquad x, y \in X, \qquad (2.6)$$

where $k < 1$ (we say that h is a contraction, or contraction mapping, on X): Then h has a unique fixed point in X.

This result can be generalised in various ways (see Exercises 2.1 and 2.2). It is also valid for complete metric spaces, where we simply replace the contraction condition (2.6) with $d_X(h(x), h(y)) \le k\, d_X(x, y)$ [see Sutherland (1975), for example].

Proof. We show that the iterates of any point $x \in X$ under h form a Cauchy sequence in X. We write $h^n(x)$ for h applied n times to x:

$$h^2(x) = h(h(x)), \qquad h^n(x) = \underbrace{h(\cdots(h(x))\cdots)}_{n \text{ times}}.$$

Now,

$$\|h^{n+1}(x) - h^n(x)\| \le k\|h^n(x) - h^{n-1}(x)\|$$
$$\le k^n \|h(x) - x\|,$$

and so, if $m > n$,

$$\|h^m(x) - h^n(x)\| \le \sum_{j=n}^{m-1} k^j \|h(x) - x\|$$
$$\le \sum_{j=n}^{\infty} k^j \|h(x) - x\|$$
$$= \frac{k^n}{1 - k} \|h(x) - x\|.$$

Thus $\{h^n(x)\}$ is a Cauchy sequence, and so it converges to some $x^* \in X$, since Y is complete and X is a closed subset of Y.

Now, $h^n(x) \to x^*$ as $n \to \infty$, and so does $h^{n+1}(x)$; since h is continuous,

$$h(x^*) = h\left(\lim_{n\to\infty} h^n(x)\right) = \lim_{n\to\infty} h(h^n(x)) = \lim_{n\to\infty} h^{n+1}(x) = x^*,$$

and so x^* is a fixed point of h.

Finally, if x^* and y^* are both fixed points of h, then

$$|x^* - y^*| = |h(x^*) - h(y^*)| \le k|x^* - y^*|,$$

which is a contradiction unless $x^* = y^*$. So the fixed point is unique. ∎

Despite its simplicity, this is a very powerful theorem, and we will use it now to prove the existence of unique solutions for ODEs when $f(x)$ is a Lipschitz function. We will take f to be a locally Lipschitz, which means that

$$|f(x) - f(y)| \le L(B)|x - y| \tag{2.7}$$

for x, y in any bounded set B. [We could write "$f \in C^{0,1}_{\text{loc}}(\mathbb{R}^m)$" (cf. the definition of $L^p_{\text{loc}}(\Omega)$ in Section 1.4.4). However, this seems much less clear than (2.7).]

2.1.2 Local Existence for Lipschitz f

We will show existence and uniqueness for a "small" time interval $[0, T]$, where we allow T to depend on the initial condition x_0. Since this ensures existence only near to $t = 0$, we call such a result a *local* existence result. We show how to prove "global" existence (i.e. for all $t \ge 0$) in the next section.

Theorem 2.3 (Local Existence). *Suppose that f satisfies*

$$|f(x) - f(y)| \le L(B)|x - y|$$

for x, y in any bounded set B. Then there exists a $T = T(x_0)$ such that the equation

$$dx/dt = f(x), \qquad x(0) = x_0 \tag{2.8}$$

has a unique solution on $[0, T]$.

Proof. We use the integral form of (2.8) provided by Lemma 2.1, and we consider the integral operator $\mathcal{J}$ given by

$$(\mathcal{J}x)(t) = x_0 + \int_0^t f(x(s))\,ds.$$

If we take a trial solution $x(t)$ that is an element of $C([0, T]; \mathbb{R}^m)$ (Lemma 2.1 shows that this is the appropriate space in which we expect the solution to lie), then $[\mathcal{J}x](t)$ is another function of t. We will show that $\mathcal{J}$ is a contraction mapping from an appropriate subset of $C([0, T]; \mathbb{R}^m)$ into itself.

Ideally we would like to make sure that all the "trial solutions" $x(t)$ used in our proof lie in the same bounded subset of $\mathbb{R}^m$, so that we can then treat f as if it were uniformly Lipschitz: we know that if $|x|, |y| \leq k$ then

$$|f(x) - f(y)| \leq L|x - y| \qquad \text{with} \qquad L = L(\{|x| \leq k\}). \qquad (2.9)$$

To do this we will show that $\mathcal{J}$ is a contraction mapping on the subspace X of $C([0, T]; \mathbb{R}^m)$, which consists of functions that satisfy a uniform bound,

$$X = \{x : x \in C^0([0, T]; \mathbb{R}^m); \|x\|_\infty \leq k\}.$$

The appropriate choices of T and k will be specified below. Note that for any choice of T and k, the space X is a closed subset of $C^0([0, T]; \mathbb{R}^m)$ equipped with the standard supremum norm,

$$\|x\|_\infty = \sup_{t \in [0, T]} |x(t)|.$$

We choose

$$T = \frac{1}{|f(x_0)| + 2Lk}. \qquad (2.10)$$

The reason for this choice of T will become obvious from the algebra that now follows, which will also tell us the correct choice of k.

To show that $\mathcal{J}$ maps X into itself we first show that

$$\|x\|_\infty \leq k \qquad \Rightarrow \qquad \|\mathcal{J}x\|_\infty \leq k. \qquad (2.11)$$

Write

$$|(\mathcal{J}x)(t)| = \left| x_0 + \int_0^t f(x(s)) \, ds \right|$$

$$\leq |x_0| + \int_0^t |f(x_0)| + L|x(t) - x_0| \, ds$$

$$\leq |x_0| + T(|f(x_0)| + 2L\|x\|_\infty)$$

$$\leq |x_0| + T(|f(x_0)| + 2Lk)$$

$$= |x_0| + 1.$$

In the light of this, we choose $k = |x_0| + 1$ to ensure (2.11).

We now show in addition that $\mathcal{J}$ maps $C([0, T]; \mathbb{R}^m)$ into itself to guarantee that $\mathcal{J}$ is a map from X into itself. Take $x \in X$ (so we have $\|x\|_\infty \leq k$), fix $t \in [0, T]$, and let τ vary in $[0, T]$; then, since

$$\sup_{\{x : |x| \leq k\}} |f(x)| \leq M$$

for some M, we have

$$|(\mathcal{J}x)(t) - (\mathcal{J}x)(\tau)| = \left| \int_0^t f(x(s)) \, ds - \int_0^\tau f(x(s)) \, ds \right|$$

$$= \left| \int_\tau^t f(x(s)) \, ds \right|$$

$$\leq M|t - \tau|,$$

so that $\mathcal{J}x$ is continuous at every $t \in [0, T]$.

We now show that $\mathcal{J}$ is a contraction mapping. We can use L as defined in (2.9) for the Lipschitz constant of f, uniform over $\{|x| \leq k\}$. Indeed,

$$\left| (\mathcal{J}x)(t) - (\mathcal{J}y)(t) \right| = \left| \int_0^t f(x(s)) - f(y(s)) \, ds \right|$$

$$\leq \int_0^t L|x(s) - y(s)| \, ds$$

$$\leq \int_0^t L\|x - y\|_\infty \, ds$$

$$= Lt\|x - y\|_\infty.$$

Thus

$$\|\mathcal{J}x - \mathcal{J}y\|_\infty \leq LT \|x - y\|_\infty.$$

Now, by the choice of T,

$$LT = \frac{L}{|f(x_0)| + 2L(1 + |x_0|)} \leq \frac{1}{2},$$

and so $\mathcal{J}$ is a contraction on X. Theorem 2.2 therefore applies, ensuring a unique fixed point for $\mathcal{J}$ in X. Lemma 2.1 then ensures a unique solution of the differential equation (2.8). ■

Note that the proof is also valid when $f(x)$ is replaced by the time-dependent function $f(x, t)$, provided that

$$|f(x, t) - f(y, t)| \leq L(B)|x - y| \qquad \text{for all} \qquad x, y \in B, \text{ bounded,}$$

with the Lipschitz constant independent of t.

We now give conditions under which we can prove that the solution exists for all time.

2.2 Global Existence

Theorem 2.3 is limited in that it ensures existence for only a bounded time interval $[0, T]$, where T in fact depends (via (2.10)) on the initial condition. To discuss whether or not solutions exist for all time, we make use of the notion of the *maximal interval of existence* of a solution of a differential equation.

We say that $[0, T^*)$ is the maximal interval of existence for a solution $x(t)$ of (2.1) if there is no solution $y(t)$ of (2.1) on a longer time interval $[0, T^+)$, $T^+ > T^*$, with $x(t) = y(t)$ on $[0, T^*)$. In other words, we cannot extend the solution $x(t)$ beyond the time T^* so that it remains a solution of (2.1). If the solution exists for all time then its maximal interval of existence is $[0, \infty)$.

To show existence for all time ("global existence"), it is enough to ensure that the solution does not "blow up" in finite time.

Lemma 2.4. *A solution $x(t)$ of Equation (2.1) has a finite maximal interval of existence $[0, T^*)$ iff $|x(t)| \to \infty$ as $t \to T^*$.*

Proof. Clearly, if $|x(t)| \to \infty$ as $t \to T^*$ then there can be no continuous extension of $x(t)$ to an interval containing T^*. Conversely, suppose that the solution is bounded on $[0, T^*)$, say

$$\sup_{t \in [0, T^*)} |x(t)| \le k.$$

It follows, since f is continuous on $\mathbb{R}^m$, that $f(x(t))$ is bounded on $[0, T^*)$. The integral form of the equation

$$x(t) = x_0 + \int_0^t f(x(s)) \, ds$$

then shows that the limit of $x(t)$ as $t \to T^*$ exists. Clearly this limit must satisfy

$$\lim_{t \to T^*} |x(t)| \le k.$$

Now we can use Theorem 2.3 to extend the interval of existence to $[0, T^* + \delta]$ for some $\delta > 0$, contradicting the maximality of T^*. ∎

Note that it does not follow that a solution that exists for all time must be bounded (e.g. the solution $x(t) = e^t$ of $dx/dt = x$ with $x(0) = 1$). Nor is global existence automatic. The solution of

$$dx/dt = x^2, \qquad x(0) = x_0 > 0$$

is

$$x(t) = \frac{x_0}{1 - x_0 t},$$

and so $x(t) \to \infty$ as $t \to x_0^{-1}$.

We now discuss what happens when the nonlinearity is known only to be continuous. We will be able to prove existence, but not uniqueness.

2.3 Existence but No Uniqueness – An Approximation Method

When the nonlinearity $f(x)$ is continuous but not locally Lipschitz, our approach will be to approximate the problem

$$dx/dt = f(x), \qquad x(0) = x_0 \tag{2.12}$$

by a sequence of problems

$$dx_n/dt = f_n(x_n), \qquad x_n(0) = x_0,$$

where every $f_n(x)$ is Lipschitz. We can already find a unique solution x_n to each member of this sequence by using Theorem 2.3. The idea is to show that some subsequence of these solutions converges to give a solution of (2.12). We will use a very similar method to construct solutions of PDEs in Chapters 7–9.

2.3.1 The Arzelà–Ascoli Theorem

The result we will use to extract this subsequence is known as the Arzelà–Ascoli theorem. This characterises the compact subsets of $C^0(X, \mathbb{R}^m)$ as sets of bounded equicontinuous functions. If $\mathcal{S}$ is such a subset and $\{f_n\}$ is a sequence in $\mathcal{S}$, then there is a subsequence of $\{f_n\}$ that is uniformly convergent.

Theorem 2.5 (Arzelà–Ascoli). *Let X be a compact subset of $\mathbb{R}^{m_1}$, and let $\{f_n\}$ be a sequence of continuous functions from X into $\mathbb{R}^{m_2}$. If f_n is uniformly bounded, that is, there exists an M such that*

$$\|f_n\|_\infty \leq M \qquad \text{for all } n,$$

and equicontinuous ("uniformly uniformly continuous"), that is, for every $\epsilon > 0$

there exists a $\delta > 0$, independent of n, such that

$$|x - y| \leq \delta \qquad \text{implies that} \qquad |f_n(x) - f_n(y)| \leq \epsilon,$$

then $\{f_n\}$ has a subsequence that converges uniformly on X.

A prime example of a family of equicontinuous functions is the space $C^{0,1}(\overline{\Omega})$ of Lipschitz functions on $\overline{\Omega}$.

Proof. First we use the compactness of X to find a countable set of points $\{x_i\}$ and an increasing sequence of integers $\{N_j\}$ such that, for each n,

$$|x - x_i| \leq 2^{-n} \qquad \text{for some} \qquad 1 \leq i \leq N_n$$

(see Exercise 2.3).

Now, since $f_n(x_1)$ is a uniformly bounded sequence, the Bolzano–Weierstrass theorem guarantees that $\{f_n\}$ has a subsequence $\{f_{n_{1j}}\}$ such that $f_{n_{1j}}(x_1)$ converges.

Since $f_{n_{1j}}(x_2)$ must be uniformly bounded, there is a further subsequence $\{f_{n_{2j}}\}$ such that $f_{n_{2j}}(x_2)$ also converges. One can continue in this way to select subsequences $\{f_{n_{kj}}\}$ such that

$$f_{n_{kj}}(x_i) \qquad \text{converges for all} \qquad 1 \leq i \leq k.$$

The trick now, which we will see again later in other proofs of compactness properties, is to take the "diagonal sequence" $g_j = f_{n_{jj}}$. For any k, $\{g_j\}$ is a subsequence of $\{f_{n_{kj}}\}$ from the kth term on, and so $g_j(x_k)$ converges as $j \rightarrow \infty$ at every x_k.

We now use this convergence over the dense subset $\{x_k\}$ along with the equicontinuity of the functions $\{f_n\}$ to show that $\{g_j\}$ is in fact a Cauchy sequence in the supremum norm and hence converges uniformly on X. Indeed, we know that given an $\epsilon > 0$, there is a $\delta > 0$ such that

$$|x - y| \leq \delta \qquad \text{implies that} \qquad |g_j(x) - g_j(y)| \leq \epsilon/3.$$

By construction of the x_i, there exists an M such that for every $x \in X$ there is an x_i with $i \leq M$ such that

$$|x - x_i| \leq \delta. \tag{2.13}$$

Now take N large enough such that

$$|g_m(x_i) - g_n(x_i)| \leq \epsilon/3 \qquad \text{for} \qquad m, n > N, \qquad 1 \leq i \leq M,$$

which we can do since $\{g_j(x_k)\}$ converges for each individual x_k. For any given x, choose the x_i that satisfies (2.13), and then

$$|g_m(x) - g_n(x)|$$
$$\leq |g_m(x) - g_m(x_i)| + |g_m(x_i) - g_n(x_i)| + |g_n(x_i) - g_n(x)|$$
$$\leq \epsilon/3 + \epsilon/3 + \epsilon/3 = \epsilon.$$

Thus $\{g_j\}$ is a Cauchy sequence in the sup norm and hence is uniformly convergent. ■

2.3.2 Local Existence for Continuous f

We now use this theorem to give an existence result for the case when we know only that f is continuous.

Theorem 2.6. *Let $f(x)$ be a continuous function. Then there exists a $T > 0$ such that the equation*

$$dx/dt = f(x), \qquad x(0) = x_0 \tag{2.14}$$

has at least one solution on $[0, T]$.

Proof. We approximate a solution of (2.14) by a sequence of solutions of uniformly Lipschitz equations, for which Theorem 2.3 guarantees a solution. The only complication is ensuring that solutions are all defined on the same interval of existence. To this end, note that for $|x - x_0| \leq R$, one can guarantee that $|f(x)| \leq M(R)$. Choose some $R > 0$, and set

$$T = \frac{R}{2M(R)}; \tag{2.15}$$

we could choose R to maximise the time T, but any choice will do.

Now, using Proposition 1.6, we approximate $f(x)$ by a sequence of Lipschitz functions f_n so that

$$\sup_{|x-x_0|\leq R} \|f_n - f\|_\infty \to 0.$$

Note that for some n_0 we have

$$\sup_{|x-x_0|\leq R} |f_n(x)| \leq 2M(R) \qquad \text{for all} \qquad n \geq n_0.$$

Now consider the solutions $x_n(t)$ of the equations

$$dx_n/dt = f_n(x_n), \qquad x_n(0) = x_0; \qquad (2.16)$$

since f_n is Lipschitz, the solutions of (2.16) exist locally, by Theorem 2.3.

Using the integral formulation from Lemma 2.1, we have, given our choice of T in (2.15),

$$\sup_{t\in[0,T]} |x_n(t) - x_0| \le R,$$

and so Lemma 2.4 shows that all the solutions exist on $[0, T]$. Since we can also show similarly that

$$|x_n(s) - x_n(t)| \le 2M(R)|s - t|,$$

the $\{x_n\}$ is a uniformly bounded equicontinuous sequence of functions from $[0, T]$ into $\mathbb{R}^m$. As such, the Arzelà–Ascoli theorem guarantees that they have a uniformly convergent subsequence, $\{x_{n_j}\}$, with

$$\sup_{t\in[0,T]} |x_{n_j}(t) - x^*(t)| \to x^*$$

as $j \to \infty$.

The function $x^*(t)$ is clearly a candidate for a solution of (2.14); to show that it is one, consider the integral form of (2.14) from Lemma 2.1:

$$x_{n_j}(t) = x_0 + \int_0^t f_{n_j}(x_{n_j}(s))\,ds.$$

Since x_{n_j} converges uniformly to x^* and f_{n_j} converges uniformly to f, all the terms converge uniformly on $[0, T]$ to give

$$x^*(t) = x_0 + \int_0^t f(x^*(s))\,ds,$$

so that x^* is indeed a solution of (2.14) on $[0, T]$. ∎

The lack of uniqueness is not just an artefact of the proof, as the following simple example shows. The equation

$$dx/dt = x^{1/2}, \qquad x(0) = 0$$

has, for $t \ge 0$, an infinite family of solutions that can be parametrised by $c \ge 0$,

$$x_c(t) = \begin{cases} 0, & 0 \le t \le c, \\ (t - c)^2/4, & t > c. \end{cases}$$

One property of the set of all possible solutions is discussed in Exercise 2.4.

2.4 Differential Inequalities

To apply Lemma 2.4 to show that the solutions we obtained in Theorem 2.3 or 2.6 in fact exist for all time, we have to obtain bounds on the solution. This is exactly what we will try to do for the PDE case. To show that such a blowup cannot happen, we need to be able to control $|x(t)|$. To do this, the usual approach is to use the original differential equation for $x(t)$ to produce a differential *inequality* for $|x(t)|$.

A differential inequality is just a differential equation involving an inequality rather than an equality. Since we often need to obtain estimates on the size of solutions of differential equations, the ability to "solve" differential inequalities is one of our main tools.

To cover a wide range of possible applications, we treat the case of a differential inequality given in terms of a right-hand derivative. The right-hand derivative $(d/dt)_+ x$ of a function $x : [0, T] \to \mathbb{R}$ is

$$\frac{d}{dt}_+ x = \lim_{h \downarrow 0^+} \frac{x(t+h) - x(t)}{h}.$$

We consider the differential inequality

$$\frac{d}{dt}_+ x \le f(x, t). \tag{2.17}$$

Clearly if we can treat this case then we can treat the inequality with $(d/dt)_+$ replaced by the standard derivative. However, the ability to treat (2.17) will occasionally be useful in what follows, mainly because of the results of Lemma 2.9 and Exercise 2.5.

We will show that if x satisfies (2.17) then it is bounded by the solution of the corresponding differential equality.

Lemma 2.7. *Let $x(t) \in \mathbb{R}$ satisfy the differential inequality*

$$\frac{d}{dt}_+ x \le f(x, t) \tag{2.18}$$

and let $y(t) \in \mathbb{R}$ be the unique solution of the differential equation

$$dy/dt = f(y, t), \tag{2.19}$$

with $x(0) \le y(0)$. Then $x(t) \le y(t)$ on $[0, T]$.

Note that the lemma contains an implicit assumption on the term $f(x, t)$, since the solution of (2.19) is required to be unique. We can circumvent this if we require $y(t)$ to be the maximal solution of (2.19) (see Hartman (1973) for a discussion of maximal solutions).

Proof. Consider the equations

$$dy_n/dt = f(y_n, t) + \frac{1}{n}, \qquad y_n(0) = y(0).$$

The first step is to show that, for each n, $x(t) \leq y_n(t)$ on $[0, T]$. If this is not the case then there must exist some $t_1 \in (0, T)$ such that $x(t_1) > y_n(t_1)$, and so there is a largest $t_2 \in [0, t_1)$ where $x(t_2) = y_n(t_2)$. Since we must then have $x(t) > y_n(t)$ for all $t \in (t_2, t_1)$, it follows that

$$\frac{x(t_2 + h) - x(t_2)}{h} > \frac{y_n(t_2 + h) - y_n(t_2)}{h}$$

for all $h > 0$ and, in particular, that

$$\frac{d}{dt_+} x(t_2) > f(y_n(t_2), t_2) + \frac{1}{n} = f(x(t_2), t_2) + \frac{1}{n},$$

contradicting (2.18). Thus, for each n, $x(t) \leq y_n(t)$ on $[0, T]$.

Now, observe that the functions $y_n(t)$ satisfy the conditions of the Arzelà–Ascoli theorem (Theorem 2.5), and so they have a subsequence that converges uniformly on $[0, T]$ to some $y^*(t)$, which must be a solution of (2.19) using the argument in the proof of Theorem 2.6. Since the only solution of (2.19) is $y(t)$ (by assumption), we must have $x(t) \leq y(t)$ on $[0, T]$, as claimed. ■

We can use this lemma to deduce the following result, known as Gronwall's inequality. This is central to the estimates we derive in what follows. In particular we will make use of the simplest case (the estimate in (2.21)) repeatedly.

Lemma 2.8 (Gronwall's Inequality). *Let $x(t) \in \mathbb{R}$ satisfy the differential inequality*

$$\frac{d}{dt_+} x \leq g(t)x + h(t).$$

Then

$$x(t) \leq x(0) \exp[G(t)] + \int_0^t \exp[G(t) - G(s)]h(s)\, ds, \qquad (2.20)$$

where

$$G(t) = \int_0^t g(r)\, dr.$$

In particular, if a and b are constants and

$$\frac{d}{dt}_+ x \le ax + b,$$

then

$$x(t) \le \left(x_0 + \frac{b}{a}\right)e^{at} - \frac{b}{a}. \tag{2.21}$$

Proof. The right-hand side of (2.20) is simply the solution of the differential equation

$$dy/dt = g(t)y + h(t),$$

solved using the integrating factor $\exp[-G(t)]$. The result follows from Lemma 2.7, and Equation (2.21) is simply a special case. ∎

The result of the following lemma allows us to estimate $(d/dt)_+|x|$ very easily.

Lemma 2.9. *If $x \in C^1([0, T]; \mathbb{R}^m)$ then $(d/dt)_+|x|$ exists, and*

$$\left|\frac{d}{dt}_+ |x|\right| \le |\dot{x}|. \tag{2.22}$$

Proof. First note that if $x, u \in \mathbb{R}^m, 0 < \theta \le 1$, and $h > 0$ then

$$|x + \theta hu| - |\theta x + \theta hu| \le (1 - \theta)|x|.$$

It follows that

$$\frac{|x + \theta hu| - |x|}{\theta h} \le \frac{|x + hu| - |x|}{h},$$

and so the expression

$$\frac{|x + hu| - |x|}{h} \tag{2.23}$$

is nondecreasing in h. Since it is bounded below by $-|u|$, the limit of (2.23) as $h \to 0$ must exist. If $x(t)$ is C^1 and we set $u = \dot{x}(t)$, then this implies that

$$\lim_{h \downarrow 0^+} \frac{|x(t) + h\dot{x}(t)| - |x(t)|}{h}$$

exists and is bounded above by $|\dot{x}(t)|$. Now,

$$\left|[|x(t+h)|-|x(t)|]-[|x(t)+h\dot{x}(t)|-|x(t)|]\right|$$
$$=\left||x(t+h)|-|x(t)+h\dot{x}(t)|\right|$$
$$\le|x(t+h)-x(t)-h\dot{x}(t)|$$
$$=o(h),$$

where $o(h)$ is such that $o(h)/h \to 0$ as $h \to 0$. Therefore

$$\frac{d}{dt_+}x(t)=\lim_{h\downarrow 0^+}\frac{|x(t)+h\dot{x}(t)|-|x(t)|}{h},$$

and inequality (2.22) follows immediately. ∎

As an example, suppose that $f(x)$ is a globally Lipschitz function, so that $|f(x)| \le |f(0)| + L|x|$. Then it follows from Lemma 2.9 that

$$\frac{d}{dt_+}|x| \le |\dot{x}| = |f(x)| \le |f(0)| + L|x|.$$

The simpler version of Gronwall's inequality ((2.21) from Lemma 2.8) now shows that

$$|x(t)| \le \left(|x(0)|+\frac{|f(0)|}{L}\right)e^{Lt}.$$

In particular, we see that $x(t)$ is bounded on any bounded interval $[0, T]$, and so it follows from Lemma 2.4 that the solution must exist for all time. Thus all equations with globally Lipschitz nonlinearities enjoy global existence.

2.5 Continuous Dependence on Initial Conditions

Finally, we investigate whether or not the solutions of the equation depend in a continuous way on the initial condition. The first, simple, result deals with equations that are globally Lipschitz, and it uses the Gronwall lemma. We will then give a much more general result that relies on the Arzelà–Ascoli theorem.

Proposition 2.10. *Let $f(x)$ be a Lipschitz function with*

$$|f(x)-f(y)| \le L|x-y|.$$

Then if x(t) and y(t) are solutions of $\dot{x} = f(x)$, with initial conditions x_0 and y_0 respectively, we have

$$|x(t) - y(t)| \leq |x_0 - y_0|e^{Lt}. \tag{2.24}$$

Exercise 2.7 provides a similar estimate when we keep x_0 fixed but change f. Clearly we could also combine both perturbations into one result.

Proof. Write $z(t) = x(t) - y(t)$, and then

$$\frac{dz}{dt} = f(x) - f(y),$$

so that, by Lemma 2.9,

$$\frac{d}{dt_+}|z| \leq |f(x) - f(y)| \leq L|x - y| = L|z|.$$

Gronwall's inequality (Lemma 2.8) now gives (2.24). ∎

Note that the argument of this proposition implies the uniqueness result of Theorem 2.3, since if $x(t)$ and $y(t)$ both solve

$$dx/dt = f(x), \qquad x(0) = 0$$

on $[0, T]$, then $x(t)$ and $y(t)$ are bounded on $[0, T]$, and so

$$|f(x(t)) - f(y(t))| \leq L|x(t) - y(t)| \qquad \text{for all} \qquad t \in [0, T]$$

for some $L = L(T)$. The above argument then shows that $x(0) = y(0)$ implies $x(t) = y(t)$ for all $t \in [0, T]$, which is uniqueness.

In fact we can show that if $f(x)$ is continuous and solutions are unique then they *must* depend continuously on their initial conditions. Although this result will work for all continuous right-hand sides and so includes the case of Proposition 2.10, the result there gives an explicit estimate of the rate of separation of solutions (2.24).

To treat existence for general continuous right-hand sides we used the Arzelà–Ascoli theorem; it is therefore no surprise that it is a related result that allows us to treat continuous dependence for such equations.

Lemma 2.11. *Let the assumptions of the Arzelà–Ascoli theorem (Theorem 2.5) hold. If all convergent subsequences converge to the same limit, then the sequence itself is uniformly convergent to that limit.*

Proof. Suppose that all convergent subsequences converge to $g(x)$, but the sequence itself does not converge to $g(x)$. Then there must exist an $\epsilon > 0$ and n_j with $n_j \to \infty$ such that

$$\|f_{n_j} - g\|_\infty \geq \epsilon. \tag{2.25}$$

However, $\{f_{n_j}\}$ is a sequence satisfying the conditions of the Arzelà–Ascoli theorem, and so it *must* have a convergent subsequence. By assumption this converges to g, contradicting (2.25). ∎

We can now see in the final theorem of this chapter that uniqueness and continuous dependence are intimately related. For simplicity, we assume that f is globally bounded, although such an assumption can be removed.

Theorem 2.12. *Suppose that $f(x)$ is continuous and globally bounded and that the solutions of*

$$dx/dt = f(x), \qquad x(0) = x_0$$

are unique and defined for all t. Then the solutions $x(t; x_0)$ depend continuously on x_0, and the convergence is uniform on bounded intervals of time:

$$\sup_{t\in[0,T]} |x(t; y_0) - x(t; x_0)| \to 0 \qquad as \qquad y_0 \to x_0.$$

Proof. Choose some $T > 0$ and consider the solutions to a set of problems

$$dx_n/dt = f(x_n), \qquad x(0) = x_0^{(n)},$$

where $x_0^{(n)} \to x_0$. For large enough n we know that $|x_0^{(n)}| \leq |x_0| + 1$, and since $f(x)$ is uniformly bounded (by M, say), we have

$$|x_n(t)| \leq |x_0| + 1 + MT \qquad \text{and} \qquad |x_n(t) - x_n(s)| \leq M|t - s|$$

for all $s, t \in [0, T]$. The solutions $x_n(t)$ satisfy the conditions of the Arzelà–Ascoli Theorem 2.5, and we know that any convergent subsequence will give a solution of

$$dx/dt = f(x), \qquad x(0) = x_0 \tag{2.26}$$

on $[0, T]$. However, the solution of (2.26) is unique by assumption, and so every convergent subsequence must have the same limit. By Lemma 2.11, this implies that

$$x_n(t) \to x(t)$$

uniformly for $t \in [0, T]$. ∎

2.6 Conclusion

We have shown (Theorem 2.3) that locally Lipschitz equations possess, locally, unique solutions. If we can show that these do not blow up in finite time, then in fact the solutions are globally defined (i.e. for all time). Proposition 2.10 guarantees that these solutions depend continuously on their initial conditions. In this case, we can use the solutions of the equation to define a dynamical system on $\mathbb{R}^m$, as outlined in (2.2) and (2.3).

Proving the corresponding existence and uniqueness results for PDEs is a much more involved task. As discussed in the introduction, there is no general theory that can treat these questions for all possible choices of PDE. We will have to tailor our treatment of the two examples in Chapters 8 and 9 to each particular equation.

Before we start to discuss existence and uniqueness for PDEs, we need to develop more functional analytic tools to formulate the problem precisely.

Exercises

2.1 Show that the conclusion of the contraction mapping Theorem 2.2 (a unique fixed point for h) remains true if the condition that h is a contraction on X is replaced with the assumption that h^n is a contraction on X for some $n > 1$.

2.2 Show that one cannot replace the condition of the contraction mapping theorem with

$$\|h(x) - h(y)\| < \|x - y\|$$

unless X is compact.

2.3 If X is compact, use a repeated application of the result in Exercise 1.2 to show that one can find a countable set $\{x_i\}$ and an increasing sequence of integers $\{N_i\}$ such that

$$|x - x_i| \le 2^{-n} \qquad \text{for some} \qquad 1 \le i \le N_n.$$

2.4 Assuming for simplicity that f is globally bounded, use the Arzelà–Ascoli theorem to show that even if the solutions of $dx/dt = f(x)$ with $x(0) = x_0$

are not unique then the set of all possible $\{x(\tau)\}$,

$$X_\tau = \{y : \text{there is a solution } x(t) \text{ with } x(\tau) = y\},$$

is closed.

2.5   Suppose that

$$\tfrac{1}{2} \frac{d}{dt} |x|^2 \leq C(t)|x|,$$

where $C(t)$ is continuous. Show that

$$\frac{d}{dt_+} |x| \leq C(t). \tag{2.27}$$

2.6   Prove that if $a(t)$ is increasing and $x(t) \geq 0$ satisfies

$$x(t) \leq a(t) + \int_0^t b(t)x(t)\,dt \tag{2.28}$$

then

$$x(t) \leq a(t) \exp\left(\int_0^t b(s)\,ds \right).$$

[Hint: Consider the new variable $y(t) = \int_0^t b(s)x(s)\,ds$, and integrate the equation for dy/dt.]

2.7   Suppose that f is a globally Lipschitz function,

$$|f(x) - f(y)| \leq L|x - y|,$$

and that $g(x)$ is a continuous function with $\|f - g\|_\infty < \infty$. If $x(t)$ is the solution of

$$dx/dt = f(x), \qquad x(0) = x_0$$

and $y(t)$ is any one of the solutions of

$$dy/dt = g(y), \qquad y(0) = x_0$$

(they may not be unique unless g is Lipschitz) use Lemma 2.9 and Gronwall's inequality (Lemma 2.8) to show that

$$|x(t) - y(t)| \leq \frac{\|f - g\|_\infty}{L} e^{Lt}. \tag{2.29}$$

[This shows that solutions of $\dot{x} = f(x)$ depend continuously on the nonlinearity $f(x)$.]

Notes

The problems discussed in this chapter are treated in significantly more detail in most advanced textbooks on ODEs and in less detail (usually relegated to an appendix) in more elementary texts. Many (Hartman, 1973; Coddington & Levinson, 1955; Hale, 1969; Doering & Gibbon, 1995) treat existence for Lipschitz equations by using the "method of successive approximations," setting

$$x_{n+1}(t) = x_0 + \int_0^t f(x_n(s)) \, ds.$$

This is (clearly) equivalent to the contraction mapping approach but can be done without the formal statement of the contraction mapping theorem.

Coddington & Levinson (1955) and Hartman (1973), two classic texts, both treat equations without uniqueness in some detail, as do Piccinini *et al.* (1984). Some further results are contained in the research monograph by Bhatia & Szegö (1967). In our brief treatment we obtained a solution as a limit of solutions of the approximating problems (2.16); it is in fact possible to show that, provided $m \neq 2$, for any given solution of (2.14) there is a corresponding sequence of approximating functions f_n such that the solutions of (2.16) converge to $x(t)$ (Robinson, 1999a).

The treatment of differential inequalities in Section 2.4 is adapted from Hale (1969) and Hartman (1973). In our treatment of PDEs we will usually obtain differential inequalities for norms in L^2 and related spaces by taking inner products, thus eliminating the need for right-hand derivatives.

We have chosen here to concentrate on autonomous equations with no explicit time dependence, since we can then use the solutions to define a dynamical system. Traditional treatments of the problems of existence and uniqueness, such as those to be found in the above texts, tend to treat the nonautonomous problem $dx/dt = f(x, t)$.

Equations that do not have unique solutions can behave much more wildly than our relatively well-behaved example $\dot{x} = x^{1/2}$. Hartman (1973) gives an example of a scalar equation $\dot{x} = f(x, t)$ that exhibits nonuniqueness for every possible "initial" condition $x(t_0) = x_0$.

3

Linear Operators

This chapter covers various aspects of the theory of linear operators. We introduce bounded operators, compact operators, and symmetric operators, and we show that a compact symmetric operator behaves very much like a real symmetric matrix. In particular this will enable us to find a basis consisting entirely of eigenfunctions of some given linear operator. Finally we show how similar results can be obtained for an unbounded linear operator.

An operator A on a vector space V is *linear* if

$$A(x + \lambda y) = Ax + \lambda Ay$$

for all $x, y \in V$ and $\lambda \in \mathbb{R}$ (or $\mathbb{C}$).

3.1 Bounded Linear Operators on Banach Spaces

We say that a linear operator A from a normed space $(X, \|\cdot\|_X)$ into another normed space $(Y, \|\cdot\|_Y)$ is bounded if there exists a constant M such that

$$\|Ax\|_Y \leq M \|x\|_X \qquad \text{for all} \qquad x \in X. \tag{3.1}$$

We write $\mathcal{L}(X, Y)$ for the space of all bounded linear maps from X into Y. The operator norm of an operator A (from X into Y) is the smallest value of M such that (3.1) holds:

$$\|A\|_{\mathcal{L}(X,Y)} = \inf\{M : (3.1) \text{ holds}\}. \tag{3.2}$$

Another equivalent definition (see Exercise 3.1) is

$$\|A\|_{\mathcal{L}(X,Y)} = \sup_{x \neq 0} \frac{\|Ax\|_Y}{\|x\|_X} = \sup_{\|x\|_X = 1} \|Ax\|_Y. \tag{3.3}$$

When there is no room for confusion we will omit the $\mathcal{L}(X, Y)$ subscript on the norm, sometimes adding the subscript "op" (for "operator") to make things clearer ($\|\cdot\|_{op}$).

The space $\mathcal{L}(X, Y)$ is a Banach space whenever Y is a Banach space. Remarkably this does not depend on whether the space X is complete or not.

Proposition 3.1. *Let X be a normed space and Y a Banach space. Then $\mathcal{L}(X, Y)$ is a Banach space.*

Proof. Let $\{A_n\}$ be a Cauchy sequence in $\mathcal{L}(X, Y)$. We need to show that $A_n \to A$ for some $A \in \mathcal{L}(X, Y)$. Since $\{A_n\}$ is Cauchy, given $\epsilon > 0$ there exists an N such that

$$\|A_n - A_m\|_{op} \leq \epsilon \qquad \text{for all} \qquad n, m \geq N. \tag{3.4}$$

We now show that for every fixed $x \in X$ the sequence $\{A_n x\}$ is Cauchy in Y. This follows since

$$\|A_n x - A_m x\|_Y = \|(A_n - A_m)x\|_Y \leq \|A_n - A_m\|_{op}\|x\|_X, \tag{3.5}$$

and $\{A_n\}$ is Cauchy in $\mathcal{L}(X, Y)$. Since Y is complete, it follows that

$$A_n x \to y,$$

where y depends on x. We can therefore define a mapping $A : X \to Y$ by $Ax = y$. We still need to show, however, that $A \in \mathcal{L}(X, Y)$ and that $A_n \to A$ in the operator norm.

First, A is linear since

$$A(x + \lambda y) = \lim_{n\to\infty} A_n(x + \lambda y) = \lim_{n\to\infty} A_n x + \lambda \lim_{n\to\infty} A_n y = Ax + \lambda Ay.$$

To show that A is bounded, take $n, m \geq N$ (from (3.4)) in (3.5) and let $m \to \infty$. Since $A_m x \to Ax$ this shows that

$$\|A_n x - Ax\|_Y \leq \epsilon\|x\|_X. \tag{3.6}$$

Since (3.6) holds for every x it follows that

$$\|A_n - A\|_{op} \leq \epsilon, \tag{3.7}$$

and so $A_n - A \in \mathcal{L}(X, Y)$. Since $\mathcal{L}(X, Y)$ is a vector space and $A_n \in \mathcal{L}(X, Y)$, it follows that $A \in \mathcal{L}(X, Y)$, and (3.7) shows that $A_n \to A$ in $\mathcal{L}(X, Y)$. ∎

One nice property of bounded linear operators is that they are automatically continuous.

Proposition 3.2. *Let $L : X \to Y$ be a linear map. Then L is continuous iff it is bounded.*

Proof. If L is bounded then

$$\|L(x_n - x)\|_Y \le \|L\|_{op} \|x_n - x\|_X,$$

which gives continuity. If L is continuous but unbounded then for every n there exists a y_n such that $\|Ly_n\|_Y > n^2 \|y_n\|_X$. Then

$$x_n = y_n / (n\|y_n\|_X) \to 0,$$

but $\|Lx_n\|_Y > n$, so that L is not continuous at the origin, a contradiction. Thus continuity implies boundedness. ∎

We now give an example of a bounded operator to which we will return repeatedly in the rest of this chapter.

Lemma 3.3. *Let $k(x, y) \in L^2(\Omega \times \Omega)$:*

$$\int_\Omega \int_\Omega |k(x, y)|^2 \, dx \, dy = C^2 < \infty.$$

Then the integral operator $K : L^2(\Omega) \to L^2(\Omega)$ defined by

$$[Ku](x) = \int_\Omega k(x, y)u(y) \, dy \qquad (3.8)$$

is bounded.

Proof. We have

$$|Ku|^2 = \int_\Omega \left(\int_\Omega k(x, y)u(y) \, dy \right)^2 dx,$$

and using the Cauchy–Schwarz inequality we obtain

$$|Ku|^2 \le \int_\Omega \left(\int_\Omega |k(x, y)|^2 \, dy \right) \left(\int_\Omega |u(y)|^2 \, dy \right) dx$$

$$= \left(\int_\Omega \int_\Omega |k(x, y)|^2 \, dx \, dy \right) \left(\int_\Omega |u(y)|^2 \, dy \right)$$

$$= C^2 |u|^2,$$

which shows that $\|K\|_{op} \le C$. ∎

3.2 Domain, Range, Kernel, and the Inverse Operator

It may be the case that an operator is not defined on the whole of X. For example, in Exercise 3.2 you are asked to show that the integration operator

$$I[f](x) = \int_0^x f(s)\, ds, \qquad x \in [0, 1]$$

is bounded from $C^0([0, 1])$ into $L^2(0, 1)$, where both spaces are equipped with the $L^2(0, 1)$ norm (i.e. we consider $C^0([0, 1])$ as a subspace of $L^2(0, 1)$). In other words we define $I[f]$ for functions that are continuous, but we consider questions of continuity by using the norm in L^2. We call this subspace on which the operator is defined its *domain*, and for a general linear operator A we write it as $D(A)$.

If $A : X \to Y$, then the image of the domain of A under application of A is called the range of A, written $R(A)$:

$$R(A) = \{v \in Y : v = Au, u \in D(A)\}.$$

This may well be a proper subspace of Y, as in the example of the integration operator, where the range is

$$\{g \in C^1([0, 1]) : g(0) = 0\}.$$

In general we could write, perhaps uninstructively,

$$X \supset D(A) \ni x \mapsto Ax \in R(A) \subset Y.$$

A bounded linear operator defined on a linear subspace of X can be extended to a bounded linear operator defined on the whole of X. (This is not obvious. We will study a related result, the Hahn–Banach theorem, in the next chapter.) Because of this it is somewhat artificial to restrict the domain of definition of a bounded linear operator, and in the rest of our discussion we will assume that $D(A) = X$. However, when we come to discuss unbounded operators in the final sections of this chapter, the domain will form an intrinsic part of the definition of the operator.

As with the theory of matrices, the concept of the inverse of a general linear operator is extremely useful. We say that A is *invertible* if the equation

$$Ax = y$$

has a unique solution for every $y \in R(A)$ (i.e. if A is injective). In this case we define the *inverse of A, A^{-1}* by $A^{-1}y = x$. It is easy to check that

$$AA^{-1}u = u \quad \text{for all} \quad u \in R(A) \quad \text{and} \quad A^{-1}Au = u \quad \text{for all} \quad u \in D(A).$$

If A is linear and A^{-1} exists then it is linear too (see Exercise 3.3).

Another important concept is the *kernel* of A, $\mathrm{Ker}(A)$, the space of all elements of $D(A)$ that A sends to zero:

$$\mathrm{Ker}(A) = \{u \in D(A) : Au = 0\}.$$

The invertibility of A is equivalent to the triviality of its kernel.

Lemma 3.4. *A is invertible iff* $\mathrm{Ker}(A) = \{0\}$.

Proof. Suppose that A is invertible. Then the equation $Ax = y$ has a unique solution for any $y \in R(A)$. However, if $\mathrm{Ker}(A)$ contains some nonzero element z then $A(x + z) = y$ also, so $\mathrm{Ker}(A)$ must be $\{0\}$. Conversely, if A is not invertible then for some $y \in R(A)$ there are two distinct solutions, x_1 and x_2, of $Ax = y$, and so $A(x_1 - x_2) = 0$, giving a nonzero element of $\mathrm{Ker}(A)$. ∎

This characterisation of invertibility will prove useful later.

3.3 The Baire Category Theorem

To prove one of the fundamental results in the theory of linear operators, the uniform boundedness principle, we will need to use a basic result of topology, the Baire category theorem. This is also useful in proving the existence of certain classes of functions (see Chapter 16). We give the result in two equivalent forms.

Theorem 3.5 (Baire Category Theorem). *If G_i is a countable family of dense open subsets of a Banach space X then*

$$G = \bigcap_{n=1}^{\infty} G_n$$

is dense in X.

The theorem also holds for complete metric spaces.

Proof. Take $x \in X$, and $r > 0$; we need to show that $B(x, r) \cap G$ is not empty.*
Now, since each G_n is dense and open, for some $y \in G_n$ and $s > 0$,

$$B(x, r) \cap G_n \supset B(y, 2s) \supset \overline{B}(y, s).$$

---

* Recall that $B(x, \epsilon)$ is an open ball with centre x and radius ϵ, and $\overline{B}(x, \epsilon)$ is the closed ball with centre x and radius ϵ.

First take $x_1 \in X$ and $r_1 < 1/2$ such that $\overline{B}(x_1, r_1) \subset G_1 \cap B(x, r)$, then take $x_2 \in X$ and $r_2 < 2^{-2}$ such that $\overline{B}(x_2, r_2) \subset G_2 \cap B(x_1, r_1)$, and in general take $x_n \in X$ and $r_n < 2^{-n}$ such that $\overline{B}(x_n, r_n) \subset G_n \cap B(x_{n-1}, r_{n-1})$.

In this way we obtain a nested sequence of closed sets

$$\overline{B}(x_1, r_1) \supset \overline{B}(x_2, r_2) \supset \cdots. \tag{3.9}$$

Since the space is complete,

$$\bigcap_{n=1}^{\infty} \overline{B}(x_n, r_n) = x_0.$$

Indeed, the points $\{x_j\}$ form a Cauchy sequence due to (3.9), and they converge to x_0, which is the intersection of all the nested sets. Now, $x_0 \in \overline{B}(x_1, r_1) \subset B(x, r)$, and also $x_0 \in \overline{B}(x_n, r_n) \subset G_n$ for all n. So $x_0 \in G \cap B(x, r)$ as required. ∎

By taking $G_j = X \setminus \overline{F_j}$ in the previous theorem we can obtain the following equivalent formulation. A set W is *nowhere dense* if its closure $\overline{W}$ contains no nonempty open set.*

Corollary 3.6. *Let X be a Banach space and F_j a countable sequence of nowhere dense subsets. Then*

$$\bigcup_{j=1}^{\infty} F_j \neq X.$$

We can now use this result to prove the Banach–Steinhaus uniform boundedness principle. Essentially this allows us to deduce a uniform bound on the operator norm of a sequence of operators from X into Y, given uniform bounds on how they act on each individual element of X. Note that we can in fact treat an uncountable collection of operators in this way.

Theorem 3.7 (Uniform Boundedness Principle). *Let X be a Banach space and Y a normed space. Let $S \subset \mathcal{L}(X, Y)$, and let*

$$\sup_{T \in S} \|Tx\|_Y < \infty \qquad \text{for all} \qquad x \in X.$$

---

* The name of the theorem derives from its original formulation, in which a countable union of nowhere dense sets was said to be "of the first category" and any other set "of the second category." Corollary 3.6 is then "no Banach space is of the first category." However, this unhelpful terminology seems now (thankfully) to be falling into disuse.

Then

$$\sup_{T \in S} \|T\|_{\mathcal{L}(X,Y)} < \infty.$$

Proof. Consider the sets

$$F_j = \{x \in X : \|Tx\|_Y \leq j \text{ for all } T \in S\}.$$

Then $\cup_j F_j = X$, and so, since F_j is closed, Corollary 3.6 shows that at least one of the F_j must have a nonempty interior, call it F_n, say. Then there exist $y \in X, r > 0$ such that

$$B(y, r) \subset F_n.$$

Therefore, if $\|x\|_X \leq r$ (so that $y + x \in F_n$),

$$\|Tx\|_Y = \|T(y + x) + T(-y)\| \leq n + \|Ty\|_Y \leq R$$

for some $R > 0$, since $\sup_{T \in S} \|Ty\|_Y$ is bounded. In particular, for any x with $\|x\|_X = r$ we have

$$\|Tx\|_Y \leq \frac{R}{r}\|x\|_X, \qquad \text{for all} \qquad T \in S,$$

and since T is linear, this follows for all x, proving the theorem. ∎

For one application see Exercise 3.4.

3.4 Compact Operators

We now introduce another class of operators, which behave more simply than general bounded operators. The compactness involved in their definition makes their analysis much more straightforward. Happily, we will find that the differential operators we consider later have inverses that are compact.

We start with a definition that can be applied to both linear and nonlinear operators.

Definition 3.8. *An operator $K : X \to Y$ is compact if the image of any set W, which is bounded in X, has compact closure in Y:*

$$\overline{K(W)} \qquad \text{is compact in } Y \text{ for all bounded} \qquad W \subset X.$$

Most of the time we will apply this definition to the image under K of a sequence $\{x_n\}$ contained in a bounded set of X. It follows that the sequence

$\{Kx_n\}$ lies in a compact subset of Y and so has a convergent subsequence. Indeed, this is an equivalent definition of a compact operator.

Although a bounded linear operator need not be compact (for example, the image of the unit ball in a Hilbert space under the identity is still the unit ball, which is not compact (Proposition 1.25)), any compact operator is bounded.

Lemma 3.9. *A compact operator is bounded.*

Proof. Since $B = B_X(0, 1)$, the unit ball in X, is bounded and K is compact, $\overline{K(B)}$ is compact and so bounded, contained in $B_Y(0, R)$, say. Thus

$$\sup_{\|x\|_X \le 1} \|K(x)\|_Y \le R,$$

and so by (3.3) $\|K\|_{\text{op}} \le R$ and K is bounded. ∎

The space of all compact linear operators from X into Y, $\mathcal{K}(X, Y)$, is a Banach space when equipped with the operator norm, provided that Y is a Banach space (cf. Proposition 3.1). The main component of this is the following (completeness) result.

Theorem 3.10. *Suppose that X is a normed space and Y is a Banach space. If $\{K_n\}$ is a sequence of compact (linear) operators in $\mathcal{L}(X, Y)$ converging to some $K \in \mathcal{L}(X, Y)$ in the operator norm, i.e.*

$$\sup_{\|x\|_X=1} \|K_n x - K x\|_Y \to 0 \qquad as \qquad n \to \infty,$$

then K is compact.

Proof. Let $\{x_n\}$ be a bounded sequence in X. Then since K_1 is compact, $K_1(x_n)$ has a convergent subsequence, $K_1(x_{n_{1j}})$. Since $x_{n_{1j}}$ is bounded, $K_2(x_{n_{1j}})$ has a convergent subsequence, $K_2(x_{n_{2j}})$. Repeat this process to get a family of nested subsequences, $x_{n_{kj}}$, with $K_l(x_{n_{kj}})$ convergent for all $l \le k$.

Now, as in the Arzelá–Ascoli theorem (Theorem 2.5), consider the diagonal sequence $y_j = x_{n_{jj}}$. This is a subsequence of the original $\{x_n\}$, and we now show that $K(y_j)$ is Cauchy, and hence convergent, to complete the proof. Choose $\epsilon > 0$, and use the triangle inequality to write

$$\|K(y_i) - K(y_j)\|_Y$$
$$\le \|K(y_i) - K_n(y_i)\|_Y + \|K_n(y_i) - K_n(y_j)\|_Y + \|K_n(y_j) - K(y_j)\|_Y.$$

Since $\{y_j\}$ is bounded and $K_n \to K$ in the operator norm, pick n large enough that

$$\|K(y_j) - K_n(y_j)\|_Y \leq \epsilon/3$$

for all y_j in the sequence. For such an n, the sequence $K_n(y_j)$ is Cauchy, and so there exists an N such that for $i, j \geq N$ we can guarantee

$$\|K_n(y_i) - K_n(y_j)\|_Y \leq \epsilon/3.$$

So now

$$\|K(y_i) - K(y_j)\|_Y \leq \epsilon \qquad \text{for all} \qquad i, j \geq N,$$

and $\{K(y_n)\}$ is a Cauchy sequence. ∎

Corollary 3.11. *If X is a normed space and Y is a Banach space then $\mathcal{K}(X, Y)$ is a Banach space.*

Proof. If $\{K_n\}$ is a Cauchy sequence of elements in $\mathcal{K}(X, Y)$ then $\{K_n\}$ is a Cauchy sequence in $\mathcal{L}(X, Y)$. Since $\mathcal{L}(X, Y)$ is a Banach space by Proposition 3.1, it follows that $K_n \to K$ for some $K \in \mathcal{L}(X, Y)$. Theorem 3.10 now shows that K is compact, and hence $K \in \mathcal{K}(X, Y)$. ∎

Theorem 3.10 can be particularly useful when applied to a sequence of finite-dimensional approximations of an operator, since any bounded operator with a finite-dimensional range is compact.

Lemma 3.12. *Let A be a bounded operator (not necessarily linear) from X into Y. If A has finite-dimensional range then A is compact.*

Proof. Let $u_n \in D(A)$ be a given bounded sequence. Since A is bounded, then $Au_n \in R(A)$ is also bounded. Since $R(A)$ is finite-dimensional, the Bolzano–Weierstrass theorem implies that Au_n has a convergent subsequence, and so A is compact. ∎

We now use Theorem 3.10 and Lemma 3.12 to show that the linear operator of Lemma 3.3 is compact.

Proposition 3.13. *The integral operator $K : L^2(\Omega) \to L^2(\Omega)$ given by*

$$[Ku](x) = \int_\Omega k(x, y)u(y)\,dy,$$

where $k \in L^2(\Omega \times \Omega)$ is compact.

Proof. Let $\{\phi_j\}$ be an orthonormal basis for $L^2(\Omega)$. It follows (see Exercise 3.5) that $\{\phi_i(x)\phi_j(y)\}$ is an orthonormal basis for $L^2(\Omega \times \Omega)$. If we write $k(x, y)$ in terms of this basis we have

$$k(x, y) = \sum_{j,k=1}^{\infty} k_{ij}\phi_i(x)\phi_j(y),$$

where the coefficients k_{ij} are as in (3.27) and the sum converges in $L^2(\Omega \times \Omega)$. Since $\{\phi_i(x)\phi_j(y)\}$ is a basis we have (cf. Proposition 1.23)

$$\|k\|_{L^2(\Omega \times \Omega)}^2 = \int_\Omega \int_\Omega |k(x, y)|^2\,dx\,dy = \sum_{i,j=1}^{\infty} |k_{ij}|^2. \qquad (3.10)$$

We now approximate K by operators derived from finite truncations of the expansion of $k(x, y)$. We set

$$k_n(x, y) = \sum_{i,j=1}^{n} k_{ij}\phi_i(x)\phi_j(y),$$

and

$$[K_n u](x) = \int_\Omega k_n(x, y)u(y)\,dy.$$

If $u \in L^2(\Omega)$ is given by $u = \sum_{l=1}^{\infty} c_l\phi_l$, then

$$K_n u = \sum_{i,j=1}^{n} k_{ij}c_j\phi_i,$$

and so K_n has rank n. It follows from Lemma 3.12 that K_n is compact for each n. If we can show that $K_n \to K$ in the operator norm then we can use Theorem 3.10 to show that K is compact.

The argument of Lemma 3.3 shows that

$$\|K - K_n\|^2 \le \int_\Omega \int_\Omega |k(x, y) - k_n(x, y)|^2\,dx\,dy = \sum_{i,j=n+1}^{\infty} |k_{ij}|^2,$$

by use of the expansion of k and k_n. Convergence of K_n to K follows since the sum in (3.10) is finite. ∎

In the next part of this chapter we will restrict attention further to compact linear maps from Hilbert spaces into themselves and show how – with the additional constraint of symmetry – this leads to a basis of eigenvalues for the space.

3.5 Compact Symmetric Operators on Hilbert Spaces

The eigenvalue theory of matrices is simplest in the case in which they are symmetric, which ensures that the eigenvalues are real and that the eigenvectors are mutually orthogonal. We can make a similar definition when H is a Hilbert space and $A \in \mathcal{L}(H, H)$.

Definition 3.14. *A linear operator $A \in \mathcal{L}(H, H)$ is* symmetric *if*

$$(u, Av) = (Au, v) \qquad \text{for all} \qquad u, v \in H.$$

For such symmetric operators there is yet another way to obtain the operator norm of A (cf. (3.2) and (3.3)) that will be useful later.

Proposition 3.15. *If A is a symmetric operator then*

$$\|A\|_{\mathcal{L}(H,H)} = \sup_{u \,:\, \|u\|=1} |(Au, u)|. \tag{3.11}$$

Proof. Using the Cauchy–Schwarz inequality it is clear that, for $\|u\| = 1$,

$$|(Au, u)| \le \|Au\| \|u\| \le \|A\|_{\mathcal{L}(H,H)}.$$

The work is in showing the opposite inequality.

If we write α for the right-hand side of (3.11) then it follows that

$$|(Au, u)| \le \alpha \|u\|^2 \qquad \text{for all} \qquad u \in H. \tag{3.12}$$

For any u and v in H we have

$$\begin{aligned}
4(Au, v) &= (A(u + v), u + v) - (A(u - v), u - v) \\
&\le \alpha(\|u + v\|^2 + \|u - v\|^2) \\
&= 2\alpha(\|u\|^2 + \|v\|^2),
\end{aligned}$$

by using (3.12) and the parallelogram law (1.27). Now if $Au \neq 0$ choose

$$v = \|u\| \frac{Au}{\|Au\|},$$

to obtain, since $\|v\| = \|u\|$,

$$\|u\| \|Au\| \leq \alpha \|u\|^2.$$

Therefore

$$\|Au\| \leq \alpha \|u\|,$$

which clearly holds as well if $Au = 0$, and so we have

$$\|A\|_{\mathcal{L}(H,H)} \leq \alpha,$$

which gives equality in (3.11). ∎

We now show that if $k(x, y)$ is symmetric then so is the integral operator K used as an example above.

Lemma 3.16. *If $k(x, y) = k(y, x)$ then the integral operator (3.8) is symmetric.*

Proof. The inner product (Ku, v) is given by

$$\int_\Omega [Ku](x)v(x)\, dx = \int_\Omega \left(\int_\Omega k(x, y)u(y)\, dy \right) v(x)\, dx$$

$$= \int_\Omega \int_\Omega k(x, y)u(y)v(x)\, dx\, dy$$

$$= \int_\Omega \int_\Omega k(y, x)u(y)v(x)\, dx\, dy$$

$$= \int_\Omega \left(\int_\Omega k(y, x)v(x)\, dx \right) u(y)\, dy$$

$$= \int_\Omega u(y)[Kv](y)\, dy,$$

which is (u, Kv). ∎

3.6 Obtaining an Eigenbasis from a Compact Symmetric Operator

An *eigenvalue* of A is a complex number λ such that there exists a nonzero $u \in H$ (the *eigenvector*) with

$$Au = \lambda u.$$

For a compact symmetric operator A we can show that at least one of the two values $\|A\|_{op}$ or $-\|A\|_{op}$ is an eigenvalue.

Lemma 3.17. *If A is a compact symmetric operator then at least one of $\pm \|A\|_{op}$ is an eigenvalue of A.*

Proof. We assume that $A \neq 0$; otherwise the result is trivial. From Proposition 3.15,

$$\|A\|_{op} = \sup_{\{|x|=1\}} |(Ax, x)|.$$

Thus there exists a sequence x_n of unit vectors such that

$$(Ax_n, x_n) \to \pm \|A\|_{op} = \alpha.$$

Since A is compact there is a subsequence x_{n_j} such that Ax_{n_j} is convergent to some y. Relabel x_{n_j} as x_n again.

Now consider

$$\|Ax_n - \alpha x_n\|^2 = \|Ax_n\|^2 + \alpha^2 - 2\alpha(Ax_n, x_n)$$
$$\leq 2\alpha^2 - 2\alpha(Ax_n, x_n);$$

by the choice of x_n, the right-hand side tends to zero as $n \to \infty$. It follows, since $Ax_n \to y$, that

$$\alpha x_n \to y,$$

and since $\alpha \neq 0$ is fixed we must have $x_n \to x$ for some $x \in H$. Therefore $Ax_n \to Ax = \alpha x$. It follows that

$$Ax = \alpha x$$

and clearly $x \neq 0$, since $\|y\| = |\alpha| \|x\| = \|A\|_{op} \neq 0$. ∎

We can apply this lemma repeatedly to obtain a basis for $R(A)$ that consists entirely of eigenvectors.

Theorem 3.18 (Hilbert–Schmidt Theorem). *Let A be a linear, symmetric, compact operator acting on an infinite-dimensional Hilbert space H. Then all eigenvalues λ_j of A are real, and if they are ordered so that*

$$|\lambda_{n+1}| \le |\lambda_n|,$$

one has

$$\lim_{n \to \infty} \lambda_n = 0.$$

Furthermore, the eigenvectors w_j can be chosen so that they form an orthonormal basis for $R(A)$, and the action of A on any $u \in H$ is given by

$$Au = \sum_{j=1}^{\infty} \lambda_j (u, w_j) w_j.$$

Proof. By Lemma 3.17 there exists a w_1 such that $\|w_1\| = 1$ and $Aw_1 = \pm \|A\| w_1$. Consider the subspace of H perpendicular to w_1:

$$H_1 = w_1^{\perp}.$$

A leaves H_1 invariant, since if $u \perp w_1$ then

$$(Au, w_1) = (u, Aw_1) = \lambda_1 (u, w_1) = 0.$$

If we consider $A_1 = A|_{H_1}$ we have another compact symmetric operator whose norm $\|A_1\|_{\text{op}} \le \|A\|_{\text{op}}$, since the supremum in (3.11) is taken over a smaller set (H_1 rather than H).

Thus we can apply the same argument to H_1 to obtain an eigenvalue $\lambda_2 = \pm \|A_1\|_{\text{op}}$, and its corresponding eigenvector w_2, by construction perpendicular to w_1. Continuing inductively in this way, we obtain a sequence of orthonormal eigenvectors w_j, with $Aw_j = \lambda_j w_j$ and $|\lambda_{n+1}| \le |\lambda_n|$.

Now suppose that $\lambda_j \not\to 0$, so that $|\lambda_j| \ge \eta$ for some $\eta > 0$. Thus, for all j, $\eta w_j \in A(B(0, 1))$; but clearly ηw_j has no convergent subsequence since the $\{w_j\}$ are orthonormal (cf. Proposition 1.25), contradicting the compactness of A.

Finally, if x is orthogonal to the linear span of the $\{w_j\}$ then

$$\|Ax\| \le |\lambda_j| \|x\|$$

for all j, so that $Ax = 0$. So we must have $x \in \text{Ker } A$. Thus there are no more (and in particular no complex) nonzero eigenvalues of A.

If $\{k_j\}_{j \in \mathcal{I}}$ is an orthonormal basis for the kernel of A, then $\{k_j\} \cup \{w_j\}$ is an orthonormal basis for H. So, writing

$$u = \sum_{j=1}^{\infty}(u, w_j)w_j + \sum_{j \in \mathcal{I}}(u, k_j)k_j,$$

we obtain

$$Au = \sum_{j=1}^{\infty}\lambda_j(u, w_j)w_j.$$

Finally, we note from this final equality that the $\{w_j\}$ form a basis for the range of A, and the proof is complete. ∎

It is natural to ask when the eigenvectors actually form a basis for all of H. We can see from the proof of Theorem 3.18 that this happens whenever $\operatorname{Ker} A = \{0\}$. Now, we saw above (Lemma 3.4) that an operator A is invertible iff $\operatorname{Ker} A = \{0\}$. We therefore have the following corollary of Theorem 3.18.

Corollary 3.19. *If A is invertible and satisfies the conditions of Theorem 3.18 then there is a basis of H consisting entirely of eigenfunctions of A.*

As one immediate and important application of these results, we consider the Sturm–Liouville boundary value problem. If L is the linear differential operator given by

$$L[u] = -\frac{d}{dx}\left(p(x)\frac{du}{dx}\right) + q(x)u(x), \tag{3.13}$$

where p and q are continuous strictly positive functions on $[a, b]$, the Sturm–Liouville eigenvalue problem is to find all those λ such that the equation

$$L[u] = \lambda w(x)u(x) \tag{3.14}$$

has a corresponding solution $u(x)$. The function $w(x)$, the "weight function," is also assumed to be strictly positive on $[a, b]$. For simplicity we study (3.14) with the boundary conditions

$$u(a) = u(b) = 0,$$

although a more general treatment is possible.

It was one of the major concerns of applied mathematics throughout the nineteenth century to show that the solutions $\{u_n(x)\}$ of (3.14) form a complete basis for some appropriate space of functions (generalising the use of Fourier series as a basis for L^2). We can do this easily using the theory developed in the last section.

Theorem 3.20. *The eigenfunctions of problem (3.14) form a complete orthonormal basis for $L_w^2(a, b)$:*

$$L_w^2(a, b) = \left\{ u : \int_a^b w(x)|u(x)|^2 \, dx < \infty \right\}. \tag{3.15}$$

On L_w^2 we use the inner product that corresponds to the norm in definition (3.15):

$$(u, v)_w = \int_a^b w(x)u(x)v(x) \, dx.$$

Proof. We show first that any λ for which (3.14) holds must be positive. Indeed,

$$
\begin{aligned}
(Lu, u) &= \int_a^b -(pu')'u + q|u|^2 \, dx \\
&= \int_a^b p|u'|^2 + q|u|^2 \, dx - {}_a^b[p(x)u'(x)u(x)] \\
&\geq \int_a^b q(x)|u(x)|^2 \, dx \\
&\geq \frac{\sup_{x \in [a,b]} q(x)}{\inf_{x \in [a,b]} w(x)} \int_a^b w(x)|u(x)|^2 \, dx \\
&= \alpha \int_a^b w(x)|u(x)|^2 \, dx = \alpha(wu, u),
\end{aligned}
$$

with $\alpha > 0$, and so $\lambda \geq \alpha > 0$.

The Green's function $G(x, y)$, the solution of the equation

$$L[G] = \delta(x - y), \qquad G(x, a) = G(x, b) = 0,$$

is symmetric in x and y (see Exercise 3.7). Since we can use $G(x, y)$ to express the solution of

$$L[u] = f(x), \qquad u(a) = u(b) = 0$$

as

$$u(x) = \int_a^b G(x, y) f(y) \, dy,$$

our original Sturm–Liouville problem is equivalent to the integral equation

$$\int_a^b G(x, y) w(y) u(y) \, dy = \frac{1}{\lambda} u(x). \qquad (3.16)$$

Therefore if we define a new function $v(x) = \sqrt{w(x)} u(x)$ and set

$$k(x, y) = G(x, y) \sqrt{w(x) w(y)},$$

then we can rewrite (3.16) as

$$Kv = \frac{1}{\lambda} v, \qquad (3.17)$$

where K is the integral operator given by

$$Kv(x) = \int_a^b k(x, y) v(y) \, dy.$$

Since $k(x, y)$ is symmetric and bounded, it follows from Lemma 3.16 and Proposition 3.13 that K is compact and symmetric, and so, using the Hilbert–Schmidt theorem (Theorem 3.18), K has a set of orthonormal eigenfunctions $\{\phi_j\}$,

$$(\phi_j, \phi_k) = \delta_{jk}.$$

Furthermore the kernel of K is $\{0\}$, since any solution of (3.14) and hence of (3.16) and (3.17) has λ finite. It follows from Corollary 3.19 that the eigenfunctions $\{\phi_j\}$ of (3.17) form a basis for $L^2(a, b)$.

Each eigenfunction ϕ_j gives rise to an eigenfunction $\varphi_j = \phi_j w^{-1/2}$ of the original Sturm–Liouville problem (3.14), and these eigenfunctions are orthonormal in $L_w^2(a, b)$:

$$(\varphi_j, \varphi_k)_w = \int_a^b w(x) \varphi_j(x) \varphi_k(x) \, dx = (\phi_j, \phi_k) = \delta_{jk}.$$

Now suppose that $f \in L_w^2(a, b)$. Then the function $w^{1/2} f$ is an element of $L^2(a, b)$ and hence can be expanded in terms of the $\{\phi_j\}$:

$$w^{1/2} f = \sum_{j=1}^{\infty} (w^{1/2} f, \phi_j) \phi_j.$$

It follows that

$$f = \sum_{j=1}^{\infty} \left(w^{1/2} f, w^{1/2} \varphi_j \right) \varphi_j$$

or

$$f = \sum_{j=1}^{\infty} (f, \varphi_j)_w \varphi_j,$$

and so the $\{\varphi_j\}$ form an orthonormal basis for $L^2_w(a, b)$, as claimed. ∎

For a related example see Exercise 3.8.

3.7 Unbounded Operators

We will now discuss unbounded operators; the main task will be to adapt the
Hilbert–Schmidt theorem (Theorem 3.18) and its corollary (Corollary 3.19) to
this case.

A linear operator is (unsurprisingly) called unbounded if it is not bounded.
Most of the linear operators that arise in the study of PDEs are unbounded
operators. This is to be expected, since the differential operator d/dx is (in
general) an unbounded operator. Let us consider the action of d/dx on the
domain $C^1([0, 1])$, considered as a subset of $L^2(0, 1)$. Now, for all $k \in \mathbb{R}$ the
function $e^{-kx} \in C^1([0, 1])$, and

$$\frac{d}{dx} e^{-kx} = -k e^{-kx}.$$

Clearly

$$\left\| \frac{d}{dx} e^{-kx} \right\|_{L^2} = k \| e^{-kx} \|_{L^2},$$

and so

$$\left\| \frac{d}{dx} \right\|_{\mathcal{L}(L^2, L^2)} \geq k$$

for all k, showing that the operator is unbounded.

Although it is artificial to restrict the domain of definition of a bounded
operator to some subspace of X, this is most certainly not the case for unbounded
operators, for which the domain is an integral part of the definition: strictly
speaking, an unbounded operator is not specified until we have also given its

domain, so we should speak about the pair $(A, D(A))$. As with Banach spaces
(where we usually avoid the clumsy $(X, \|\cdot\|_X)$ notation and just talk about "the
space X") we will often refer just to "the operator A" in cases when the domain
has been specified.

3.8 Extensions and Closable Operators

Suppose that we want to consider the differential operator as a linear map from
$L^2(0, 1)$ into $L^2(0, 1)$, which would enable us to take advantage of the Hilbert
space structure of L^2. The correct choice of the domain of A is not immediately
clear. For example, we could choose $D(A)$ to be $C^j([0, 1])$ for any $j \geq 1$.
Somehow we would like to choose $D(A)$ to be the "natural" domain consistent
with the definition of A with which we are familiar (in this case $C^1([0, 1])$).

What we want is a canonical way to extend d/dx to a larger (and in some
ways optimal) domain in L^2.

Definition 3.21. *An operator* $(\hat{A}, D(\hat{A}))$ *is an* extension *of* $(A, D(A))$ *if*
$D(\hat{A}) \supset D(A)$ *and* $\hat{A} = A$ *on* $D(A)$.

We can find one such domain in a standard way called "closing the operator."
Suppose that we start with an operator $(A, D(A))$ with $A : X \to Y$. Consider
a sequence $\{x_n\}$ in X such that

$$x_n \to x \quad \text{in} \quad X \quad \text{and} \quad Ax_n \to y \quad \text{in} \quad Y. \qquad (3.18)$$

Now, if $x \in D(A)$ then we would like to have $y = Ax$, but this does not follow
automatically, since A is unbounded and so is not continuous. To guarantee that
$y = Ax$, it is sufficient that for every sequence $x_n \to 0$, either $Ax_n \to 0$ or
Ax_n does not converge.

Under this assumption we can define an extension of A, $\hat{A}$, taking $D(\hat{A})$ to
be all those $x \in X$ for which there are sequences satisfying (3.18) and defining
$\hat{A}x = y$. For then, if there are two sequences $\{x_n\}$ and $\{\bar{x}_n\}$ that satisfy

$$x_n \to x \quad \text{in} \quad X \quad \text{and} \quad Ax_n \to y \quad \text{in} \quad Y$$

and

$$x_n \to x \quad \text{in} \quad X \quad \text{and} \quad Ax_n \to \bar{y} \quad \text{in} \quad Y,$$

it follows, since $x_n - \bar{x}_n \to 0$, that $y = \bar{y}$, and so $\hat{A}x = y$ is a well-defined
extension of A. This gives us a *closed* operator $\hat{A}$.

Definition 3.22. *An operator A is closed if, whenever $\{x_n\} \in D(A)$ obeys (3.18), then $x \in D(A)$ with $Ax = y$.*

The next result shows that a symmetric operator whose domain is dense can always be closed. Such operators occur in many applications, as we will see in what follows.

Proposition 3.23. *If A is a symmetric operator on a Hilbert space H with dense domain, then it has a closed extension $\bar{A}$ that is also symmetric.*

Proof. Suppose that $x_n \to 0$ and that $Ax_n \to f$. Then for any $u \in D(A)$ we have

$$(f, u) = \lim_{n \to \infty} (Ax_n, u) = \lim_{n \to \infty} (x_n, Au) = 0.$$

Since $D(A)$ is dense in H it follows that $(f, u) = 0$ for all $u \in H$, and so, choosing $u = f$, we have that $f = 0$. We can therefore define an extension as outlined above, taking $\hat{A}$ to be all those $x \in X$ for which there are sequences satisfying (3.18). The symmetry is clear. ∎

Because of this result we will not distinguish in what follows between such an operator and its closed extension. We will return briefly to the example of the differentiation operator in Chapter 5 (Section 5.2.1).

3.9 Spectral Theory for Unbounded Symmetric Operators

In Section 3.6 we developed an eigenvalue theory for compact symmetric operators on a Hilbert space. In this section we want to do something similar for unbounded operators, using our previous results.

Since an unbounded operator is not necessarily defined on the whole of H, we have to adjust slightly our definition of a symmetric operator.

Definition 3.24. *A linear operator $(A, D(A))$ is* symmetric *if*

$$(Ax, y) = (x, Ay) \qquad \text{for all} \qquad x, y \in D(A).$$

The result we will use to translate eigenvalue problems for unbounded A back to problems for bounded operators is the following simple lemma.

Lemma 3.25. *If A is an unbounded symmetric operator whose range is the whole of H and whose inverse is well defined, then A^{-1} is bounded and symmetric.*

One condition that ensures that A^{-1} is well defined is that A is bounded below (see Exercise 3.9).

Proof. If A^{-1} is not bounded then there exists a sequence $\{y_n\} \in D(A)$ such that

$$|Ay_n| = 1 \qquad \text{and} \qquad |y_n| \to \infty.$$

For any $f \in R(A) = H$, there exists an $x \in D(A)$ such that $Ax = f$. Therefore

$$|(y_n, f)| = |(y_n, Ax)| = |(Ay_n, x)| \leq |x|.$$

This shows that the linear functionals L_n on H given by

$$f \mapsto (y_n, f)$$

are bounded for each $f \in H$ and for each n. It follows from the principle of uniform boundedness (Theorem 3.7) that there is a uniform bound on the norm of these functionals L_n. Now, since clearly

$$\|L_n\|_{\text{op}} \leq |y_n| \qquad \text{and} \qquad L_n(y_n) = |y_n|^2,$$

we have

$$\|L_n\|_{\text{op}} = |y_n|,$$

and so the $\{y_n\}$ are bounded, which is a contradiction.

To show that A^{-1} is symmetric, take $x, y \in H$, with $x = Au$ and $y = Av$, where $u, v \in D(A)$. Then

$$(A^{-1}x, y) = (A^{-1}Au, Av) = (u, Av) = (Au, v) = (x, A^{-1}y). \qquad \blacksquare$$

Now, any eigenfunction for A is an eigenfunction for A^{-1} and vice versa, since

$$Aw_n = \lambda_n w_n \qquad \Longleftrightarrow \qquad A^{-1}w_n = \lambda_n^{-1} w_n.$$

Thus we have the following result from Lemma 3.25, Theorem 3.18, and its corollary (Corollary 3.19).

Corollary 3.26. *Let A be a symmetric linear operator on H whose range is all of H, and suppose further that its inverse is defined and compact. Then A has* *an infinite set of real eigenvalues λ_n with corresponding eigenfunctions w_n:*

$$Aw_n = \lambda_n w_n.$$

If the eigenvalues are ordered so that

$$|\lambda_{n+1}| \geq |\lambda_n|,$$

one has

$$\lim_{n \to \infty} \lambda_n = \infty.$$

Furthermore, the w_n can be chosen so that they form an orthonormal basis for H, and in terms of this basis the operator A can be represented by

$$Au = \sum_{j=1}^{\infty} \lambda_j(u, w_j)w_j. \tag{3.19}$$

We will apply this result to eigenfunctions of the Laplacian operator in Chapter 6. Note that one consequence of Corollary 3.26 is that the domain of A can be represented as

$$D(A) = \left\{ u : u = \sum_{j=1}^{\infty} c_j w_j, \ \sum_{j=1}^{\infty} |c_j|^2 \lambda_j^2 < \infty \right\}. \tag{3.20}$$

It follows from this representation that $D(A)$ is a Hilbert space when equipped with the inner product

$$((u, v))_{D(A)} = (Au, Av) \tag{3.21}$$

and corresponding norm

$$\|u\|_{D(A)} = \|Au\|. \tag{3.22}$$

3.10 Positive Operators and Their Fractional Powers

A positive operator is one that satisfies

$$(Au, u) \geq k\|u\|^2 \tag{3.23}$$

for some $k > 0$. It follows immediately that if the operator in Corollary 3.26 is positive then all its eigenvalues λ_j are positive, and $\lambda_j \to \infty$ as $j \to \infty$.

In this case, since we have a representation of A by (3.19), we can use this to define its *fractional powers* by the natural expression

$$A^\alpha u = \sum_{j=1}^{\infty} \lambda_j^\alpha (u, w_j) w_j. \tag{3.24}$$

The domain of A^α in H, $D(A^\alpha)$, is

$$D(A^\alpha) = \{ u : \|A^\alpha u\| < \infty \}$$

(i.e. such that the right-hand side of (3.24) converges), which is the same as (cf. (3.20))

$$D(A^\alpha) = \left\{ u : u = \sum_{j=1}^{\infty} c_j w_j, \ \sum_{j=1}^{\infty} |c_j|^{2\alpha} \lambda_j^2 < \infty \right\}. \tag{3.25}$$

We can make $D(A^\alpha)$ a Hilbert space by using the inner product (cf. (3.21))

$$(u, v)_{D(A^\alpha)} = (A^\alpha u, A^\alpha v),$$

which of course gives rise to a corresponding norm (cf. (3.22))

$$\|u\|_{D(A^\alpha)} = \|A^\alpha u\|.$$

We will make only occasional use of these fractional power spaces in what follows, although they can be used as the basis of a general theory of existence and uniqueness (see Notes). However, the following interpolation inequality will prove useful in Chapter 6. The proof, which uses the eigenvalue expansion and Hölder's inequality, is set as an exercise (Exercise 3.11).

Lemma 3.27. *If $u \in D(A^k)$ then*

$$\|A^s u\| \leq \|A^l u\|^{(k-s)/(k-l)} \|A^k u\|^{(s-l)/(k-l)} \tag{3.26}$$

for $0 \leq l < s < k$.

We can define other functions of operators. In particular, the exponential e^{-At} is much used in certain approaches to PDEs (see Exercise 3.12).

Exercises

3.1 Show that Definitions (3.2) and (3.3) of the operator norm are equivalent.

3.2 Show that the integration operator

$$I[f](x) = \int_0^x f(s)\,ds, \qquad x \in [0, 1]$$

is a bounded operator from $C^0([0, 1])$ into itself. Show also that it is a bounded operator acting on $C^0([0, 1])$ as a subset of $L^2(0, 1)$ into $L^2(0, 1)$.

3.3 If A is linear and invertible, show that A^{-1} is also linear.

3.4 Suppose that $\{e_j\}_{j=1}^\infty$ is a basis for a Banach space X, such that any $x \in X$ can be written uniquely as

$$x = \sum_{j=1}^\infty c_j e_j,$$

where the sum converges in X. Define a linear map P_n by

$$P_n\left(\sum_{j=1}^\infty c_j e_j\right) = \sum_{j=1}^n c_j e_j.$$

Use the principle of uniform boundedness (Theorem 3.7) to show that there exists a constant C such that

$$\|P_n\|_{\mathrm{op}} \le C \qquad \text{for all} \qquad n \in \mathbb{Z}^+.$$

3.5 Suppose that $\{\phi_j(x)\}$ is an orthonormal basis for $L^2(\Omega)$. Show that $\{\phi_i(x)\phi_j(y)\}$ is an orthonormal basis for $L^2(\Omega \times \Omega)$ and hence that, if $k \in L^2(\Omega \times \Omega)$, it can be written in the form (3.10) with

$$k_{ij} = \int_{\Omega \times \Omega} k(x, y)\phi_i(x)\phi_j(y)\,dx\,dy. \qquad (3.27)$$

3.6 This is a partial converse of the Hilbert–Schmidt theorem (Theorem 3.18). Show that if A can be expressed in the form

$$Au = \sum_{n=1}^\infty \lambda_n(u, w_n)w_n$$

where $\lambda_n \to 0$ and $(w_n, w_m) = \delta_{mn}$ then A is compact and symmetric. (Hint: Use the result of Lemma 3.12 and Theorem 3.10.)

3.7   Let L be defined by (3.13), and suppose that $u_1(x)$ and $u_2(x)$ are nonzero functions satisfying $L[u] = 0$, along with the boundary conditions

$$u_1(a) = 0 \quad \text{and} \quad u_2(b) = 0.$$

Use these to find the Green's function $G(x, y)$, which is the solution of the problem

$$L_x[G] = 0 \quad \text{for all} \quad x \in [a, y) \cup (y, b],$$

along with the conditions at $x = y$:

$$G(x, y) \text{ continuous at } x = y \quad \text{and} \quad [G'(x, y)]_{y-}^{y+} = p(y)^{-1}.$$

In particular, show that

$$G(x, y) = \begin{cases} Cu_1(x)u_2(y), & a \le x < y, \\ Cu_1(y)u_2(x), & y \le x \le b, \end{cases} \tag{3.28}$$

for some constant C, ensuring that $G(x, y)$ is symmetric in x and y.

3.8   If $k \in L^2(\Omega \times \Omega)$ and $k(x, y) = k(y, x)$ show that the solution of the integral equation

$$\int_\Omega k(x, y)u(y)\,dy = f(x), \qquad f \in L^2(\Omega) \tag{3.29}$$

is given in terms of the eigenvalues and eigenfunctions of the equation

$$\int_\Omega k(x, y)u(y)\,dy = \lambda u(x) \tag{3.30}$$

by

$$u(x) = \sum_{j=1}^{\infty} \frac{(f, u_j)}{\lambda_j} u_j(x).$$

[Hint: Consider the integral operator K defined by

$$[Ku](x) = \int_\Omega k(x, y)u(y)\,dy.$$

You may assume that $\lambda = 0$ is not an eigenvalue of (3.30).]

3.9   Show that if A is a linear operator that is bounded below (i.e. if there exists a constant $k > 0$ such that $\|Ax\| \ge k\|x\|$), then A^{-1} is well defined. Show also that if A is bounded, then $A^{-1} : R(A) \to D(A)$ is bounded.

3.10 One can define the fractional powers of a larger class of operators than those treated in Section 3.10. These are the so-called sectorial operators, which satisfy certain conditions on their spectrum. In this situation, the appropriate definition is

$$A^{-\alpha} = \frac{1}{\Gamma(\alpha)} \int_0^\infty t^{\alpha-1} e^{-At} \, dt.$$

Show that this agrees with (3.24) when H has a basis of eigenfunctions of A. [Hint: Apply $A^{-\alpha}$ to each eigenfunction, and use the definition $\Gamma(x) = \int_0^\infty t^{x-1} e^{-t} \, dt$.]

3.11 Prove (3.26) by using the eigenfunction expansion $u = \sum c_j w_j$ and the corresponding expression for $\|A^s u\|$,

$$\|A^s u\|^2 = \sum_{j=1}^\infty \lambda_j^{2s} |c_j|^2,$$

along with the Hölder inequality.

3.12 If A is an operator then we can define e^A by the standard exponential series

$$e^A = \sum_{n=1}^\infty \frac{A^n}{n!}.$$

When A is a positive symmetric operator with compact inverse, as in Section 3.10, this definition implies that

$$e^{-At} u = \sum_{j=1}^\infty e^{-\lambda_j t} (u, w_j) w_j.$$

Using this equality, show that if $x \in D(A)$ then $e^{-At} x$ is differentiable on $[0, \infty)$, and

$$\frac{d}{dt} e^{-At} x = -A e^{-At} x. \tag{3.31}$$

[Hint: You need check only (3.31) at $t = 0$, and then it holds for all $t \geq 0$ by linearity.]

Notes

By aiming directly for the eigenbasis expansion of H as provided by Corollary 3.26 we have bypassed much of the spectral theory of linear operators. Indeed, we have not even discussed the spectrum itself, the starting point for any more detailed investigation of the subject. The spectrum of a linear operator $A : X \to Y$, denoted $\sigma(A)$, is the complement of all those complex numbers λ for which $(A - \lambda)$ has a bounded inverse with dense domain in X. Various properties that we introduced cut down on the possible complexities of the spectrum.

For example, we have avoided introducing the definition of a self-adjoint operator, which allows spectral theory to be applied directly to unbounded operators, rather than using the roundabout route via Lemma 3.25 adopted here. Suppose that $(A, D(A))$ is a densely defined operator from a Hilbert space H into itself. Then we define $D(A^*)$ to be the set of all $v \in H$ such that there exists a $w \in H$ so that

$$(Au, v) = (u, w)$$

for all $u \in D(A)$. The *adjoint* of A is the operator $(A^*, D(A^*))$ defined by

$$A^* v = w.$$

We say that A is self-adjoint if $(A, D(A)) = (A^*, D(A^*))$. Since A^* is an extension of A, this is a stronger requirement than symmetry, and it allows us to conclude more about the spectrum of A.

Detailed treatments of spectral theory can be found in Bollobás (1990), Meise & Vogt (1997), Renardy & Rogers (1992), Davies (1995), and Young (1988), among many other texts.

Fractional powers of operators are used extensively by some authors, in particular Henry (1984) and Cazenave & Haraux (1998). We will concentrate instead on the Sobolev space formalism, which is introduced in Chapter 5.

4

Dual Spaces

In this chapter we study spaces of bounded linear maps from X into $\mathbb{R}$. It turns out that such maps play a large part in the theory of partial differential equations (PDEs). In particular, we will be able to prove a result that goes some way towards circumventing the lack of compactness of bounded sets in infinite-dimensional spaces.

A bounded linear map from a Banach space X into $\mathbb{R}$ (an element of $\mathcal{L}(X, \mathbb{R})$) is called a *linear functional* on X – it associates a real number with every element of X. Functional analysis was originally solely concerned with the study of such functionals. Because of its importance, we adopt a special notation for $\mathcal{L}(X, \mathbb{R})$.

Definition 4.1. *Let X be a real Banach space. The space $\mathcal{L}(X, \mathbb{R})$ of all linear functionals on X is denoted X^* and is called the* dual space *of X.*

It follows from Proposition 3.1 that X^* is itself a Banach space when equipped with the operator norm. Since this is the natural norm on X^*, we will denote

$$\|f\|_{X^*} = \|f\|_{\mathcal{L}(X,\mathbb{R})}.$$

When no confusion can arise, the simpler notation $\|f\|_*$ will sometimes be used.

4.1 The Hahn–Banach Theorem

We will now prove one of the fundamental results in the theory of linear functionals, the Hahn–Banach theorem. This result guarantees that any linear functional defined on a linear subspace of X can be extended to a similarly bounded functional on the whole of X.

The proof relies on Zorn's lemma, which is in fact more of an axiom than a lemma since it can be shown to be equivalent to the axiom of choice. Despite its somewhat abstract formulation, Zorn's lemma is a very powerful tool.

Zorn's lemma guarantees the existence of an element in a partially ordered set P that is maximal with respect to some order relation; to state the lemma precisely we need to define these terms. A *partial order* on a set P is an operation $\leq$ that satisfies

 (i) $a \leq a$ for all $a \in P$,
 (ii) $a \leq b$ and $b \leq c$ implies that $a \leq c$, and
(iii) if $a \leq b$ and $b \leq a$ then $a = b$.

A subset C of P is called a *chain* if any two elements of C can be compared: for all pairs $a, b \in C$ either $a \leq b$ or $b \leq a$ (or both if $a = b$). An element b is an *upper bound* for $S \subset P$ if $s \leq b$ for all $s \in S$. Finally, we say that m is a maximal element of P if $m \leq a$ implies that $a = m$.

Lemma 4.2 (Zorn's Lemma). *If P is a partially ordered set and all chains in P have an upper bound then P has a maximal element.*

For another application of Zorn's lemma, see Exercise 4.1.

Theorem 4.3 (Hahn–Banach Theorem). *Let X be a Banach space and M a linear subspace of X. Suppose that f is a linear functional on M, such that*

$$|f(x)| \leq k\|x\| \qquad \text{for all} \qquad x \in M. \tag{4.1}$$

Then there exists an extension F of f to all of X, such that

$$|F(x)| \leq k\|x\| \qquad \text{for all} \qquad x \in X. \tag{4.2}$$

In the case when X is a Hilbert space we do not need to resort to Zorn's lemma and the proof is much simpler (see Exercise 4.4).

Proof. We will apply Zorn's lemma to all possible extensions of f satisfying the bound (4.1). Indeed, consider the set E of pairs (g, G), where $g : G \to \mathbb{R}$ is a linear functional on G that extends f and satisfies

$$|g(x)| \leq k\|x\| \qquad \text{for all} \qquad x \in G. \tag{4.3}$$

Clearly E is nonempty, as $(f, M) \subset E$. We define an order on E by $(g, G) \geq (h, H)$ if g is an extension of h, that is, if $G \supseteq H$ and $g = h$ on H. Since any chain $C = \{(g_i, G_i)\}$ has an upper bound, namely the functional

$$u(x) = g_i(x), \qquad x \in G_i,$$

defined on

$$U = \bigcup_i G_i,$$

an application of Zorn's lemma guarantees a maximal element (F, W) of E. All that now remains is to show that $W = X$.

We argue by contradiction and show that if $W \neq X$ then F can be further extended. So, if $W \neq X$ there exists a $z \in X \setminus W$. Any element of the space Z spanned by z and W can be uniquely written as

$$x = w + \alpha z,$$

since, if not, there would be two representations

$$w_1 + \alpha_1 z = w_2 + \alpha_2 z \qquad \Rightarrow \qquad (\alpha_1 - \alpha_2)z = w_1 - w_2;$$

since $z \notin W$ and $w_1 - w_2 \in W$, this implies that $\alpha_1 = \alpha_2$ and $w_1 = w_2$. So for any c, we can define a linear functional on Z by

$$g(w + \alpha z) = F(w) + \alpha c.$$

We now have to show that we can find a value of c that ensures that the functional g satisfies the bound (4.1) on Z.

If this were so, we would need

$$|F(w) + \alpha c| \leq k\|w + \alpha z\| \tag{4.4}$$

for all $w \in W$ and all $\alpha \in \mathbb{R}$. Now, since both sides are linear, this holds provided that

$$F(w) + \alpha c \leq k\|w + \alpha z\|. \tag{4.5}$$

(If the quantity in the left-hand modulus sign in (4.4) is negative then we want $-F(w) - \alpha c \leq k\|w + \alpha z\|$, which is

$$F(-w) + (-\alpha)c \leq k\|-w + (-\alpha)z\|,$$

ensured by (4.5).) This is equivalent to

$$c \leq \frac{1}{\alpha}(k\|w + \alpha z\| - F(w)),$$

$$-c \leq \frac{1}{\alpha}(k\|w - \alpha z\| - F(w)),$$

with $w \in W$ and $\alpha \geq 0$. So we need to find a constant c that satisfies

$$\frac{1}{\alpha_1}(F(w_1) - k\|w_1 - \alpha_1 z\|) \leq c \leq \frac{1}{\alpha_2}(k\|w_2 + \alpha_2 z\| - F(w_2)) \qquad (4.6)$$

for all $\alpha_1, \alpha_2 \geq 0$, $w_1, w_2 \in W$. Such a c will exist provided that

$$\sup_{w_1 \in W, \alpha_1 > 0} \frac{1}{\alpha_1}(F(w_1) - k\|w_1 - \alpha_1 z\|)$$

$$\leq \inf_{w_2 \in W, \alpha_2 > 0} \frac{1}{\alpha_2}(k\|w_2 + \alpha_2 z\| - F(w_2)).$$

So we can perform the extension provided that, for all $w_1, w_2 \in W$ and $\alpha_1, \alpha_2 > 0$, we have

$$\alpha_2(F(w_1) - k\|w_1 - \alpha_1 z\|) \leq \alpha_1(k\|w_2 + \alpha_2 z\| - F(w_2)),$$

which is

$$F(\alpha_2 w_1 + \alpha_1 w_2) \leq k\|\alpha_1 w_2 + \alpha_1 \alpha_2 z\| + k\|\alpha_2 w_1 - \alpha_1 \alpha_2 z\|. \qquad (4.7)$$

Now, we know from (4.3) that

$$|F(\alpha_2 w_1 + \alpha_1 w_2)| \leq k\|\alpha_2 w_1 + \alpha_1 w_2\|;$$

using the triangle inequality gives (4.7), and so we can find the constant c needed in (4.6).

So we have shown that F can be further extended. This contradicts the maximality of (F, W) in E, and hence $W = X$ and the theorem is proved. ∎

As an indication of why studying dual spaces can be expected to yield interesting results, we apply the Hahn–Banach theorem to show that if $f(x) = f(y)$ for all $f \in X^*$ then $x = y$.

Lemma 4.4. *Let X be a Banach space. If $x, y \in X$ and $f(x) = f(y)$ for every $f \in X^*$ then $x = y$.*

Proof. If $x = y = 0$ then we are done, so without loss of generality assume that $x \neq 0$. Let Z be the linear space spanned by x and y. If x and y are linearly dependent then set

$$f(\alpha x) = \alpha |x|; \tag{4.8}$$

otherwise set

$$f(\alpha x + \beta y) = \alpha |x| \qquad \text{for all} \qquad \beta \in \mathbb{R}. \tag{4.9}$$

In both cases $f(x) \neq f(y)$. Now extend f to an element $F \in X^*$ by using the Hahn–Banach theorem; clearly $F(x) \neq F(y)$. ∎

The proof of this lemma included the additional factor of $|x|$ in Definitions (4.8) and (4.9) so as to provide the following immediate corollary (f in the statement below is just F from the proof of Lemma 4.4).

Corollary 4.5. *Let $x \in X$, a nontrivial Banach space. Then there exists an $f \in X^*$ such that $f(x) = |x|$ and $\|f\|_* = 1$.*

4.2 Examples of Dual Spaces

Before we turn to how dual spaces can be used in analysis, we will first give several examples.

4.2.1 The Dual Space of L^p, $1 < p < \infty$

We first consider the Lebesgue L^p spaces, with $1 < p < \infty$, and observe that if $f \in L^q(\Omega)$, with (p, q) conjugate indices, we can define a linear functional L_f on L^p via

$$L_f(g) = \int_\Omega f(x)g(x)\,dx, \tag{4.10}$$

since Hölder's inequality gives

$$|L_f(g)| \leq \|f\|_{L^q}\|g\|_{L^p}.$$

Certainly, then, $\|L_f\|_{(L^p)^*} \leq \|f\|_{L^q}$, and if we consider

$$g(x) = |f(x)|^{q-2}f(x)$$

then we have

$$\|g\|_{L^p} = \left(\int_\Omega |f(x)|^{(q-1)p} \, dx \right)^{1/p} = \left(\int_\Omega |f(x)|^q \, dx \right)^{1/p} = \|f\|_{L^q}^{q/p},$$

and

$$|L_f(g)| = \left| \int_\Omega |f(x)|^q \, dx \right| = \|f\|_{L^q}^q.$$

In this case, therefore,

$$|L_f(g)| = \|f\|_{L^q} \|g\|_{L^p},$$

showing that in fact

$$\|L_f\|_{(L^p)^*} = \|f\|_{L^q}.$$

This shows that the map $f \mapsto L_f$ is an isometry from L^q *into* $(L^p)^*$.

The question remains whether every element of $(L^p)^*$ can be realised as L_f for some $f \in L^q$, i.e. whether the map $f \mapsto L_f$ is onto. One can show that this is the case by using results from measure theory that are beyond the scope of this discussion. It follows that L^q and $(L^p)^*$ are isometrically isomorphic, which we will write as $L^q \simeq (L^p)^*$. Since it is therefore natural to identify $(L^p)^*$ with L^q via (4.10), you will often see $L^q = (L^p)^*$ (a convenient abuse of notation). [For details of the proof see Adams (1975), Meise & Vogt (1997), or Yosida (1980).]

4.2.2 The Dual Space of l^p, $1 < p < \infty$

For the l^p sequence spaces (see Definition 1.19) it is much easier to show that every element of $(l^p)^*$ can be derived from an element of l^q, and we now give this simple proof in full. In this case, we associate an element L_ϕ of $(l^p)^*$ with an element $\phi \in l^q$ via

$$L_\phi(x) = \sum_{k=1}^\infty \phi_k x_k. \tag{4.11}$$

Theorem 4.6. *If* $1 < p < \infty$, *then* $(l^p)^* \simeq l^q$, *where* (p, q) *are conjugate indices.*

Proof. First it is clear that if $\phi \in l^q$ and L_ϕ is defined by (4.11) then

$$|L_\phi(x)| \le \|\phi\|_{l^q} \|x\|_{l^p},$$

so that $L_\phi \in (l^p)^*$ with norm

$$\|L_\phi\|_{(l^p)^*} \leq \|\phi\|_{l^q}. \tag{4.12}$$

We now show the converse, that with any element $L \in (l^p)^*$ we can associate an element ϕ of l^q such that $L = L_\phi$ as defined in (4.11), with

$$\|\phi\|_{l^q} \leq \|L\|_{(l^p)^*}.$$

Since $L = 0$ corresponds trivially to $\phi_k = 0$ for all k, we may assume that $L \neq 0$.

If $\{e_k\}$ is the basis for l^p given by $e_k = \{\delta_{jk}\}$, we set $\phi_k = L(e_k)$. Now, if $x \in l^p$ is given by

$$x = \sum_{k=1}^{\infty} x_k e_k,$$

then, since L is linear and continuous,

$$L(x) = \sum_{k=1}^{\infty} x_k \phi_k.$$

Thus $L = L_\phi$. It remains to show that ϕ is in fact an element of l^q with $\|\phi\|_{l^q} = \|L\|_{(l^p)^*}$.

To this end, consider the sequence $\phi^{(m)}$, where

$$\phi_n^{(m)} = \begin{cases} |\phi_n|^q/\phi_n, & n \leq m \text{ and } \phi_n \neq 0, \\ 0, & n > m \text{ or } \phi_n = 0. \end{cases}$$

Then

$$L\left(\phi^{(m)}\right) = \sum_{n=1}^{\infty} \phi_n^{(m)} \phi_n = \sum_{n=1}^{m} |\phi_n|^q.$$

Also, by definition of $\|L\|_{(l^p)^*}$,

$$|L\left(\phi^{(m)}\right)| \leq \|L\|_{(l^p)^*} \|\phi^{(m)}\|_{l^p} = \|L\|_{(l^p)^*} \left(\sum_{n=1}^{\infty} |\phi_n^{(m)}|^p\right)^{1/p}$$

$$= \|L\|_{(l^p)^*} \left(\sum_{n=1}^{m} |\phi_n|^{(q-1)p}\right)^{1/p}.$$

Now, since $(q-1)p = q$, we have

$$|L(\phi^{(m)})| = \sum_{n=1}^{m} |\phi_n|^q \leq \|L\|_{(l^p)^*} \left(\sum_{n=1}^{m} |\phi_n|^q \right)^{1/p};$$

this yields (on dividing), for m large enough,

$$\left(\sum_{n=1}^{m} |\phi_n|^q \right)^{1/q} \leq \|L\|_{(l^p)^*},$$

and so

$$\|\phi\|_{l^q} \leq \|L\|_{(l^p)^*}. \qquad (4.13)$$

Thus $\phi \in l^q$.

Combining (4.12) and (4.13) shows that in fact $\|\phi\|_{l^q} = \|L\|_{(l^p)^*}$, and we are done. ■

4.2.3 The Dual Spaces of L^1 and L^∞

We are still left with the spaces $L^1(\Omega)$ and $L^\infty(\Omega)$. It is easy to see that every element of L^∞ can be used to define an element of $(L^1)^*$ via (4.10), and one can check, by using Proposition 1.16 and arguments similar to those in Section 4.2.1 (see Exercise 4.3), that

$$\|L_f\|_{(L^1)^*} = \|f\|_{L^\infty}.$$

In this case, as with the L^p spaces above, every element of the dual space of $L^1(\Omega)$ can be obtained, via (4.10), from an element of $L^\infty(\Omega)$. (Since $L^\infty \simeq L^1$, one way of defining L^∞ is as the dual space of L^1.)

However, $L^1(\Omega)$ is *not* the whole dual space of $L^\infty(\Omega)$, which is much more complicated to characterise (see Yosida (1980) for example). We will illustrate this lack of symmetry in more detail, using the simpler example of the sequence spaces l^1 and l^∞.

4.2.4 The Dual Space of l^1 and l^∞

In the last section we claimed that $(L^1)^* \simeq L^\infty$ but that L^1 is not itself the dual space of L^∞. We can see this more clearly with the sequence spaces. First we show that $(l^1)^* \simeq l^\infty$.

Theorem 4.7. $(l^1)^* \simeq l^\infty$.

Proof. Take $\phi \in l^\infty$, $\phi = \{\phi_k\}$. Then ϕ corresponds to a linear functional L_ϕ on l^1 given by

$$L_\phi(x) = \sum_{k=1}^\infty \phi_k x_k.$$

Clearly we have

$$|L_\phi(x)| \leq \sum_{k=1}^\infty |\phi_k x_k| \leq \sup_k |\phi_k| \sum_{k=1}^\infty |x_k| = \|\phi\|_{l^\infty} \|x\|_{l^1},$$

so that

$$\|L_\phi\|_{(l^1)^*} \leq \|\phi\|_{l^\infty}.$$

Now take an $L \in (l^1)^*$, and consider $x \in l^1$ with $x = \sum_{k=1}^\infty x_k e_k$, where $\{e_k\}$ is a basis for l^1, as used in the proof of Theorem 4.6. Then

$$L(x) = \sum_{k=1}^\infty x_k \phi_k,$$

with $\phi_k = L(e_k)$, and so in particular $L_\phi = L$. Now we have to show that in fact $\phi \in l^\infty$ with $\|\phi\|_{l^\infty} = \|L\|_{(l^1)^*}$. This follows since

$$|\phi_k| \leq \|L\|_{(l^1)^*} \|e_k\|_{l^1} = \|L\|_{(l^1)^*},$$

which implies that

$$\|\phi\|_{l^\infty} \leq \|L\|_{(l^1)^*}.$$

Thus $\|L\|_{(l^1)^*} = \|\phi\|_{l^\infty}$, and the theorem is proved. ∎

Now, l^1 is *not* the dual space of l^∞ – it is in fact the dual space of a subspace of l^∞, the space c_0, which is the space of all sequences that converge to zero:

$$c_0 = \{\{x_j\} : x_j \to 0\}.$$

This is a Banach space when equipped with the l^∞ norm. (Note that $c_0 \subset l^\infty$ with no further assumptions, since a convergent sequence is bounded.)

Theorem 4.8. $(c_0)^* \simeq l^1$.

Note that the proof is a simpler version of that of Theorem 4.6.

Proof. If $\phi \in l^1$, $\phi = \{\phi_k\}$ then this gives a linear functional on c_0 by setting

$$L_\phi(x) = \sum_{k=1}^{\infty} \phi_k x_k,$$

when $x = \{x_k\}$. Indeed,

$$|L_\phi(x)| \le \sum_{k=1}^{\infty} |\phi_k x_k| \le \sup_k |x_k| \sum_{k=1}^{\infty} |\phi_k| = \|\phi\|_{l^1} \|x\|_{l^\infty},$$

so that

$$\|L_\phi\|_{(c_0)^*} \le \|\phi\|_{l^1}.$$

Now take $L \in (c_0)^*$ with $L \ne 0$ ($L = 0$ corresponds trivially to $\phi_k = 0$ for all k), and let $\phi_k = L(e_k)$. Then

$$L(x) = L\left(\sum_{k=1}^{\infty} x_k e_k\right) = \sum_{k=1}^{\infty} x_k L(e_k) = \sum_{k=1}^{\infty} x_k \phi_k,$$

so that $L_\phi = L$.

Now consider the elements $x^{(m)}$ of c^0 with

$$x_n^{(m)} = \begin{cases} \text{sign } \phi_n, & n \le m, \\ 0, & n > m, \end{cases}$$

and so

$$L(x^{(m)}) = \sum_{k=1}^{m} |\phi_k| \le \|L\|_{(c_0)^*} \|x^{(m)}\|_{l^\infty} = \|L\|_{(c_0)^*}.$$

Thus

$$\sum_{k=1}^{m} |\phi_k| \le \|L\|_{(c_0)^*}$$

for all m sufficiently large (so that $x^{(m)} \ne 0$), and so $\phi \in l^1$, with

$$\|\phi\|_{l^1} \le \|L\|_{(c_0)^*}.$$

This shows that $\|L\|_{(c_0)^*} = \|\phi\|_{l^1}$, and the theorem is proved. ∎

We will see at the end of Section 4.4 that it follows that the dual space of l^∞ is *not* l^1.

4.3 Dual Spaces of Hilbert Spaces

Since $(L^p)^* \simeq L^q$, if $p = q = 2$ this shows that the dual of L^2 is identifiable with L^2 itself. This is a particular case of a more general result valid for any Hilbert space, the Riesz representation theorem. This shows that any linear functional l on H can be represented as an inner product with some appropriate element x_l of H itself (4.14) and that the norms of l (in H^*) and x_l (in H) are the same.

Theorem 4.9 (Riesz Representation Theorem). *For any Hilbert space H, $H^* \simeq H$. In particular, for every $x \in H$,*

$$l_x(y) \equiv (x, y) \tag{4.14}$$

is bounded and has norm $\|l_x\|_{H^*} = \|x\|$. Furthermore, for every bounded linear functional $l \in H^*$ there exists a unique $x_l \in H$ such that

$$l(y) = (x_l, y) \qquad \text{for all} \qquad y \in H,$$

and $\|x_l\|_H = \|l\|_{H^*}$. It follows that $l \mapsto x_l$ is continuous.

Proof. The first part of the theorem is straightforward, since (4.14) provides a linear functional l_x associated with x. The Cauchy–Schwarz inequality gives

$$|l_x(y)| = |(x, y)| \le \|x\| \|y\|$$

so that $l_x \in H^*$ with $\|l_x\|_{H^*} \le \|x\|$, and the choice $y = x$ shows that in fact $\|l_x\|_{H^*} = \|x\|$.

The second part of the theorem involves a little more work. Suppose that $l \in H^*$. Then the kernel of l, $K = \{y \in H : l(y) = 0\}$, is a closed subspace of H. The subspace $K^\perp$ of vectors orthogonal to K is a one-dimensional subspace of H, since if $u, v \in K^\perp$ we have

$$l[l(u)v - l(v)u] = 0,$$

and so $l(u)v - l(v)u \in K$. Since u and v are orthogonal to all elements of K, so is $l(u)v - l(v)u$, which implies that $l(u)v - l(v)u = 0$, and so u and v are proportional.

Now choose a unit vector z in $K^\perp$. We can decompose every $y \in H$ as $y = (z, y)z + w$, where $w \in K$. Then $l(y) = (z, y)l(z)$, and so if we set $x_l = l(z)z$, we have

$$(x_l, y) = (l(z)z, y) = l(y),$$

and the norm equality is clear. ∎

4.4 Reflexive Spaces

If H is a Hilbert space the Riesz theorem shows that H^* is isometrically isomorphic to H. The situation in a general Banach space X is somewhat more complicated. For an element $x \in X$ one can define a linear functional G_x on X^* (i.e. an element of $X^{**} \equiv (X^*)^*$) by

$$G_x(f) = f(x) \qquad \text{for all} \qquad f \in X^*. \tag{4.15}$$

If we set $Ax = G_x$ this gives a linear map $A : X \to X^{**}$. Noting that

$$|G_x(f)| \le \|f\|_{X^*} \|x\|,$$

one has

$$\|G_x\|_{X^{**}} \le \|x\|.$$

We saw in Corollary 4.5 that given an $x \in X$ there is an $f \in X^*$ with $\|f\|_{X*} = 1$ and $f(x) = |x|$; this shows that in fact

$$\|G_x\|_{X^{**}} = \|x\|, \tag{4.16}$$

and so A is an isometry from X onto a subspace of X^{**}. When this isometry is onto (i.e. when $X \simeq X^{**}$), then X is called *reflexive*.

The Riesz representation theorem shows that all Hilbert spaces are reflexive, and Theorem 4.6 shows that the l^p spaces, $1 < p < \infty$, are reflexive, as are the Lebesgue spaces $L^p(\Omega)$, with $1 < p < \infty$. Theorems 4.7 and 4.8 show that the sequence space c_0 is not reflexive, since $(c_0)^{**} = (l^1)^* = l^\infty$, and c_0 is a proper subset of l^∞.

We will now see that the sequence spaces l^1 and l^∞ are not reflexive, using the following result.

Proposition 4.10. *E is reflexive iff E^* is reflexive.*

Proof If E is reflexive, then $E \simeq E^{**}$, and so

$$E^* = (E^{**})^* = ((E^*)^*)^* = (E^*)^{**},$$

which shows that E^* is reflexive.

Now suppose that E^* is reflexive, so that $E^* \simeq (E^*)^{**} = (E^{**})^*$. Then for any element $\Psi \in (E^{**})^*$ there is an $f \in E^*$ such that

$$\Psi(G) = G(f) \qquad \text{for all} \qquad G \in E^{**}. \tag{4.17}$$

Now, E can be identified with a subset of E^{**} via (4.15), so that for each $x \in E$ there is a corresponding element $G_x \in E^{**}$ given by

$$G_x(f) = f(x) \qquad \text{for all} \qquad f \in E^*. \tag{4.18}$$

Suppose that $\Psi(G_x) = 0$ for any such G_x ("$\Psi = 0$ on E"). Then it follows that $f(x) = 0$ for any $x \in E$. Since $f \in E^*$, it follows that $f = 0$. Since Ψ and f are related as in (4.17), it follows that Ψ itself must also be zero.

Now suppose that E is a proper subspace of E^{**} (under identification (4.18)). Then the result of Exercise 4.2 (cf. Lemma 4.4) shows that there exists a non-zero linear functional on E^{**} that vanishes on E. But we have just shown that if $\Psi = 0$ on E then $\Psi = 0$, and so this is impossible. We conclude that $E \simeq E^{**}$. ∎

We now know that since c_0 is not reflexive, then neither are $(c_0)^* \simeq l^1$ nor $(l^1)^* \simeq l^\infty$. In particular this shows that $(l^\infty)^* \not\simeq l^1$, since otherwise we would have $(l^1)^{**} \simeq l^1$.

It follows similarly that the Lebesgue spaces $L^1(\Omega)$ and $L^\infty(\Omega)$ are not reflexive and that $(L^\infty)^* \not\simeq L^1$.

4.5 Notions of Weak Convergence

As the final topic of this chapter we introduce the extremely powerful notion of weak convergence, which will enable us – to some extent – to circumvent the problem that the unit ball is not compact in an infinite-dimensional space.

4.5.1 Weak Convergence

Since we know from Lemma 4.4 that linear functionals can distinguish elements of X, we will define a notion of convergence based on the application of linear functionals.

Definition 4.11. *Let X be a Banach space. A sequence $x_n \in X$ converges weakly to x (in X),*

$$x_n \rightharpoonup x \qquad in \qquad X,$$

if $f(x_n) \to f(x)$ for every $f \in X^*$.

First we will motivate the terminology. Another term used for the standard convergence in the norm of the space ($x_n \to x$) is strong convergence of x_n to

x – we will show that this "strong" convergence implies weak convergence, but not vice versa.

Lemma 4.12. *If $x_n \to x$ (strong convergence) then $x_n \rightharpoonup x$ (weak convergence).*

For some more properties of weak limits, see Exercise 4.5.

Proof Every element of X^* is a bounded linear functional and hence continuous; thus $f(x_n) \to f(x)$ for every $f \in X^*$, which is precisely $x_n \rightharpoonup x$. ∎

However, there are simple examples of weakly convergent subsequences that do not converge strongly. For example, let $\{e_j\}$ be an orthonormal basis in a separable Hilbert space. Then we know that the $\{e_j\}$ have no convergent subsequence (Proposition 1.25), but we can show that $e_j \rightharpoonup 0$. Indeed, using the Riesz representation theorem (Theorem 4.9) every element $l \in H^*$ has a representation as $(x_l, \cdot)$ for some $x_l \in H$, and so it suffices to consider sequences (x_l, e_j) for each $x_l \in H$. Since $\{e_j\}$ is an orthonormal basis,

$$\|x_l\|^2 = \sum_{j=1}^{\infty} |(x_l, e_j)|^2,$$

and so $|(x_l, e_j)| \to 0$ as $j \to \infty$. But this is exactly $l(e_j) \to 0$ for every $l \in H^*$, and so $e_j \rightharpoonup 0$. For another example, see Exercise 4.6.

We now want to investigate further some important properties of weak limits. Although the two properties in the following proposition, uniqueness of weak limits and boundedness of weakly convergent sequences, are elementary, the proofs rely on the Hahn–Banach theorem (Theorem 4.3) and the uniform boundedness principle (Theorem 3.7), both powerful results.

Proposition 4.13. *Weak limits are unique, and weakly convergent sequences are bounded.*

Proof If $x_n \rightharpoonup x$ and $x_n \rightharpoonup y$ then from the definition we must have $f(x) = f(y)$ for all $f \in X^*$. Lemma 4.4 shows that this implies that $x = y$. To show boundedness, observe that, for each $f \in X^*$, $\{f(x_n)\}$ is a convergent sequence of real numbers, and so bounded:

$$|f(x_n)| \leq C_f \qquad \text{for all } n.$$

Now, we can define, as in (4.15), an element $G_n \in X^{**}$ that corresponds to x_n,

$$G_n(f) = f(x_n), \qquad \text{for all} \quad f \in X^*,$$

and then it follows that

$$|G_n(f)| \le C_f \qquad \text{for all } n:$$

$\{G_n(f)\}$ is a bounded sequence for every $f \in X^*$. Since X^* is complete (Proposition 3.1) the uniform boundedness principle (Theorem 3.7) shows that the sequence $\{\|G_n\|_{X^{**}}\}$ is bounded. We showed in (4.16) that $\|G_n\|_{X^{**}} = \|x_n\|$, and so $\{x_n\}$ is bounded. ∎

The example of the orthonormal basis $\{e_j\}$ with $|e_j| = 1$ for all j but $e_j \rightharpoonup 0$ shows that taking weak limits can decrease the norm. However, the following result shows that it can never increase it.

Lemma 4.14. *If* $x_n \rightharpoonup x$ *in* X *then*

$$\|x\| \le \liminf_{n \to \infty} \|x_n\|. \tag{4.19}$$

Proof If $x_n \rightharpoonup x$, choose $f \in X^*$ such that $f(x) = \|x\|$ and $\|f\|_* = 1$ (see Corollary 4.5). Then

$$\|x\| = \|f(x)\| = \lim_{n \to \infty} \|f(x_n)\| \le \liminf_{n \to \infty} \|f\|_* \|x_n\| = \liminf_{n \to \infty} \|x_n\|,$$

and (4.19) follows. ∎

One important case in which weak convergence becomes strong convergence is if we look at the image of a weakly convergent sequence under a compact operator (see also Exercise 4.7).

Theorem 4.15. *Suppose that* $A : X \to Y$ *is compact and that* $x_n \rightharpoonup x$ *in* X. *Then* $Ax_n \to Ax$ *in* Y.

Proof The first step is to show that $Ax_n \rightharpoonup Ax$ in Y, i.e. if we take $f \in Y^*$ then $f(Ax_n) \to f(Ax)$. The composition $g = f \circ A$ is a linear map from X into $\mathbb{R}$, and it is bounded since

$$|g(x)| \le \|f\|_{Y^*} \|A\|_{\mathcal{L}(X,Y)} \|x\|_X.$$

Thus $g \in X^*$, and since $x_n \rightharpoonup x$, $g(x_n) \to g(x)$ by definition. So $f[Ax_n] \to f[Ax]$ for every $f \in Y^*$ which is $Ax_n \rightharpoonup Ax$ in Y.

Now, if $Ax_n \not\to Ax$ then there is an $\epsilon > 0$ and a sequence of n_k such that

$$\|Ax_{n_k} - Ax\|_Y \geq \epsilon. \tag{4.20}$$

Clearly, $x_{n_k} \rightharpoonup x$. Since A is compact, there is a further subsequence of x_{n_k}, x_j, such that $Ax_j \to z$ for some z. Since strong convergence implies weak convergence and weak limits are unique (Proposition 4.13), we must have $z = Ax$. But this contradicts (4.20), proving the result. ∎

4.5.2 Weak-* Convergence

We could also apply the definition of weak convergence to elements of X^*. By definition, $f_n \rightharpoonup f$ in X^* provided that

$$G(f_n) \to G(f) \tag{4.21}$$

for every element $G \in X^{**}$. However, a more useful type of weak convergence for a sequence $\{f_n\}$ in X^* is weak-* ("weak star") convergence, defined in terms of the action of the f_n on elements of X.

Definition 4.16. *A sequence $f_n \in X^*$ converges weakly-* to f, written*

$$f_n \overset{*}{\rightharpoonup} f,$$

if $f_n(x) \to f(x)$ for every $x \in X$.

We now prove some properties of weak-* convergence in line with Proposition 4.13, and we also discuss its relationship to weak convergence.

Proposition 4.17
 (i) *Weak-* limits are unique, and weakly-* convergent sequences are bounded,*
 (ii) *weak convergence implies weak-* convergence, and*
(iii) *weak-* convergence implies weak convergence if X is reflexive.*

Proof
 (i) Uniqueness for weak-* limits follows immediately from the definition, for if $f, g \in X^*$ with $f(x) = g(x)$ for all $x \in X$, then $f = g$. The principle of uniform boundedness (Theorem 3.7) can be applied immediately to the sequence $\{f_n\}$ to show that it is bounded.

(ii) In general (see above, (4.15)) we know that $X \subset X^{**}$, using the identification

$$G_x(f) = f(x) \qquad \text{for all} \qquad f \in X^*. \tag{4.22}$$

Weak convergence ensures that

$$G(f_n) \to G(f) \qquad \text{for all} \qquad G \in X^{**}, \tag{4.23}$$

and in particular we have (4.23) for those G of the form G_x for some $x \in X$, which gives, by the definition of G_x,

$$f_n(x) \to f(x) \qquad \text{for all} \qquad x \in X,$$

which is weak-* convergence.

(iii) If X is reflexive then any element G of X^{**} can be written as G_x, as in (4.22), for some $x \in X$. For each such G we have

$$G_x(f_n) = f_n(x) \to f(x) = G_x(f), \tag{4.24}$$

using the weak-* convergence to take the limit. Weak convergence now follows, since (4.23) holds for all $G \in X^{**}$. ∎

4.6 The Alaoglu Weak-* Compactness Theorem

The real power of the theory of dual spaces and weak convergence comes from the following compactness theorem. Recall that the unit ball in an infinite-dimensional space is not compact (Proposition 1.25). However, the following result and its corollary show that these balls are weakly compact. In particular, we will see that a bounded sequence in a reflexive Banach space has a *weakly* convergent subsequence.

Theorem 4.18 (Alaoglu weak-* compactness). *Let X be a separable Banach space and let f_n be a bounded sequence in X^*. Then f_n has a weakly-* convergent subsequence.*

In fact this theorem is valid in any Banach space [see Yosida (1980)].

Proof. Let x_k, $k \in \mathbb{N}$, be a sequence that is dense in X. Note that if $f_n(x_k)$ converges for each x_k then $f_n(x)$ converges for every $x \in X$. Indeed, if $\|f_n\|_{X^*} \leq M$ then, given $\epsilon > 0$ and $x \in X$, first choose k with

$$\|x - x_k\|_X \leq \epsilon/3M.$$

Then

$$|f_n(x) - f_m(x)|$$

$$\leq |f_n(x) - f_n(x_k)| + |f_n(x_k) - f_m(x_k)| + |f_m(x_k) - f_m(x)|$$

$$\leq \epsilon/3 + |f_n(x_k) - f_m(x_k)| + \epsilon/3,$$

with the remaining term bounded by $\epsilon/3$ if m and n are large enough. The rest of the proof follows the diagonal argument used in Theorems 2.5 and 3.9. ∎

The following simple corollary will be extremely useful.

Corollary 4.19 (Reflexive weak compactness). *Let X be a reflexive Banach space and x_n a bounded sequence in X. Then x_n has a subsequence that converges weakly in X.*

Proof. Corresponding to each x_n there exists an element G_n of X^{**} given as in (4.15) by the definition

$$G_n(f) = f(x_n) \qquad \text{for all} \qquad f \in X^*.$$

Since we saw in (4.16) that $\|G_n\|_{X^{**}} = \|x_n\|$, $\{G_n\}$ is a bounded sequence in $X^{**} = (X^*)^*$. We can therefore use the result of Theorem 4.18 to find a subsequence G_{n_j} that converges weakly-* to some $G \in X^{**}$. Since X is reflexive, there exists an $x \in X$ such that

$$G(f) = f(x) \qquad \text{for all} \qquad f \in X^*.$$

The weak-* convergence of G_{n_j} to G,

$$G_{n_j}(f) \to G(f) \qquad \text{for all} \qquad f \in X^*,$$

therefore implies that

$$f(x_{n_j}) \to f(x) \qquad \text{for all} \qquad f \in X,$$

which is weak convergence of x_{n_j} to x in X. ∎

Since any Hilbert space H is reflexive, a bounded sequence in H has a weakly convergent subsequence.

In the next chapter we introduce the Sobolev spaces of functions whose derivatives lie in L^p. These form the basis of our existence and uniqueness analysis in Chapters 6–9.

Exercises

4.1 Use Zorn's lemma to show that every Hilbert space has a basis. (Hint: Find a maximal orthonormal set.)

4.2 Show that if Y is a proper linear subspace of X then there exists a nonzero element of X^* that vanishes on Y.

4.3 For Ω of finite volume, show that if $f \in L^\infty(\Omega)$ then the norm of the linear functional L_f defined on $L^1(\Omega)$ by

$$L_f(g) = \int_\Omega f(x)g(x)\,dx \qquad \text{for all} \qquad g \in L^1(\Omega)$$

satisfies

$$\|L_f\|_{(L^1)^*} = \|f\|_{L^\infty}.$$

[Hint: Consider the sequence of functions $g_p(x) = |f(x)|^{p-2} f(x)$, and use Proposition 1.16.]

4.4 Use the Riesz representation theorem to prove the Hahn–Banach theorem for a Hilbert space.

4.5 Suppose that M is a linear subspace of a Banach space X and that $\{x_n\}$ is a sequence of elements of M that converges weakly in X to some x. Show that $x \in M$. Deduce that

$$x = \sum_{j=1}^\infty c_j x_j$$

for some coefficients $\{c_j\}$. [Hint: As a first step show that if $f(x) = 0$ for every $f \in X^*$ such that $f|_M = 0$, then $x \in M$.]

4.6 If $x_n \in C^0([a, b])$ and $x_n \rightharpoonup x$ in $C^0([a, b])$, show that $\{x_n\}$ is pointwise convergent on $[a, b]$, i.e. that $x_n(t)$ converges for every $t \in [a, b]$.

4.7 Let H be a Hilbert space. Show that if $x_n \rightharpoonup x$ in H and $\|x_n\| \to \|x\|$ then $x_n \to x$.

Notes

Various generalisations of the Hahn–Banach theorem are possible. In particular, one can replace the right-hand side of both (4.1) and (4.2) with $p(x)$, where $p(x)$ is any positive convex function,

$$p(x + y) \le p(x) + p(y);$$

see Bollobás (1990), for example.

The ideas of weak convergence here are treated to a greater or lesser extent in the various texts mentioned in the notes of the previous chapter. They are mentioned briefly in the appendix of Evans (1998) as part of the functional analytic background he assumes. Renardy & Rogers (1992) give a good overview of the theory, while Kreyszig (1978) has a lengthy (and readable) discussion including the detailed examples of the l^p sequence spaces used above. Bollobás (1990) discusses linear functionals at length, and Meise & Vogt (1997) give a high-level account of dual spaces (Proposition 4.10 is adapted from there and gives a good idea of the general level of the text).

A proper characterisation of the dual of L^∞ is given in Yosida (1980).

5

Sobolev Spaces

The subject of this chapter could easily be the subject of a whole book. Indeed, it is, and Adams's 1975 book, *Sobolev Spaces*, is a very nicely presented and detailed survey of the subject. We will give a much simplified presentation here, mainly restricting attention to those Sobolev spaces that are also Hilbert spaces, and by considering only domains with smooth boundaries. Nonetheless, the results we will obtain are typical of the more general theory.

Sobolev spaces are a family of spaces of functions akin to $C^r(\overline{\Omega})$, in that they consist of functions with conditions on their derivatives. Rather than requiring the derivatives to be in $C^0(\overline{\Omega})$, we require them to be elements of $L^p(\Omega)$. First we introduce a generalised notion of the derivative that allows us to make such a definition sensibly.

5.1 Generalised Notions of Derivatives

We saw briefly in the introduction that, owing to considerations of energy, the natural phase space for many problems modelled by PDEs will be L^2 or some related space. However, since we are treating differential equations we need some way to assign a meaning to the "derivative" of an L^2 function, unless we are always to treat ungainly spaces such as $L^2(\Omega) \cap C^1(\Omega)$.

5.1.1 The Weak Derivative

We first introduce the "weak derivative," which has to be an element of $L^1_{\text{loc}}(\Omega)$. Suppose that u *is* differentiable. Then for any "test function" $\phi \in C_c^\infty(\Omega)$, the identity

$$\int_\Omega \frac{du}{dx_j}\phi\, dx = -\int_\Omega u \frac{d\phi}{dx_j}\, dx \qquad (5.1)$$

holds, after an integration by parts in the x_j variable (the boundary terms vanish since ϕ has compact support in Ω). Repeating this process $|\alpha|$ times, we have similarly

$$\int_\Omega D^\alpha u \phi \, dx = (-1)^{|\alpha|} \int_\Omega u D^\alpha \phi \, dx \tag{5.2}$$

for any multi-index α.

The weak derivative is defined by analogy with (5.1): for a function $u \in L^1_{\text{loc}}(\Omega)$, we say that v is the *weak derivative* of u with respect to x_j, written $v = D_j u$, if $v \in L^1_{\text{loc}}(\Omega)$ and

$$\int_\Omega v \phi \, dx = - \int_\Omega u \frac{d\phi}{dx_j} \, dx \tag{5.3}$$

for all $\phi \in C^\infty_c(\Omega)$. The requirement that u and v are elements of $L^1_{\text{loc}}(\Omega)$ is natural, since the support of ϕ will be contained in some compact subdomain K of Ω, and then $u, v \in L^1(K)$ ensures that both integrals make sense.

We can also define, inductively, higher-order weak derivatives: if u, $v \in L^1_{\text{loc}}(\Omega)$ then v is the αth weak derivative of u, $v = D^\alpha u$ if

$$\int_\Omega v \phi \, dx = (-1)^{|\alpha|} \int_\Omega u D^\alpha \phi \, dx.$$

As an example, consider the $L^1(0, 2)$ function

$$u(x) = \begin{cases} x, & 0 < x \le 1, \\ 1, & 1 < x < 2. \end{cases} \tag{5.4}$$

For any $\phi \in C^\infty_c(0, 2)$ we have

$$-\int_0^2 u \phi' \, dx = -\int_0^1 x \phi' \, dx - \int_1^2 \phi' \, dx$$

$$= -[x\phi]_0^1 + \int_0^1 \phi \, dx - \phi(2) + \phi(1)$$

$$= \int_0^1 \phi \, dx$$

(since $\phi(2) = 0$), and so u has weak derivative

$$v(x) = \begin{cases} 1, & 0 < x \le 1, \\ 0, & 1 < x < 2. \end{cases} \tag{5.5}$$

However, even being differentiable almost everywhere is not enough to have a weak derivative. The function

$$u(x) = \begin{cases} x, & 0 < x \le 1, \\ 2, & 1 < x < 2 \end{cases} \tag{5.6}$$

does *not* have a weak derivative. Following the above computations, we obtain

$$-\int_0^2 u\phi' \, dx = \int_0^1 \phi \, dx + \phi(1),$$

but there is no $L^1_{\text{loc}}(0, 2)$ function v such that

$$\int_0^2 v\phi \, dx = \int_0^1 \phi \, dx + \phi(1). \tag{5.7}$$

The theory of distributions, along with the definition of the distribution derivative, gives a way around this problem.

5.1.2 The Distribution Derivative

The main idea behind the theory of distributions is that for each continuous function f on Ω, and indeed for each function $f \in L^p_{\text{loc}}(\Omega)$, there is a linear functional L_f on $C_c^\infty(\Omega)$ defined by

$$\langle L_f, \phi \rangle = \int_\Omega f(x)\phi(x) \, dx,$$

where we use the notation $\langle \rho, \phi \rangle$ to denote the action of the linear functional ρ on the test function ϕ.

Furthermore, if f, g are locally integrable and $L_f = L_g$ then we must have $f = g$ almost everywhere. This result, which is the content of the next lemma, gives the theory its considerable power.

Lemma 5.1. *If $f, g \in L^1_{\text{loc}}(\Omega)$ and $L_f = L_g$, then $f = g$ almost everywhere.*

Proof. It clearly suffices to show that if $u \in L^1_{\text{loc}}(\Omega)$ and

$$\int_\Omega u\phi \, dx = 0 \tag{5.8}$$

for all test functions ϕ, then $u = 0$ a.e. For a given $\Omega' \subset\subset \Omega$, we know that $C_c^\infty(\Omega')$ is dense in $L^1(\Omega')$ by using Corollary 1.14, and so, for a given $\epsilon > 0$, choose $\psi_\epsilon \in C_c^\infty(\Omega')$ such that

$$\|u - \psi_\epsilon\|_{L^1(\Omega')} \leq \epsilon.$$

Now, set $g = \mathrm{sgn}(\psi_\epsilon)$. Then $\|g\|_\infty \leq 1$, and since Ω' has finite volume, $g \in L^1(\Omega')$. We now approximate g by its mollification g_h, choosing h small enough that

$$\|g - g_h\|_{L^1(\Omega')} \leq \delta.$$

It follows from Proposition 1.18 that $g_h \in C_c^\infty(\Omega')$, and (using (1.7) and (1.8)) we obtain

$$|g_h(x)| = \left| h^{-m} \int_{\Omega'} \rho\left(\frac{x-y}{h}\right) g(y)\, dy \right|$$

$$\leq \|g\|_\infty \left| h^{-m} \int_{\Omega'} \rho\left(\frac{x-y}{h}\right) dy \right|$$

$$= \|g\|_\infty \leq 1.$$

Then, setting $\phi = g_h$ in (5.8) we have

$$0 = \int_{\Omega'} u g_h\, dx \geq \int_{\Omega'} \psi_\epsilon g\, dx - \left| \int_{\Omega'} \psi_\epsilon(g_h - g) + (u - \psi_\epsilon)g_h\, dx \right|$$

$$\geq |\psi_\epsilon|_{L^1(\Omega')} - \|\psi_\epsilon\|_\infty \|g_h - g\|_{L^1} - \|u - \psi_\epsilon\|_{L^1}\|g_h\|_\infty$$

$$\geq \|u\|_{L^1(\Omega')} - \|\psi_\epsilon\|_\infty \delta - 2\epsilon.$$

Choosing first ϵ, then δ, both sufficiently small, we obtain a contradiction unless $\|u\|_{L^1(\Omega')} = 0$. Since this is valid for each $\Omega' \subset\subset \Omega$ we must have $u = 0$ a.e. in Ω. ∎

We would like to define the space of distributions, written $\mathcal{D}'(\Omega)$, as the space of all continuous linear functionals on $\mathcal{D}(\Omega) \equiv C_c^\infty(\Omega)$. However, there are some subtleties, since $C_c^\infty(\Omega)$ is not a Banach space, and without a norm we cannot immediately give meaning to the idea that a distribution has to be a *continuous* linear functional. We get around this by defining an appropriate notion of convergence in $\mathcal{D}(\Omega)$.

Definition 5.2. *A sequence $\{\phi_n\}$ in $\mathcal{D}(\Omega)$ is said to* converge *to ϕ in $\mathcal{D}(\Omega)$ if there is a compact set K with $\mathrm{supp}(\phi_n) \subset K$ for all n and if ϕ_n and all its derivatives converge uniformly to ϕ on K.*

We can now define $\mathcal{D}'(\Omega)$ as the set of all linear functionals on $\mathcal{D}(\Omega)$, which are sequentially continuous; that is, if $\phi_n \to \phi$ in $\mathcal{D}(\Omega)$ then

$$\langle u, \phi_n \rangle \to \langle u, \phi \rangle. \tag{5.9}$$

Often we blur the distinction between a distribution and a function, as in the case when $u = L_f$,

$$\langle u, \phi \rangle = \int_\Omega f\phi \, dx,$$

for some $f \in L^1_{\text{loc}}$. In this way we might say $u \in L^1_{\text{loc}}(\Omega)$ as shorthand for "u can be written in the form L_f for some $f \in L^1_{\text{loc}}(\Omega)$." [This is the same sense in which we might have claimed in Chapter 4 that $(L^p)^* = L^q$ for (p, q) conjugate.]

We can now define the distribution derivative in a way that generalises the weak derivative and includes many more possibilities. If $u \in \mathcal{D}'(\Omega)$ we define the distribution derivative $D^\alpha u$ in line with (5.2) as the linear functional given by

$$\langle D^\alpha u, \phi \rangle = (-1)^{|\alpha|} \langle u, D^\alpha \phi \rangle.$$

It is easy to check that $D^\alpha u$ is sequentially continuous (as in (5.9)) and so is an element of $\mathcal{D}'(\Omega)$ (see Exercises 5.1 and 5.2).

This circumvents the problem we found with (5.6), since (5.7) immediately defines the distribution derivative of u as the linear functional

$$\phi \mapsto \int_0^1 \phi \, dx + \phi(1).$$

In fact, if we define the Dirac delta "function" (which is really a distribution) δ_y by

$$\langle \delta_y, \phi \rangle = \phi(y),$$

then we can write

$$Du = v + \delta_1,$$

where v is the weak derivative (5.5) of the function in (5.4). (Strictly this is $Du = L_v + \delta_1$, but this kind of pedantry is unhelpful.)

Note that Lemma 5.1 shows that weak derivatives are unique and that when they exist they must coincide with the distribution derivative (in the sense discussed above).

5.2 General Sobolev Spaces

In the first chapter we considered various examples of Banach and Hilbert spaces. One was the family $C^r(\overline{\Omega})$ of functions classified according to their degree of differentiability,

$$C^r(\overline{\Omega}) = \{u : D^\alpha u \in C^0(\overline{\Omega}), \text{ for all } 0 \le |\alpha| \le r\}.$$

The main class of spaces we will use in the study of PDEs is based instead on integrability properties of the weak derivative. The Sobolev space $W^{k,p}(\Omega)$ consists of functions whose weak derivatives of all orders up to k lie in L^p.

Definition 5.3. *The* Sobolev space $W^{k,p}(\Omega)$ *is defined as*

$$W^{k,p}(\Omega) = \{u : D^\alpha u \in L^p(\Omega), \text{ for all } 0 \le |\alpha| \le k\},$$

with norm

$$\|u\|_{W^{k,p}} = \left\{ \sum_{0 \le |\alpha| \le k} \|D^\alpha u\|_{L^p}^p \right\}^{1/p}.$$

Our main task in the remainder of this chapter is to investigate various properties of the Sobolev spaces $W^{k,2}$. For example, we will show that in certain cases all elements of $W^{k,2}(\Omega)$ also lie in $C^r(\overline{\Omega})$ [for some appropriate $r \ge 0$], which means that even if we work with Sobolev spaces we can then deduce that the functions are in fact classically differentiable and smooth.

First, we show that these Sobolev spaces are Banach spaces.

Theorem 5.4. $W^{k,p}(\Omega)$ *is a separable Banach space.*

Proof. If $\{u_n\}$ is a Cauchy sequence in $W^{k,p}$, then $\{D^\alpha u_n\}$ is a Cauchy sequence in L^p for all $0 \le |\alpha| \le k$. Since the L^p spaces are complete, there exist functions u and u_α such that $u_n \to u$ and $D^\alpha u_n \to u_\alpha$ in L^p. We now have to show that $u \in W^{k,p}$ and that $u_\alpha = D^\alpha u$.

Now, for any function $\phi \in C_c^\infty(\Omega)$,

$$\int_\Omega D^\alpha u \phi \, dx = (-1)^{|\alpha|} \int_\Omega u D^\alpha \phi \, dx = (-1)^{|\alpha|} \lim_{n \to \infty} \int_\Omega u_n D^\alpha \phi \, dx$$

$$= \lim_{n \to \infty} \int_\Omega D^\alpha u_n \phi \, dx$$

$$= \int_\Omega u_\alpha \phi \, dx, \qquad (5.10)$$

and so $D^\alpha u = u_\alpha$. Since $D^\alpha u_n \to D^\alpha u$ in L^p, it follows that $u \in W^{k,p}$.

For the separability, observe that the map $u \mapsto (u_\alpha)_{|\alpha| \le k}$ is an isometry into the product space $\Pi_{|\alpha| \le k} L^p(\Omega)$. Since a linear subspace of a separable space is itself separable and a finite product of separable spaces is separable (see Exercise 1.1), $W^{k,p}$ is separable. ∎

In the course of Theorem 5.3 we used equality (5.10), and this gives us the following useful result. We say that $u_n \to u$ in $\mathcal{D}'(\Omega)$ if

$$\langle u_n, \phi \rangle \to \langle u, \phi \rangle \qquad \text{for all} \qquad \phi \in \mathcal{D}(\Omega).$$

Corollary 5.5. *If $u_n \to u$ in $\mathcal{D}'(\Omega)$, then $D^\alpha u_n \to D^\alpha u$ in $\mathcal{D}'$ for any multi-index α.*

See Exercise 5.3 for one useful consequence of convergence in $\mathcal{D}'$.

5.2.1 Sobolev Spaces and the Closure of Differential Operators

Note that, given the result of Theorem 5.4, the definitions of the weak derivative and of the Sobolev spaces $W^{k,p}$ provide the solution to the problem of closing differential operators discussed in Section 3.8. For example, suppose that we have a sequence $u_n \in C^1(\Omega)$ with $u_n \to u$ in $L^2(\Omega)$. Now, if $Du_n \to v$ in L^2 then u_n must be a Cauchy sequence in $W^{1,2}$, and since $W^{1,2}$ is a Banach space we must have $u \in W^{1,2}$ and so $v = Du$. In this way the weak derivative gives the natural generalisation of d/dx to a closed domain in $L^2(\Omega)$, namely $W^{1,2}(\Omega)$.

5.2.2 The Hilbert Space $H^k(\Omega)$

It is the case $p = 2$ that arises most naturally in many applications. Because of this, we will be dealing in what follows mainly with the spaces $W^{k,2}$ based on L^2. Because this particular case is used so much more frequently than the other Sobolev spaces, it has its own notation, $W^{k,2} = H^k$. This notation is chosen because H^k is a Hilbert space when equipped with the inner product

$$((u, v))_{H^k} = \sum_{0 \le |\alpha| \le k} (D^\alpha u, D^\alpha v).$$

The H^k norm corresponding to this inner product is

$$\|u\|_{H^k} = \left(\sum_{0 \le |\alpha| \le k} |D^\alpha u|^2 \right)^{1/2}.$$

Definition 5.6. *The* Sobolev space $H^k(\Omega)$ *is defined by*

$$H^k(\Omega) = \{u : D^\alpha u \in L^2(\Omega), \text{ for all } 0 \le |\alpha| \le k\}.$$

In Chapter 1 we saw that $C_c^\infty(\Omega)$ is dense in $L^2(\Omega)$ (Corollary 1.14). However, it is not dense in $H^k(\Omega)$, $k \ge 1$, as the following example shows. Suppose that u is a nonzero solution of the differential equation

$$\sum_{|\alpha| \le k} (-1)^{|\alpha|} D^{2\alpha} u = 0$$

on some parallelepiped containing Ω. By choice of suitable boundary conditions a smooth solution can be found using separation of variables, and $u|_\Omega$ is then an element of $H^k(\Omega)$.

Now consider the linear functional L on $H^k(\Omega)$ given by

$$L(\phi) = ((u, \phi))_{H^k}.$$

Since $u \ne 0$, we must have $L(u) = \|u\|_{H^k}^2 \ne 0$. However, if $\phi \in C_c^\infty(\Omega)$ then

$$
\begin{aligned}
L(\phi) &= ((u, \phi))_{H^k} \\
&= \sum_{|\alpha| \le k} (D^\alpha u, D^\alpha \phi) \\
&= \left(\sum_{|\alpha| \le k} (-1)^{|\alpha|} D^{2\alpha} u, \phi \right) \\
&= 0.
\end{aligned}
$$

In other words, L is zero on all test functions. It follows that $u \in H^k$ cannot be approximated by test functions in the H^k norm.

Since $C_c^\infty(\Omega)$ is not dense in $H^k(\Omega)$, its completion with respect to the H^k norm is a different space from $H^k(\Omega)$, which deserves its own special notation.

Definition 5.7. *The space* $H_0^k(\Omega)$ *is the completion of the space* $C_c^\infty(\Omega)$ *in* $H^k(\Omega)$.

We will see later that we can interpret $H_0^1(\Omega)$ as all those functions in $H^1(\Omega)$ that "are zero on $\partial\Omega$." The inclusion of this boundary condition enables us to prove the following inequality, know as Poincaré's inequality. This shows that the L^2 norm of ∇u suffices to bound the L^2 norm of u itself. (For a slightly more general result see Exercise 5.4.)

Proposition 5.8 (Poincaré's inequality). *Let Ω be bounded in one direction so that, for example, $|x_1| \leq d < \infty$. Then there is a constant C such that*

$$|u| \leq C|Du| \qquad \text{for all} \qquad u \in H_0^1(\Omega). \qquad (5.11)$$

We will usually write Du now for the vector of partial derivatives referred to as ∇u previously,

$$Du = (D_1 u, \ldots, D_m u),$$

and then (cf. (1.5))

$$|Du|^2 = |\nabla u|^2 = \sum_{j=1}^{m} |D_j u|^2.$$

Proof. We need only show (5.11) for elements of $C_c^\infty(\Omega)$, since such functions are dense in $H_0^1(\Omega)$ by definition. In that case, integrating by parts in the x_1 variable and using the Cauchy–Schwarz inequality shows that

$$|u|^2 = \int_\Omega |u(x)|^2 \, dx = -\int_\Omega x_1 \frac{\partial}{\partial x_1} |u(x)|^2 \, dx \leq 2d|u| \left| \frac{\partial u}{\partial x_1} \right|, \qquad (5.12)$$

and so $|u| \leq 2d|\partial u/\partial x_1| \leq 2d|Du|$. ∎

When we have a Poincaré inequality we can use

$$\|u\|_{H_0^1}^2 = \sum_{|\alpha|=1} |D^\alpha u|^2 = |Du|^2 \qquad (5.13)$$

as an alternative norm on $H_0^1(\Omega)$, equivalent to the standard H^1 norm. This follows since*

$$\|u\|_{H_0^1}^2 \leq \|u\|_{H^1}^2 = |u|^2 + \|u\|_{H_0^1}^2 \leq (1 + C)\|u\|_{H_0^1}^2.$$

In many situations it is more convenient to work with the simpler norm in (5.13), and we refer to this as "the H_0^1 norm," in contrast to the "standard" H^1 norm, which includes the L^2 term. Corresponding to the H_0^1 norm we have the "H_0^1

---

* Note that we have not kept strictly to the statement of Proposition 5.8, since the constant C here is C^2 from there. In what follows we will often be cavalier about constants in our proof, changing the definition of "C" from line to line. Generally, much of our efforts will be in showing that something is bounded, and the particular value of the constant is unimportant.

inner product,"

$$((u, v))_{H_0^1} = \sum_{|\alpha|=1} (D^\alpha u, D^\alpha v). \tag{5.14}$$

In a similar way (see Exercise 5.4) we can use the "H_0^k inner product"

$$((u, v))_{H_0^k} = \sum_{|\alpha|=k} (D^\alpha u, D^\alpha v) \tag{5.15}$$

and the "H_0^k norm"

$$\|u\|_{H_0^k}^2 = \sum_{|\alpha|=k} |D^\alpha u|^2 \tag{5.16}$$

on $H_0^k(\Omega)$.

We can also consider spaces of functions that are locally in H^k:

$$H_{\text{loc}}^k(\Omega) = \{u : u \in H^k(\Omega') \text{ for all } \Omega' \subset\subset \Omega\}.$$

As before (cf. definition for convergence in $L_{\text{loc}}^p(\Omega)$; see Section 1.4.4), we say that $u_n \to u$ in $H_{\text{loc}}^k(\Omega)$ if $u_n \to u$ in $H^k(\Omega')$ for every $\Omega' \subset\subset \Omega$.

Finally we introduce Sobolev spaces H^k with negative k; H^{-k} is the dual space of H_0^k. We saw above that the space of all linear functionals on $H^k(\Omega)$ contains nonzero elements that nevertheless vanish on all test functions. By definition these cannot be distributions (i.e. elements of $\mathcal{D}'(\Omega)$). For this reason we define the dual spaces $H^{-k}(\Omega)$ as the spaces of linear functionals on $H_0^k(\Omega)$, in which test functions *are* dense.

Definition 5.9. *The space $H^{-k}(\Omega)$ is the dual space of $H_0^k(\Omega)$.*

The following simple lemma is clear from the definitions.

Lemma 5.10. *If $u \in H^k(\Omega)$, with $k \in \mathbb{Z}$, then $D^\alpha u \in H^{k-|\alpha|}(\Omega)$.*

However, this lemma has a converse that can be more useful.

Proposition 5.11. *Suppose that $f \in H^{-k}(\Omega)$. Then there exist functions $g_\alpha \in L^2(\Omega)$ such that*

$$f = \sum_{|\alpha|=k} D^\alpha g_\alpha. \tag{5.17}$$

Proof. Using the Riesz representation theorem, there exists a unique $u \in H_0^k(\Omega)$ such that

$$((u, v))_{H_0^k} = \langle f, v \rangle$$

for all $v \in H_0^k(\Omega)$. Using the definition of the H_0^k inner product (in (5.15)) and the distribution derivative, for any test function ϕ the above equality shows that

$$\langle f, \phi \rangle = \sum_{|\alpha|=k} (D^\alpha u, D^\alpha \phi) = (-1)^k \sum_{|\alpha|=k} (D^{2\alpha} u, \phi),$$

and so

$$f = (-1)^k \sum_{|\alpha|=k} D^{2\alpha} u.$$

It follows that we can take $g_\alpha = (-1)^k D^\alpha u$ and obtain (5.17). ∎

5.3 Outline of the Rest of the Chapter

In the rest of this chapter we will investigate the properties of the Sobolev spaces $H^k(\Omega)$ in detail, covering the following topics. As in Chapter 1, Ω will always denote an open subset of $\mathbb{R}^m$.

First, we investigate how elements of $H^k(\Omega)$ can be approximated by smooth functions. We have seen that this is possible for elements of $L^p(\Omega)$, and similar density results hold for Sobolev spaces, allowing us to prove results by assuming that our functions are differentiable and then taking appropriate limits. We show initially that $C^\infty(\Omega)$ is dense in $H^k(\Omega)$. By spending some time showing that it is possible to extend functions in $H^k(\Omega)$ to functions in $H^k(\mathbb{R}^m)$, we then show that $C^\infty(\overline{\Omega})$ is also dense in $H^k(\Omega)$.

Next, we explore the relationship of the spaces $H^k(\Omega)$ with the Lebesgue spaces $L^p(\Omega)$ and spaces of continuous functions $C^r(\overline{\Omega})$, when Ω is a bounded subset of $\mathbb{R}^m$. We will see that if $k < m/2$ then if $u \in H^k(\Omega)$ it also lies in $L^p(\Omega)$ (for some appropriately chosen $p > 2$), and if $k > m/2$ then u is a continuous function in $C^0(\overline{\Omega})$.

In the following section we prove an analogue of the Arzelà–Ascoli theorem for Sobolev spaces: we show that a bounded subset of H^1 forms a compact subset in L^2. In applications this will help us overcome the problem that the unit ball in an infinite-dimensional space is not compact.

In many PDEs boundary conditions are specified, and we need to be able to make sense of these in the Sobolev space context. We show that there is a natural way to extend the idea of "restricting a continuous function to the boundary" to functions in $H^1(\Omega)$.

The last, brief, section of the chapter (Section 5.10) concerns Sobolev spaces of periodic functions, which we will find useful later on when we consider the Navier–Stokes equations. (For such spaces the arguments of Sections 5.4–5.9 can be much simplified, and the easy proofs of the corresponding results for the periodic case are given in Appendix A.)

5.4 $C^\infty(\Omega)$ is Dense in $H^k(\Omega)$

We saw above that $C_c^\infty(\Omega)$ is not dense in $H^k(\Omega)$. We will now show, however, that $C^\infty(\Omega)$ is dense in $H^k(\Omega)$ (we have lost the compact support property of our approximating sequence). This will follow using the process of mollification introduced in Chapter 1. The main result will follow from the next lemma, which shows that "mollification and weak derivatives commute."

Lemma 5.12. *Suppose that $u \in L_{\text{loc}}^1(\Omega)$ and that the weak derivative $D^\alpha u$ exists. Then provided that $h < \text{dist}(x, \partial\Omega)$ we have*

$$D^\alpha u_h(x) = (D^\alpha u)_h(x).$$

Proof. We have (Equation (1.8))

$$u_h(x) = h^{-m} \int_\Omega \rho\left(\frac{x-y}{h}\right) u(y)\, dy,$$

and, differentiating under the integral sign, we obtain

$$D^\alpha u_h(x) = h^{-m} \int_\Omega \left[D_x^\alpha \rho\left(\frac{x-y}{h}\right) \right] u(y)\, dy$$

$$= (-1)^{|\alpha|} h^{-m} \int_\Omega \left[(D_y^\alpha \rho)\left(\frac{x-y}{h}\right) \right] u(y)\, dy$$

$$= h^{-m} \int_\Omega \rho\left(\frac{x-y}{h}\right) D_y^\alpha u(y)\, dy$$

by using the definition of the weak derivative. It follows that

$$D^\alpha u_h(x) = (D^\alpha u)_h(x),$$

as claimed. ∎

The following important result on the convergence of mollified functions in the Sobolev spaces $H^k_{\text{loc}}(\Omega)$ is an immediate corollary of this lemma and Proposition 1.18.

Corollary 5.13. *If* $u \in H^k_{\text{loc}}(\Omega)$ *then* $u_h \to u$ *in* $H^k_{\text{loc}}(\Omega)$. *Furthermore,* $u_h \in C^\infty(\Omega)$.

Proof. We know that, for any $v \in L^2_{\text{loc}}(\Omega)$, $v_h \to v$ in $L^2_{\text{loc}}(\Omega)$, from Proposition 1.18. Since $D^\alpha u_h = (D^\alpha u)_h$, it follows that $D^\alpha u_h \to D^\alpha u$ in $L^2_{\text{loc}}(\Omega)$ for all $|\alpha| \leq k$. Thus $\{u_h\}$ is a Cauchy sequence in $H^k(\Omega')$ for each $\Omega' \subset\subset \Omega$. Since $H^k(\Omega')$ is complete it follows that $u_h \to \tilde{u}$ in $H^k(\Omega')$, and clearly $u = \tilde{u}$. The smoothness of u_h follows as in Proposition 1.18. ∎

We can now prove the promised density result by introducing a device known as a partition of unity, which enables us to split the whole domain into a series of more manageable pieces.

For Ω an open subset of $\mathbb{R}^m$, let $\{U_j\}$ be a countable collection of bounded open subsets of $\mathbb{R}^m$ that cover Ω, with the additional property that every compact subset of Ω intersects with only finitely many of the U_j (this is called "local finiteness"). Then a *partition of unity subordinate to the covering* $\{U_j\}$ is a set of functions $\psi_j \in C^\infty_c(\mathbb{R}^m)$ such that

(i) $0 \leq \psi_j \leq 1$,
(ii) supp $\psi_j \subset U_j$, and
(iii) $\sum_{j=1}^\infty \psi_j = 1$ on a neighbourhood of Ω.

We prove the existence of such a partition of unity in the following theorem.

Theorem 5.14. *Suppose that* $\{U_j\}$ *is a collection of bounded open subsets of* Ω *as above. Then there exists a partition of unity subordinate to the covering* $\{U_j\}$.

We will denote the open ϵ neighbourhood of a set X by $N(X, \epsilon)$,

$$N(X, \epsilon) = \{z : z = x + y, \ x \in X, \ y \in B(0, \epsilon)\},$$

and the closed ϵ neighbourhood of X by $\overline{N}(X, \epsilon)$,

$$\overline{N}(X, \epsilon) = \{z : z = x + y, \ x \in X, \ y \in \overline{B}(0, \epsilon)\}.$$

Proof. We start by constructing a new covering of Ω by sets $\{V_j\}$, which has the same properties as the covering $\{U_j\}$ except that $\overline{V}_j \subset U_j$. We do this inductively. Suppose that we have already constructed $V_1, \ldots, V_k$, such that

$$\Omega = \bigcup_{j=1}^{k} V_j \cup \bigcup_{j=k+1}^{\infty} U_j.$$

The complement of

$$\bigcup_{j=1}^{k} V_j \cup \bigcup_{j=k+2}^{\infty} U_j$$

is a closed subset of U_{k+1}; we take V_{k+1} to be any open set containing this set such that $\overline{V}_{k+1} \subset U_{k+1}$. Since any point $x \in \Omega$ is contained in only finitely many of the U_j, we must have $x \notin \cup_{j=N+1}^{\infty} U_j$ for some N. It then follows that $x \in \cup_{j=1}^{N} V_j$, and since $V_j \subset U_j$ we still have the local finiteness property of $\overline{V}_j \subset \Omega$.

If W_j is an open set such that $\overline{V}_j \subset W_j$ and $\overline{W}_j \subset U_j$ we can find a $C_c^\infty(\Omega)$ function that is 1 on $\overline{V}_k$ and zero outside W_k: if the ϵ neighbourhood of $\overline{V}_k$ is contained in W_k then define

$$f_j(x) = 1 - \min\left(1, \frac{3}{\epsilon} \operatorname{dist}(x, \overline{N}(\overline{V}_k, \epsilon/3))\right)$$

and take ϕ_j to be the mollification of f_j given by $\phi_j = (f_j)_{\epsilon/3}$. If we set

$$\phi(x) = \sum_{k=1}^{\infty} \phi_k(x)$$

then for any given x the sum has only finitely many nonzero terms. It follows that $\phi(x) \in C_c^\infty(\Omega)$, and so if we define

$$\psi_j(x) = \phi_j(x)/\phi(x)$$

we obtain the required partition of unity $\{\psi_j\}$. ∎

We now use this device to prove our first density result.

Theorem 5.15. $C^\infty(\Omega) \cap H^k(\Omega)$ *is dense in* $H^k(\Omega)$.

Proof. We let $\{\Omega_j\}$ be an increasing sequence of bounded open subsets of Ω with

$$\overline{\Omega}_j \subset\subset \overline{\Omega}_{j+1}$$

and

$$\bigcup_{j=1}^{\infty} \Omega_j = \Omega$$

(cf. proof of Proposition 1.15). We take $\{\psi_j\}$ to be a partition of unity subordinate to the covering $\{\Omega_{j+1} \setminus \overline{\Omega}_{j-1}\}$, with the convention that $\Omega_0 = \Omega_{-1} = \emptyset$.

For a given $u \in H^k(\Omega)$ we choose (for each j)

$$h_j \leq \operatorname{dist}(\Omega_j, \partial\Omega_{j+1})$$

such that

$$\|(\psi_j u)_{h_j} - \psi_j u\|_{H^k(\Omega)} \leq 2^{-j}\epsilon.$$

By the choice of h_j and ψ_j, only a finite number of the functions $(\psi_j u)_{h_j}$ are nonzero on any compact subdomain of Ω, and so

$$v = \sum_{j=1}^{\infty} (\psi_j u)_{h_j}$$

is a function in $C^\infty(\Omega)$. Furthermore we have

$$\|u - v\|_{H^k(\Omega)} \leq \sum_{j=1}^{\infty} \|(\psi_j u)_{h_j} - \psi_j u\|_{H^k}$$

$$\leq \sum_{j=1}^{\infty} 2^{-j}\epsilon = \epsilon,$$

which completes the proof. ∎

If $\Omega = \mathbb{R}^m$ we can deduce a stronger result.

Corollary 5.16. $C_c^\infty(\mathbb{R}^m)$ is dense in $H^k(\mathbb{R}^m)$.

Proof. Given $u \in H^k(\mathbb{R}^m)$ and $\epsilon > 0$ we first use Theorem 5.15 to find a function $f \in C^\infty(\mathbb{R}^m) \cap H^k(\mathbb{R}^m)$ such that

$$\|u - f\|_{H^k(\mathbb{R}^m)} < \epsilon/3.$$

Now we choose a test function $\phi \in C_c^\infty(\mathbb{R}^m)$ with $\|\phi\|_\infty \leq 1$ such that

$$\phi(x) = \begin{cases} 1, & |x| \leq 1, \\ 0, & |x| \geq 2, \end{cases}$$

and consider the sequence of functions $\{f_n\}$ in $C_c^\infty(\mathbb{R}^m)$ given by

$$f_n(x) = f(x)\phi(x/n).$$

Since $f \in H^k(\mathbb{R}^m)$, for each α with $|\alpha| \leq k$ the integral

$$\int_{\mathbb{R}^m} |D^\alpha f(x)|^2 \, dx$$

is finite. In particular there exists an N such that

$$\sum_{|\alpha| \leq k} \int_{|x| > N} |D^\alpha f(x)|^2 \, dx < \epsilon/3.$$

It follows that

$$\|f - f_N\|_{H^k(\mathbb{R}^m)} < 2\epsilon/3$$

and hence that

$$\|u - f_N\|_{H^k(\mathbb{R}^m)} < \epsilon.$$

Since $f_N \in C_c^\infty(\mathbb{R}^m)$ the corollary is proved. ∎

In Section 5.6 we will show that in fact $C^\infty(\overline{\Omega})$ is dense in $H^k(\Omega)$. We outline the idea here, which is relatively simple. Filling in the gaps will take up the next two sections.

Suppose that we can extend any function u in $H^k(\Omega)$ to a function v in $H^k(\mathbb{R}^m)$ (so $v|_\Omega = u$). Then we can use Theorem 5.15 to approximate v by functions $v_n \in C^\infty(\mathbb{R}^m)$. The restrictions of v_n to $\overline{\Omega}$, $u_n = v_n|_{\overline{\Omega}}$, should then be a sequence in $C^\infty(\overline{\Omega})$ that approximates u. Clearly, however, there is some work to be done here. In particular, we need to show that we can perform such an extension.

5.5 An Extension Theorem

Many of the theorems we will prove in the remainder of this chapter are simpler to prove when $u \in H_0^k(\Omega)$. It is thus a common ploy to extend a function in $H^k(\Omega)$ to a function in $H^k(\mathbb{R}^m)$ with compact support, work in this more convenient space, and then take the restriction of the function back to its original domain. However, without some assumptions on Ω we cannot find such an extension, in general. We will show how to construct an extension in the case when $\partial\Omega$ is compact and of class C^k.

We proceed in two steps. First, we show that a function in $H^k(\mathbb{R}^m_+)$, where

$$\mathbb{R}^m_+ = \{x \in \mathbb{R}^m : x_m > 0\},$$

can be extended to a function $E_0[u]$ in $H^k(\mathbb{R}^m)$ such that

$$\|E_0[u]\|_{H^k(\mathbb{R}^m)} \le C\|u\|_{H^k(\mathbb{R}^m_+)}.$$

Then, by a suitable change of coordinates, we treat extension from general bounded domains.

5.5.1 Extending Functions in $H^k(\mathbb{R}^+_m)$

We start off extending a function defined on $\mathbb{R}^+_m$, for which the boundary is flat.

Theorem 5.17. *Let $k \ge 0$ be an integer. Then there exists a bounded linear mapping*

$$E_0 : H^k(\mathbb{R}^m_+) \to H^k(\mathbb{R}^m)$$

such that

$$E_0[u]|_{\mathbb{R}^m_+} = u \qquad \text{for all} \qquad u \in H^k(\mathbb{R}^m_+)$$

and

$$\|E_0[u]\|_{H^k(\mathbb{R}^m)} \le C\|u\|_{H^k(\mathbb{R}^m_+)}. \tag{5.18}$$

(In fact E_0 is a bounded linear mapping from $H^j(\mathbb{R}^m_+)$ into $H^j(\mathbb{R}^m)$ for all $0 \le j \le k$.)

Proof. We first consider how to extend functions in $C^k(\mathbb{R}^m_+) \cap H^k(\mathbb{R}^m_+)$, and then we use a density argument to include all of $H^k(\mathbb{R}^m_+)$. If we only had to make $E_0[u]$ continuous across $x_m = 0$, we could take

$$u(x', x_m) = u(x', -x_m), \qquad x' \in \mathbb{R}^{m-1}, \quad x_m < 0,$$

but then the derivatives would not agree in general. However, if we take

$$u(x', x_m) = 3u(x', -x_m) - 2u(x', -2x_m), \qquad x_m < 0,$$

then both u and its derivative are continuous. In general, we try the extension $E_0[u]$ given by

$$u(x', x_m) = \sum_{j=1}^{k+1} c_j u(x', -jx_m), \qquad x_m < 0.$$

To make the first k derivatives agree, we need

$$\sum_{j=1}^{k+1} (-j)^i c_j = 1, \qquad i = 0, \dots, k.$$

In other words, we need to find a vector $\mathbf{c} = (c_1, \dots, c_{k+1})^T$ such that

$$A\mathbf{c} = 1,$$

where A is a matrix with coefficients $a_{ij} = (-j)^{i-1}$. This equation has a solution provided that A is invertible.

Now, if $\det A = 0$ then there is a nonzero vector $\mathbf{d}$ such that $A\mathbf{d} = 0$. This would imply that there is a polynomial,

$$\sum_{i=0}^{k} d_{i+1} x^i,$$

of degree at most k that has $k + 1$ distinct nonzero roots (each of the negative integers between -1 and $-(k+1)$); this is clearly ridiculous, so $\det A \neq 0$ and A is invertible.

Now, writing

$$\mathbb{R}^m_- = \{x \in \mathbb{R}^m : x_m < 0\},$$

it is clear that

$$\|D^\alpha E_0[u]\|^2_{L^2(\mathbb{R}^m)} = \|D^\alpha u\|^2_{L^2(\mathbb{R}^m_+)} + \left| D^\alpha \sum_{j=0}^{k} c_j u(x', -jx_m) \right|^2_{L^2(\mathbb{R}^m_-)}$$

$$\leq \|D^\alpha u\|^2_{L^2(\mathbb{R}^m_+)} + \left| \sum_{j=0}^{k} c_j (-j)^{\alpha_m} D^\alpha u(x', -jx_m) \right|^2_{L^2(\mathbb{R}^m_-)}$$

$$\leq C(k, \alpha) \|D^\alpha u\|^2_{L^2(\mathbb{R}^m_+)}.$$

Since $C^k(\mathbb{R}^m_+) \cap H^k(\mathbb{R}^m_+)$ is dense in $H^k(\mathbb{R}^m_+)$ (Theorem 5.15) the result follows by taking limits. ∎

To extend this result to general domains Ω, we will need an appropriate theory of how Sobolev spaces behave under transformations of coordinates.

5.5.2 Coordinate Changes

We now discuss how to change coordinates in order to flatten out the boundary of Ω. This will prove useful later in the chapter when we discuss "boundary values" of functions in Sobolev spaces and in the next chapter when we discuss the regularity of solutions of Poisson's equation.

Suppose that Ω and Ω' are open subsets of $\mathbb{R}^m$. Then a bijection $\Phi : \Omega \to \Omega'$ is a k-diffeomorphism (or C^k-diffeomorphism) if

(i) Φ, Φ^{-1}, and their derivatives of orders 1 up to k are bounded and continuous on Ω and Ω', respectively, and

(ii) there are positive constants k_1 and k_2 such that

$$k_1 \leq |\det \nabla \Phi| \leq k_2$$

in Ω.

Transforming coordinates will have an effect on the definition of a function defined with respect to our initial coordinate system. The effect of coordinate transformations on functions can be represented by *pullback operators*. If $u : \Omega \to X$, then under the change of coordinates Φ, u becomes a new function $(\Phi^{-1})^* u : \Omega' \to X$,

$$[(\Phi^{-1})^* u](y) = u(\Phi^{-1}(y)),$$

that is,

$$(\Phi^{-1})^* u = u \circ \Phi^{-1}.$$

Note that whereas the coordinate transformation acts on points in Ω, the pullback operator acts on functions defined on Ω.

For a function v defined on Ω', the pullback of Φ is defined similarly,

$$\Phi^* v = v \circ \Phi.$$

Obviously, the "new" function $\Phi^* v$ is defined on Ω.

We now show that if Φ is a k-diffeomorphism then the pullback operators map continuously between $H^k(\Omega)$ and $H^k(\Omega')$.

Theorem 5.18. *If Φ is a k-diffeomorphism then Φ^* and $(\Phi^{-1})^*$ are bounded linear maps from $H^k(\Omega')$ into $H^k(\Omega)$, or vice versa, respectively.*

In the proof we use the notations $|u|_\Omega$ and $(u, v)_\Omega$ to denote, respectively, the norm and the inner product in $L^2(\Omega)$, since we will need to distinguish between estimates in $L^2(\Omega)$ and $L^2(\Omega')$.

Proof. It suffices to prove the result for Φ^*. First consider the effect of Φ^* on functions in $L^2(\Omega') \cap C^k(\Omega')$. Starting with functions u and v in $L^2(\Omega') \cap C^k(\Omega')$, the standard change of variable formula for integrals gives

$$(u, v)_{\Omega'} = (\Phi^*u, (\det\nabla\Phi)\Phi^*v)_\Omega, \tag{5.19}$$

and so in particular

$$k_1|\Phi^*u|_\Omega^2 \le |u|_{\Omega'}^2, \tag{5.20}$$

and Φ^* is a bounded linear map from $L^2(\Omega') \cap C^k(\Omega')$ (with the $L^2(\Omega')$ norm) into $L^2(\Omega)$. By taking limits we can deduce that Φ^* is a bounded linear map from $L^2(\Omega')$ into $L^2(\Omega)$.

Now, if $u \in C^k(\Omega') \cap H^k(\Omega')$, we can compute the derivatives of Φ^*u by using the standard chain and product rules. These derivatives will be of the form

$$D^\alpha(\Phi^*u)(x) = \sum_{|\beta|\le|\alpha|} \phi_\beta(x)D^\beta u(\Phi(x)),$$

where $\phi_\beta(x)$ is an expression containing various derivatives of Φ but independent of u. Now, the derivatives of Φ are bounded on Ω, and $D^\beta u \in L^2(\Omega')$ for each β with $|\beta| \le k$. It follows from (5.20) that if $|\alpha| \le k$ then $D^\alpha(\Phi^*u)$ is an element of $L^2(\Omega')$, with

$$|D^\alpha(\Phi^*u)|_\Omega \le \sum_{|\beta|\le|\alpha|} c_\beta|D^\beta u|_{\Omega'}$$
$$\le C_\alpha\|u\|_{H^k(\Omega')}.$$

We conclude that $\Phi^*u \in H^k(\Omega)$, with

$$\|\Phi^*u\|_{H^k(\Omega)} \le C\|u\|_{H^k(\Omega')},$$

as required. ∎

5.5.3 Straightening the Boundary

We will treat the problem of straightening out the boundary for bounded domains "of class C^k."

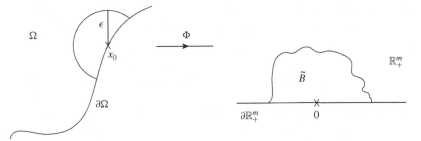

Figure 5.1. Coordinate transformation Φ from $\Omega \cap B(x_0, \epsilon)$ to $\tilde{B}$, which straightens $\partial\Omega \cap B(x_0, \epsilon)$.

Definition 5.19. *We say that $\Omega \subset \mathbb{R}^m$ is a* bounded domain of class C^k *or a* bounded C^k domain *provided that at each point $x_0 \in \partial\Omega$ there is an $\epsilon > 0$ and a C^k-diffeomorphism Φ of $B(x_0, \epsilon)$ onto a subset $\tilde{B}$ of $\mathbb{R}^m$ such that*

(i) $\Phi(x_0) = 0$,
(ii) $\Phi(B \cap \Omega) \subset \mathbb{R}^m_+$, and
(iii) $\Phi(B \cap \partial\Omega) \subset \partial\mathbb{R}^m_+$.

This is illustrated in Figure 5.1. Note that, since Ω is bounded, $\partial\Omega$ is compact, and we can find a finite cover of $\partial\Omega$ by a set of balls $\{U_j\}_{j=1}^N$ with corresponding C^k-diffeomorphisms Φ_j.

To treat problems near such boundaries, we take a partition of unity $\{\psi_j\}_{j=1}^N$ subordinate to the cover $\{U_j\}_{j=1}^N$. Then in a neighbourhood of $\partial\Omega$ we can write any function u as

$$u = \sum_{j=1}^{N}(\psi_j u).$$

We now consider, for each j, the function

$$u_j = \left(\Phi_j^{-1}\right)^*(\psi_j u),$$

which is "part" of u with the coordinates transformed so that the boundary has become flat (as in Figure 5.1). We use this technique in the next section to prove a general extension result.

5.5.4 Extending Functions in $H^k(\Omega)$

We now prove an extension theorem valid for Sobolev spaces on general bounded domains.

Theorem 5.20. *If Ω is a bounded C^k domain, then for each open set $\Omega' \supset \overline{\Omega}$ there exists a bounded linear extension operator E such that if $u \in H^k(\Omega)$ then $E[u] \in H_0^k(\Omega')$ and*

$$\|E[u]\|_{H^k(\Omega')} \leq C_{k,\Omega'} \|u\|_{H^k(\Omega)}. \tag{5.21}$$

(In fact we have (5.21) for each H^j with $0 \leq j \leq k$.)

Throughout the proof we use C to denote various constants, which may change from line to line. When the dependence of the constant on some index (as in (5.22)) or a function (as in (5.23)) is important we highlight it in the notation. We will often adopt this approach from now on.

Proof. We take the covering of $\partial\Omega$ with neighbourhoods $\{U_j\}$ as outlined above, ensuring that $\overline{U_j} \subset\subset \Omega'$, and choose a partition of unity $\{\psi_j\}$ subordinate to the cover $\{U_j\}$. We will write $\Omega_j = \Omega \cap U_j$, and $\tilde{\Omega}_j = \Phi_j(\Omega_j)$.

For $u \in H^k(\Omega)$, consider for each j the function $u_j \in H^k(\mathbb{R}_+^m)$ defined by

$$u_j = \left(\Phi_j^{-1}\right)^* \psi_j u,$$

which is zero outside $\tilde{\Omega}_j$. By the result of Theorem 5.18, we know that

$$\|u_j\|_{H^k(\mathbb{R}_+^m)} \leq C_j \|\psi_j u\|_{H^k(\Omega_j)}, \tag{5.22}$$

and since, for any $\psi \in C_c^\infty(\mathbb{R}^m)$, we have

$$\|\psi u\|_{H^k(\Omega)} \leq C(\psi) \|u\|_{H^k(\Omega)} \tag{5.23}$$

(see Exercise 5.5), it follows that

$$\|u_j\|_{H^k(\mathbb{R}_+^m)} \leq C_j \|u\|_{H^k(\Omega)}.$$

The function u_j can be extended, using Theorem 5.17, to a function $E_0[u_j] \in H^k(\mathbb{R}^m)$, which satisfies (by (5.18))

$$\|E_0[u_j]\|_{H^k(\mathbb{R}^m)} \leq C \|u_j\|_{H^k(\mathbb{R}_+^m)} \leq C_j \|u\|_{H^k(\Omega)}. \tag{5.24}$$

Multiplying $E_0[u_j]$ by a $C_c^\infty(\mathbb{R}^m)$ cutoff function θ_j, whose support is in $\Phi_j(U_j)$ and is equal to 1 on $\Phi_j(\text{supp } \psi_j)$, we obtain a function

$$\Phi_j^*(\theta_j E_0[u_j])$$

that equals $\psi_j u$ in Ω_j, has support within U_j, and satisfies

$$\|\Phi_j^*(\theta_j E_0[u_j])\|_{H^k(\mathbb{R}^m)} \leq C_j \|u\|_{H^k(\Omega)}.$$

So we can extend $\sum_j \psi_j u$ as required. What is left of the original function (defined only on Ω) is

$$\left(1 - \sum_{j=1}^N \psi_j\right) u. \tag{5.25}$$

Since the $\{U_j\}$ are a cover of $\partial\Omega$, the function in (5.25) is in $H^k(\Omega)$ with support in a compact subset of Ω; as such, this can be extended by zero, and on setting

$$E[u] = \left(1 - \sum_{j=1}^N \psi_j\right) u + \sum_{j=1}^N \Phi_j^*(\theta_j E_0[u_j]),$$

we conclude the proof by summing the estimates in (5.24). ∎

5.6 Density of $C^\infty(\overline{\Omega})$ in $H^k(\Omega)$

We are now in a position to make rigorous the discussion at the end of Section 5.4 by proving that $C^\infty(\overline{\Omega})$ is dense in $H^k(\Omega)$.

Theorem 5.21. *If Ω is of class C^k and $\partial\Omega$ is compact, then $C^\infty(\overline{\Omega})$ is dense in $H^k(\Omega)$.*

Proof. Take $u \in H^k(\Omega)$ and extend it to a function $E[u]$ in $H^k(\mathbb{R}^m)$ by using Theorem 5.20. Now, Theorem 5.15 gives a sequence of functions $v_n \in C^\infty(\mathbb{R}^m)$ such that $v_n \to E[u]$ in $H^k(\mathbb{R}^m)$. Setting $u_n = v_n|_{\overline{\Omega}}$, one clearly has $u_n \in C^\infty(\overline{\Omega})$, and

$$\|u_n - u\|_{H^k(\Omega)} \leq \|v_n - E[u]\|_{H^k(\mathbb{R}^m)},$$

so that u_n converges to u in $H^k(\Omega)$, as required. ∎

Note that if $u \in C^0(\overline{\Omega}) \cap H^k(\Omega)$ then the approximating sequence $\{u_n\}$ constructed above converges uniformly on $\overline{\Omega}$. Indeed, $u_n = (v_h)|_\Omega$, where $v \in H_0^k(\Omega')$, with $\Omega' \supset \overline{\Omega}$. It follows from Proposition 1.6 that v_h converges uniformly to u on compact subsets of Ω', and so in particular on $\overline{\Omega}$. Thus $u_n \to u$ uniformly on $\overline{\Omega}$, as claimed.

5.7 The Sobolev Embedding Theorem – H^k, C^r, and L^p

In this section we investigate how the spaces H^k, C^r, and L^p are related.

5.7.1 Integrability of Functions in Sobolev Spaces

To prevent too much notation we will use the abbreviation

$$|u|_p = \|u\|_{L^p}.$$

Exactly why we need to do this will become clear as the proofs progress!

We first prove a version of our results in the simple case of a one-dimensional domain. We show that if $u \in H^1(a, b)$ then $u \in C^0([a, b])$. However, one does have to approach this result with a little caution. What it really says is that if $u \in H^1(a, b)$, then there exists a function $v \in C^0([a, b])$ such that $u = v$ almost everywhere. Indeed, as mentioned above, an element of L^2 is not in fact a function, but an equivalence class of functions that are equal almost everywhere. As an example, the function

$$u(x) = \begin{cases} 1, & -1 < x < 0, \\ 0, & x = 0, \\ 1, & 0 < x < 1 \end{cases}$$

is in $H^1(-1, 1)$, since

$$\int |u(x)|^2 = 2$$

and its weak derivative is zero (which is clearly in L^2!). The function u is almost everywhere equal to the continuous function 1.

Proposition 5.22. *If $u \in H^1(a, b)$, with a, b finite, then*

$$\max_{x \in [a,b]} |u(x)| \le C \|u\|_{H^1(a,b)}$$

and in fact $u \in C^0([a, b])$.

Proof. First observe that if $u \in C_c^1(\mathbb{R})$ then

$$u(x) = \int_{-\infty}^{x} Du(x)\, dx, \tag{5.26}$$

which immediately implies the bound

$$|u|_\infty \le |Du|_1. \tag{5.27}$$

Now, if $u \in H^1(a, b)$ we can use Theorem 5.20 to extend u to a function $E[u] \in H_0^1(a', b')$, where $(a', b') \supset [a, b]$. If we take a sequence $u_n \in C_c^\infty(a', b')$ such that $u_n \to E[u]$ in $H^1(a', b')$ then we can use (5.27) to deduce that

$$|u_n - u_m|_\infty \le |D(u_n - u_m)|_1 \le (b' - a')^{1/2}|D(u_n - u_m)|_2,$$

which shows that $\{u_n\}$ is also a Cauchy sequence in $L^\infty(a', b')$. It follows that $E[u] \in C^0(a', b')$, with

$$|E[u]|_\infty \le C|D(E[u])|_2 \le C\|E[u]\|_{H^1(a',b')} \le C\|u\|_{H^1(a,b)},$$

by (5.21). Since $u = E[u]|_{(a,b)}$, we now have $u \in C^0([a, b])$ with the estimate

$$\max_{x \in [a,b]} |u(x)| \le |E[u]|_\infty \le C\|u\|_{H^1(a,b)},$$

as claimed. ∎

For $m > 1$ the arguments are more involved, but the ideas are essentially the same. First we consider smooth functions u on the whole of $\mathbb{R}^m$, and we prove L^p bounds for u based on the L^1 norm of Du. A simple scaling argument gives us the correct value of p to consider. Choose some $u \in C_c^\infty(\mathbb{R}^m)$ and consider the scaled functions

$$u_\lambda(x) = u(\lambda x).$$

Straightforward calculations show that

$$|u_\lambda|_p = \lambda^{-m/p}|u|_p$$

and

$$|Du_\lambda|_1 = \lambda^{1-m}|Du|_1.$$

Then if we apply our potential estimate

$$|u|_p \le C|Du|_1$$

to u_λ we obtain

$$\lambda^{-m/p}|u|_p \le C\lambda^{1-m}|Du|_1$$

or

$$|u|_p \leq C\lambda^{1-m+(m/p)}|Du|_1.$$

Unless $1 - m + (m/p) = 0$ we can take $\lambda \to 0$ or $\lambda \to \infty$ to deduce that $u = 0$ almost everywhere, which is clearly nonsense. Thus the only possible value for p is $m/(m-1)$. It is this estimate that we obtain in the following proposition.

We will then be able to use these results to show, with the help of Proposition 1.16, that appropriate bounds on Du imply that u is in fact continuous.

Proposition 5.23. *If* $u \in C_c^1(\mathbb{R}^m)$, $m > 1$, *then*

$$|u|_{m/(m-1)} \leq |Du|_1. \tag{5.28}$$

Proof. For any $1 \leq j \leq m$, we have, as in (5.26) above,

$$u(x) = \int_{-\infty}^{x_j} D_j u(x_1, \ldots, y_j, \ldots, x_m)\, dy_j,$$

and so certainly (for any j)

$$|u(x)| \leq \int_{-\infty}^{\infty} |Du(x_1, \ldots, y_j, \ldots, x_m)|\, dy_j.$$

We can therefore write

$$|u(x)|^{m/(m-1)} \leq \prod_{j=1}^{m}\left(\int_{-\infty}^{\infty} |Du(x_1, \ldots, y_j, \ldots, x_m)|\, dy_j\right)^{1/(m-1)}.$$

The idea now is to integrate this inequality with respect to each variable x_1, $x_2, \ldots, x_m$ in turn and to apply the generalised Hölder inequality from Exercise 1.8 after each integration.

Integrating with respect to x_1 gives

$$\int_{-\infty}^{\infty} |u(x)|^{m/(m-1)}\, dx_1 \leq \int_{-\infty}^{\infty} \prod_{j=1}^{m}\left(\int_{-\infty}^{\infty} |Du|\, dy_j\right)^{1/(m-1)} dx_1$$

$$= \left(\int_{-\infty}^{\infty} |Du|\, dy_1\right)^{1/(m-1)} \int_{-\infty}^{\infty} \prod_{j=2}^{m}\left(\int_{-\infty}^{\infty} |Du|\, dy_j\right)^{1/(m-1)} dx_1,$$

and using the generalised Hölder inequality

$$\left|\int \prod_{k=1}^{m-1} f_k\, dx\right| \leq \prod_{k=1}^{m-1} \|f_k\|_{L^{m-1}} \tag{5.29}$$

on the second term, we obtain

$$\int_{-\infty}^{\infty} |u(x)|^{m/(m-1)}\,dx_1$$

$$\leq \left(\int_{-\infty}^{\infty} |Du|\,dy_1\right)^{1/(m-1)} \left(\prod_{j=2}^{m}\iint_{-\infty}^{\infty} |Du|\,dx_1\,dy_j\right)^{1/(m-1)}.$$

Now we integrate this inequality with respect to x_2, which gives

$$\iint_{-\infty}^{\infty} |u(x)|^{m/(m-1)}\,dx_1\,dx_2 \leq \left(\iint_{-\infty}^{\infty} |Du|\,dx_1\,dy_2\right)^{1/(m-1)}$$

$$\times \int_{-\infty}^{\infty}\left(\int_{-\infty}^{\infty} |Du|\,dy_1\right)^{1/(m-1)} \prod_{j=3}^{m}\left(\iint_{-\infty}^{\infty} |Du|\,dx_1\,dy_j\right)^{1/(m-1)}\,dx_2.$$

If we apply the generalised Hölder inequality (5.29) once again, we now have

$$\iint_{-\infty}^{\infty} |u(x)|^{m/(m-1)}\,dx_1\,dx_2 \leq \left(\iint_{-\infty}^{\infty} |Du|\,dx_1\,dy_2\right)^{1/(m-1)}$$

$$\times \left(\iint_{-\infty}^{\infty} |Du|\,dy_1\,dx_2\right)^{1/(m-1)} \prod_{j=3}^{m}\left(\iiint_{-\infty}^{\infty} |Du|\,dx_1\,dx_2\,dy_j\right)^{1/(m-1)}.$$

Each new integration and application of Hölder's inequality adds one variable to each integration while decreasing the number of terms in the product by one. Eventually (after the integration with respect to x_m) we obtain

$$\int_{\mathbb{R}^m} |u(x)|^{m/(m-1)}\,dx \leq \prod_{j=1}^{m}\left(\int \cdots \int_{-\infty}^{\infty} |Du|\,dx_1 \ldots dy_j \ldots dx_m\right)^{1/(m-1)}$$

$$= \left(\int_{\mathbb{R}^m} |Du(x)|\,dx\right)^{m/(m-1)},$$

which is exactly (5.28). ∎

We now use this basic estimate to prove a similar result, but based on the norm of Du in L^p rather than L^1.

Proposition 5.24. *Let $m > 1$. Then if $1 \leq p < m$ then there exists a constant $C(p, m)$ such that for any $u \in C_c^1(\mathbb{R}^m)$*

$$|u|_{mp/(m-p)} \leq C|Du|_p. \tag{5.30}$$

Proof. We have already shown (5.30) when $p = 1$, so we consider only the case $p > 1$. To obtain the estimate in this case, we simply apply the estimate (5.28) to $v = |u|^\gamma$ for some $\gamma > 1$ to be choosen appropriately later. [Since $\gamma > 1$, $v \in C_c^1(\mathbb{R})$.] With this choice of v, (5.28) becomes

$$\left(\int_{\mathbb{R}^m} |u|^{\gamma m/(m-1)} \, dx \right)^{(m-1)/m} \leq \int_{\mathbb{R}^m} |D|u|^\gamma| \, dx$$

$$= \gamma \int_{\mathbb{R}^m} |u|^{\gamma-1} |Du| \, dx$$

$$\leq \gamma \left(\int_{\mathbb{R}^m} |u|^{p(\gamma-1)/(p-1)} \, dx \right)^{(p-1)/p} \left(\int_{\mathbb{R}^m} |Du|^p \, dx \right)^{1/p}, \quad (5.31)$$

where Hölder's inequality was used in the last line. If we choose γ so that

$$\frac{\gamma m}{m-1} = \frac{p(\gamma-1)}{(p-1)}$$

or

$$\gamma = \frac{p(m-1)}{m-p},$$

then this becomes

$$\left(\int_{\mathbb{R}^m} |u|^{mp/(m-p)} \, dx \right)^{(m-p)/pm} \leq C \left(\int_{\mathbb{R}^m} |Du|^p \, dx \right)^{1/p},$$

which is precisely (5.30). ■

Note that if $u \in C_c^s(\mathbb{R}^m)$ then we can apply the above result to each of the derivatives of u up to order s in turn to deduce that

$$\|u\|_{W^{s-1, pm/(m-p)}} \leq C \|u\|_{W^{s,p}}. \quad (5.32)$$

We use this simple observation to consider the particular case $u \in H^k(\mathbb{R}^m)$ [i.e. $p = 2$].

Theorem 5.25. *If $k < m/2$ and $u \in H^k(\mathbb{R}^m)$, then $u \in L^{2m/(m-2k)}(\mathbb{R}^m)$, with*

$$\|u\|_{L^{2m/(m-2k)}} \leq C(k, m) \|u\|_{H^k}.$$

Proof. Assuming initially that $u \in C_c^k(\mathbb{R}^m)$ we use the above observation (5.32) repeatedly. Starting with $u \in H^k = W^{k,2}$, we have

$$\|u\|_{W^{k-1,2m/(m-2)}} \leq C \|u\|_{W^{k,2}} = C \|u\|_{H^k},$$

and then, since

$$\frac{[2m/(m-2)]m}{(m - [2m/(m-2)])} = \frac{2m}{m-4},$$

we have, using (5.32) again,

$$\|u\|_{W^{k-2,2m/(m-4)}} \leq C \|u\|_{W^{k-1,2m/(m-2)}} \leq C \|u\|_{H^k}.$$

We can repeat this process k times, since

$$\frac{2m}{m-2s} < m$$

for $s < (m/2) - 1$, and we end up with

$$\|u\|_{W^{0,2m/(m-2k)}} = \|u\|_{L^{2m/(m-2k)}} \leq C \|u\|_{H^k}$$

as in the statement of the theorem.

The result for an arbitrary $u \in H^k(\mathbb{R}^m)$ follows using the density of $C_c^\infty(\mathbb{R}^m)$ in $H^k(\mathbb{R}^m)$, proved in Corollary 5.16. ∎

Using the extension theorem (Theorem 5.20) we can readily adapt this result to the case when $u \in H^k(\Omega)$ with Ω a bounded domain in $\mathbb{R}^m$.

Theorem 5.26. *Let $\Omega \subset \mathbb{R}^m$ be a bounded C^k domain. If $u \in H^k(\Omega)$ and $k < m/2$, then $u \in L^{2m/(m-2k)}(\Omega)$ with*

$$\|u\|_{L^{2m/(m-2k)}(\Omega)} \leq C \|u\|_{H^k(\Omega)}.$$

If $k = m/2$ then $u \in L^p(\Omega)$ for each $1 \leq p < \infty$, and there exists a constant $C(p)$ such that

$$\|u\|_{L^p} \leq C(p) \|u\|_{H^1}.$$

In particular we can see that if $m = 2$ and $u \in H^1(\Omega)$, then u is an element of $L^p(\Omega)$ for every $1 \leq p < \infty$. (For problems with the case $p = \infty$ see Exercise 5.6.)

Proof. Extend u to a function $E[u] \in H_0^k(\mathbb{R}^m)$ by using Theorem 5.20. Then for $k < m/2$ we can apply the result of Theorem 5.25 directly to obtain

$$\|u\|_{L^{2m/(m-2k)}(\Omega)} \le \|E[u]\|_{L^{2m/(m-2k)}(\mathbb{R}^m)}$$
$$\le C\|E[u]\|_{H^k(\mathbb{R}^m)}$$
$$\le C\|u\|_{H^k(\Omega)},$$

using (5.21) in the last line.

For the case $k = m/2$, observe that $E[u]$ is in fact an element of $H_0^k(\Omega')$ for some bounded open set Ω' that contains $\overline{\Omega}$. It then follows, using the Lebesgue embedding result (1.16), that $E[u] \in W^{k,q}(\mathbb{R}^m)$ for each $q < 2$, with

$$\|E[u]\|_{W^{k,q}(\mathbb{R}^m)} \le |\Omega'|^{(2-q)/2q}\|E[u]\|_{W^{k,2}(\Omega')} \le C\|u\|_{H^k(\Omega)}.$$

It follows from the argument in Theorem 5.25 that if $E[u] \in W^{k,q}(\mathbb{R}^m)$ then $E[u] \in W^{0,qm/(m-kq)}(\mathbb{R}^m) = L^{qm/(m-kq)}(\mathbb{R}^m)$, with

$$|E[u]|_{mq/(m-kq)} \le C|E[u]|_{W^{k,q}} \le C\|u\|_{H^k(\Omega)}.$$

Therefore $u = E[u]|_\Omega$ is an element of $L^{mq/(m-kq)}(\Omega)$ with

$$\|u\|_{L^{mq/(m-kq)}(\Omega)} \le C(q)\|u\|_{H^k(\Omega)}.$$

Since we can choose any $1 \le q < 2$, we can obtain any value for $p = mq/(m - kq)$ with $1 \le p < \infty$, and the proof is complete. ∎

Note that since Ω is bounded, we can use the L^p embedding inequality (1.16) to deduce that if $u \in H^k(\Omega)$ then $u \in L^p(\Omega)$ for any $p < 2m/(m - 2k)$. If we want a bound for u in one of these Lebesgue spaces then it is possible to derive inequalities for this norm that improve on the result of Theorem 5.26 by incorporating the L^2 norm. We give one example, known as Ladyzhenskaya's inequality, which will be useful later.

Lemma 5.27. *If $\Omega \subset \mathbb{R}^2$ is a bounded C^1 domain then, for $u \in H^1(\Omega)$,*

$$\|u\|_{L^4} \le C|u|^{1/2}\|u\|_{H^1}^{1/2}.$$

Recall that we use $|u| = |u|_2 = \|u\|_{L^2(\Omega)}$ when the context renders this notation unambiguous.

Proof. We prove the inequality first for $u \in C_c^1(\Omega')$, where $\overline{\Omega} \subset\subset \Omega'$. Because

$$[u(x)]^2 = 2 \int_{-\infty}^{x_j} u D_j u \, dy_j, \qquad j = 1, 2 \tag{5.33}$$

we have

$$\max_{x_j} |u(x)|^2 \le 2 \int_{-\infty}^{\infty} |u D_j u| \, dy_j.$$

Therefore

$$\iint_{-\infty}^{\infty} |u|^4 \, dx_1 \, dx_2 \le \left(\int_{-\infty}^{\infty} \max_{x_2} |u(x)|^2 \, dx_1 \right) \left(\int_{-\infty}^{\infty} \max_{x_1} |u(x)|^2 \, dx_2 \right)$$

$$\le 4 \left(\iint_{-\infty}^{\infty} |u D_2 u| \, dx_1 \, dx_2 \right) \left(\iint_{-\infty}^{\infty} |u D_1 u| \, dx_1 \, dx_2 \right).$$

Since

$$\int_{\mathbb{R}^2} |u D_j u| \, dx \le |u| |D_j u|$$

we obtain

$$\|u\|_{L^4} \le 4^{1/4} |u|^{1/2} |D_1 u|^{1/4} |D_2 u|^{1/4} \le C |u|^{1/2} |Du|^{1/2}.$$

We can use the density of $C_c^1(\Omega')$ in $H_0^1(\Omega')$ to prove the result for $H_0^1(\Omega')$, and the extension theorem, as above, provides the result for $H^1(\Omega)$. ∎

A similar inequality valid when $\Omega \subset \mathbb{R}^3$ is proved in Exercise 5.7.

5.7.2 Sobolev Spaces and Spaces of Continuous Functions

The case we avoided in Proposition 5.24 was $p > m$. We now rectify this and show that if Ω is a bounded subset of $\mathbb{R}^m$ and $Du \in L^p(\Omega)$ with $p > m$ then we can bound the sup norm of u.

Proposition 5.28. *Let $p > m > 1$, and let Ω be a bounded subset of $\mathbb{R}^m$. Then if $u \in C_c^1(\Omega)$,*

$$|u|_\infty \le C |Du|_p.$$

In the proof we will write $\tilde{m} = m/(m-1)$ and $\tilde{p} = p/(p-1)$.

Proof. We assume initially that $|\Omega| = 1$ and apply estimate (5.31), used above, to v^γ, where now we take

$$v = \frac{|u|}{|Du|_{L^p}}.$$

Then (5.31), which is

$$|v^\gamma|_{\tilde{m}} \leq \gamma |v^{\gamma-1}|_{\tilde{p}},$$

implies that

$$|v|^\gamma_{\tilde{m}\gamma} \leq \gamma |v|^{\gamma-1}_{\tilde{p}(\gamma-1)},$$

and so

$$|v|_{\tilde{m}\gamma} \leq \gamma^{1/\gamma} |v|^{1-1/\gamma}_{\tilde{p}(\gamma-1)}$$
$$\leq \gamma^{1/\gamma} |v|^{1-1/\gamma}_{\tilde{p}\gamma}, \tag{5.34}$$

where we have used the embedding result for Lebesgue spaces from (1.16),

$$|f|_s \leq |\Omega|^{(r-s)/rs} |f|_r \qquad \text{if} \qquad s < r, \tag{5.35}$$

with $|\Omega| = 1$.

We set $\delta = \tilde{m}/\tilde{p} > 1$ and choose a sequence of values for γ, $\gamma_n = \delta^n$, which tends to infinity. For this sequence $\{\gamma_n\}$ we obtain from (5.34) the sequence of estimates

$$|v|_{\tilde{m}\gamma_n} \leq \gamma_n^{1/\gamma_n} |v|^{1-1/\gamma_n}_{\tilde{m}\gamma_{n-1}}. \tag{5.36}$$

To start an iterative process by using (5.36), we need to find a bound for $|v|_{\tilde{m}}$. Now, using (5.28), we have

$$|v|_{\tilde{m}} = |v|_{m/(m-1)} = \frac{|u|_{m/(m-1)}}{|Du|_p} \leq \frac{|Du|_1}{|Du|_p},$$

and by using (5.35) once again we can deduce that $|v|_{\tilde{m}} \leq 1$.

Iterating the estimate in (5.36) now shows that

$$|v|_{\tilde{m}\gamma_n} \leq \prod_{n=1}^{\infty} \gamma_n^{1/\gamma_n}$$
$$= \prod_{n=1}^{\infty} \delta^{n\delta^{-n}}$$
$$= \delta^\sigma,$$

where

$$\sigma = \sum_{n=1}^{\infty} n\delta^{-n} < \infty.$$

It follows that

$$|v|_{\bar{m}\gamma_n} \le C,$$

where C does not depend on n. Using Proposition 1.16 we can deduce that

$$\|v\|_{\infty} \le C$$

and hence that

$$\|u\|_{\infty} \le C|Du|_{L^p}.$$

If $|\Omega| \ne 1$, set $y_j = |\Omega|^{1/m}x_j$, obtaining

$$\|u\|_{\infty} \le C|\Omega|^{1/m-1/p}|Du|_{L^p}. \qquad \blacksquare$$

We now adapt this result to show that if $u \in H^k(\Omega)$ with k large enough then in fact $u \in C^0(\overline{\Omega})$. Caution along the lines of that recommended for Proposition 5.22 is needed here – the result in fact guarantees that there is a representative of the equivalence class of u in H^k that is equal to a continuous function.

Theorem 5.29. *Let* $\Omega \subset \mathbb{R}^m$ *be a bounded* C^k *domain. If* $u \in H^k(\Omega)$ *with* $k > m/2$, *then* $u \in C^0(\overline{\Omega})$ *and there exists a constant* $C(m, k)$ *such that*

$$\|u\|_{\infty} \le C\|u\|_{H^k}. \qquad (5.37)$$

Different methods can be used to show that this result is also valid when $\Omega = \mathbb{R}^m$; see Renardy & Rogers (1992), for example.

Proof. We first prove the result for $u \in H_0^k(\Omega')$, where Ω' is an open subset of $\mathbb{R}^m$ (we will take $\Omega' \supset \overline{\Omega}$ below). Consider an approximating sequence for u, $u_n \in C_c^k(\Omega')$. If we use the iterative estimation process of Theorem 5.25, based on (5.32), for each element of this sequence, then we will eventually obtain

$$\|u_n\|_{W^{s,p}} \le C\|u_n\|_{H^k},$$

with $s \ge 1$ and $p > m$. We can now use Proposition 5.28 to deduce that (5.37) holds for each u_n. A similar argument applied to $(u_n - u_m)$ shows that $\{u_n\}$, which is Cauchy in $H_0^k(\Omega')$, is also Cauchy in $C^0(\overline{\Omega'})$, and so $u \in C^0(\overline{\Omega'})$.

Choosing $\Omega' \supset \overline{\Omega}$, for $u \in H^k(\Omega)$ we use the extension Theorem 5.20, extending u to a function $E[u] \in H_0^k(\Omega')$, and then we have

$$\|u\|_{L^\infty(\Omega)} \leq \|E[u]\|_{L^\infty(\Omega')}$$
$$\leq C\|E[u]\|_{H^k(\Omega')}$$
$$\leq C\|u\|_{H^k(\Omega)},$$

giving (5.37). ∎

For example, if $\Omega \subset \mathbb{R}$ then $H^1(\Omega) \subset C^0(\overline{\Omega})$, whereas if $\Omega \subset \mathbb{R}^2$ we need to take $u \in H^2(\Omega)$ to ensure that u is continuous.

We can also use Theorem 5.29 to show that functions with many derivatives in L^2 are continuously differentiable, provided the boundary is sufficiently smooth.

Corollary 5.30. *If $\Omega \subset \mathbb{R}^m$ is a bounded C^k domain and $k > (m/2) + j$ then each $u \in H^k(\Omega)$ is an element of $C^j(\overline{\Omega})$, with*

$$\|u\|_{C^j} \leq C\|u\|_{H^k}.$$

Proof. See Exercise 5.8. ∎

An important consequence of this corollary is that if Ω is a bounded C^∞ domain (i.e. C^k for every $k \geq 0$) and $u \in H^k(\Omega)$ for each $k \geq 0$, then in fact $u \in C^\infty(\overline{\Omega})$.

5.7.3 The Sobolev Embedding Theorem

We end this part of the chapter with a summary of the results we have obtained above.

Theorem 5.31 (Sobolev embedding theorem). *Let Ω be a bounded C^k domain in $\mathbb{R}^m$, and suppose that $u \in H^k(\Omega)$.*

(i) *If $k < m/2$ then $u \in L^{2m/(m-2k)}(\Omega)$, and there exists a constant C such that*

$$\|u\|_{L^{2m/(m-2k)}(\Omega)} \leq C\|u\|_{H^k(\Omega)}.$$

(ii) If $k = m/2$ then $u \in L^p(\Omega)$ for every $1 \leq p < \infty$, and for each p there exists a constant $C = C(p)$ such that

$$\|u\|_{L^p(\Omega)} \leq C \|u\|_{H^k(\Omega)}.$$

(iii) If $k > j + (m/2)$ then $u \in C^j(\overline{\Omega})$, and there exists a constant C such that

$$\|u(x)\|_{C^j(\overline{\Omega})} \leq C_j \|u\|_{H^k(\Omega)}.$$

Note that since Ω is bounded, it follows trivially in part (iii) that $u \in L^p(\Omega)$ for every $1 \leq p \leq \infty$.

5.8 A Compactness Theorem

The Arzelà–Ascoli theorem implies that a set of uniformly bounded continuous functions whose derivatives are uniformly bounded (so that they are equicontinuous) form a compact subset of C^0. In this section we prove a similar result for the Sobolev spaces, the Rellich–Kondrachov compactness theorem, which shows that a bounded subset of H^1 (with u and Du bounded in L^2) is a compact subset of L^2. In applications, this will help get around the fact that the unit ball in L^2 is not compact, since it implies that the unit ball in H^1 is compact in L^2.

Theorem 5.32 (Rellich–Kondrachov Compactness Theorem). *Let Ω be a bounded C^1 domain. Then $H^1(\Omega)$ is compactly embedded in $L^2(\Omega)$.*

The result of this theorem can be used to generate useful new inequalities (see for example Exercise 5.9).

Proof. Take a bounded sequence of functions $\{u_n\} \in H^1(\Omega)$ and extend them to a sequence $\{v_n\}$ bounded in $H^1(\mathbb{R}^m)$ with compact support U, using Theorem 5.20.

Now approximate this sequence by using the mollification approach to give a new sequence $(v_n)_h$, which we will assume also has its support within U. Note that

$$|v_n(x) - (v_n)_h(x)| \leq \int_{B(0,1)} \rho(z)|v_n(x) - v_n(x - hz)|\, dz$$

$$\leq \int_{B(0,1)} \rho(z) \int_0^{h|z|} \left| \frac{\partial}{\partial r} v_n(x - rs) \right| dr\, dz \qquad \text{with} \qquad s = z/|z|,$$

and so

$$\int_U |v_n(x) - (v_n)_h(x)|\, dx \le h\left(\int_U |Dv_n|\, dx\right)\left(\int_{B(0,1)} |z|\rho(z)\, dz\right)$$

$$\le h|U|^{1/2}|Dv_n|_{L^2(U)} \le Ch.$$

Thus $(v_n)_h$ converges to v_n in L^1 uniformly over all n. If we use the L^p interpolation inequality from Exercise 1.9, we have

$$|(v_n)_h - v_n| \le \|(v_n)_h - v_n\|_{L^1}^{(r-2)/2(r-1)}\|(v_n)_h - v_n\|_{L^r}^{r/2(r-1)},$$

and so by choosing $r > 2$ such that $H^1(U) \subset L^r(U)$ (Theorem 5.31) we can deduce that $(v_n)_h$ converges to v_n in L^2 uniformly over all n.

Now, for each fixed h the sequence $(v_n)_h$ is uniformly bounded and equicontinuous, since

$$|(v_n)_h(x)| \le \int_{B(0,1)} \rho(z)|v_n(x - hz)|\, dz \le h^{-m}\|\rho\|_\infty\|v_n\|_{L^1} \le Ch^{-m},$$

and

$$|D(v_n)_h(x)| \le \left|\int_{B(0,1)} \rho(z)(D_x v_n)(x - hz)\, dz\right|$$

$$= h^{-1}\left|\int_{B(0,1)} \rho(z)(D_z v_n)(x - hz)\, dz\right|$$

$$= h^{-1}\left|\int_{B(0,1)} D\rho(z)v_n(x - hz)\, dz\right|$$

$$\le h^{-(m+1)}\|D\rho\|_\infty\|v_n\|_{L^1}.$$

It follows using the Arzelà–Ascoli theorem (Theorem 2.5) that for each h there is a subsequence of $\{(v_n)_h\}$ that converges uniformly on U. It certainly converges, therefore, in $L^2(U)$.

We now use a diagonal argument, as in the proof of the Arzelà–Ascoli theorem. Pick a subsequence $v_{n_{1j}}$ such that $(v_{n_{1j}})_1$ converges in L^2. Now pick a subsequence of this, $v_{n_{2j}}$, such that $(v_{n_{2j}})_{1/2}$ also converges in L^2. Define a whole sequence of subsequences in this way, $(v_{n_{mj}})_{1/l}$, converging in L^2 for all $l \le m$. Now take the diagonal sequence $w_j = v_{n_{jj}}$, which has $(w_j)_{1/n}$ convergent in L^2 for all integers n.

Then

$$|w_n - w_m| \le |w_n - (w_n)_{1/l}| + |w_m - (w_m)_{1/l}| + |(w_n)_{1/l} - (w_m)_{1/l}|,$$

and choosing l large enough that $|w_n - (w_n)_{1/l}| < \epsilon/3$ for each n and N

large enough that $|(w_n)_{1/l} - (w_m)_{1/l}| < \epsilon/3$ for all $n, m > N$, we obtain the result. ∎

The following corollary is a straightforward consequence of the completeness of $H^k(\Omega)$.

Corollary 5.33. *Let Ω be bounded and $\partial\Omega$ C^{k+1}. Then $H^{k+1}(\Omega)$ is compactly embedded in $H^k(\Omega)$.*

We also have $L^2(\Omega)$ compactly embedded in $H^{-1}(\Omega)$ (see Exercise 5.10).

5.9 Boundary Values

A PDE is usually made up of two components: the differential equation itself and the boundary condition. For example, Dirichlet boundary conditions specify the values of u on the boundary ($u|_{\partial\Omega}$). To treat such problems in the current setting we need to able to define Sobolev spaces on manifolds such as $\partial\Omega$.

When Ω is a bounded C^k domain there is a natural way to do this. Recall (from Definition 5.19) that then every point x_0 in $\partial\Omega$ has a neighbourhood $U(x_0)$ such that $\partial\Omega \cap U(x_0)$ is given by $\Phi^{-1}(y)$ as y ranges over a subset of $\partial\mathbb{R}^m_+$. We can find a finite set of neighbourhoods $\{U_j\}_{j=1}^N$ that cover $\partial\Omega$ since $\partial\Omega$ is compact, and we will denote by Φ_j the corresponding C^k-diffeomorphisms.

Essentially the Sobolev space $H^k(\partial\Omega)$ consists of functions that are in H^k on each region $\partial\Omega \cap U_j$. To be more precise, we let $\{\psi_j\}$ be a partition of unity subordinate to the cover $\{U_j\}$, as guaranteed by Theorem 5.14. We can then define $H^k(\partial\Omega)$, identifying $\partial\mathbb{R}^m_+$ with $\mathbb{R}^{m-1}$. The definition is related to the procedure used to straighten the boundary in Section 5.5.3.

Definition 5.34. *If Ω is a bounded C^r domain then, for $k \le r$, $u \in H^k(\partial\Omega)$ if*

$$\left(\Phi_j^{-1}\right)^*(\psi_j u) \in H^k(\mathbb{R}^{m-1}) \qquad \text{for all } j.$$

One can define a norm on $H^k(\partial\Omega)$ by

$$|u|^2_{H^k(\partial\Omega)} = \sum_{j=1}^N \left|\left(\Phi_j^{-1}\right)^*(\psi_j u)\right|^2_{H^k(\mathbb{R}^{m-1})}.$$

Although the particular norm depends on the particular choice of covering $\{U_j\}$ and corresponding diffeomorphisms $\{\Phi_j\}$, we can show that all such norms are equivalent.

We now show how to consider the restrictions of functions in $H^k(\Omega)$ to the boundary of Ω.

Theorem 5.35. *Suppose that Ω is a bounded C^1 domain. Then there exists a bounded linear operator*

$$T : H^1(\Omega) \to L^2(\partial\Omega),$$

the "trace operator," such that for all $u \in H^1(\Omega) \cap C^0(\overline{\Omega})$

$$Tu = u|_{\partial\Omega}. \tag{5.38}$$

As in the proof of Theorem 5.20, we will write $\Omega_j = U_j \cap \Omega$ and $\tilde{\Omega}_j = \Phi_j(\Omega_j)$.

Proof. The first step is to consider a function in $H^1(\mathbb{R}_+^m)$ and to show that there is a bounded linear operator from $H^1(\mathbb{R}_+^m)$ into $L^2(\partial\mathbb{R}_+^m) = L^2(\mathbb{R}^{m-1})$. For any $u \in H^1(\mathbb{R}_+^m) \cap C^1(\overline{\mathbb{R}_+^m})$, such that $u(x) \to 0$ as $x_m \to \infty$, we have

$$\int_{\partial\mathbb{R}_+^m} |u(x')|^2 \, dx' = -\int_{\mathbb{R}_+^m} D_m(|u|^2) \, dx$$

$$= -\int_{\mathbb{R}_+^m} 2u \, D_m u \, dx$$

$$\leq \int_{\mathbb{R}_+^m} |u|^2 + |D_m u|^2 \, dx$$

$$\leq \|u\|_{H^1(\mathbb{R}_+^m)}^2.$$

For $u \in H^1(\Omega) \cap C^1(\overline{\Omega})$ we take the partition of unity $\{\psi_j\}$ subordinate to the cover $\{U_j\}$ that occurs in Definition 5.34, and we use the previous result to obtain

$$\int_{\mathbb{R}^{m-1}} \left|(\Phi_j^{-1})^*(\psi_j u)\right|^2 dx \leq \left\|(\Phi_j^{-1})^*(\psi_j u)\right\|_{H^1(\mathbb{R}_+^m)}^2$$

$$\leq C_j \|\psi_j u\|_{H^1(\Omega_j)}$$

$$\leq C_j \|u\|_{H^1(\Omega)}.$$

Summing over j we have

$$\|u\|_{L^2(\partial\Omega)} \leq C\|u\|_{H^1(\Omega)} \tag{5.39}$$

for all $u \in H^1(\Omega) \cap C^1(\overline{\Omega})$.

If we now define an operator T for all $u \in H^1(\Omega) \cap C^1(\overline{\Omega})$ by

$$Tu = u|_{\partial\Omega}$$

then we can rewrite (5.39) as

$$\|Tu\|_{L^2(\partial\Omega)} \le C\|u\|_{H^1(\Omega)}. \tag{5.40}$$

Now for a general $u \in H^1(\Omega)$, use Theorem 5.21 to take a sequence of functions $u_n \in C^\infty(\overline{\Omega})$ that converges to u in $H^1(\Omega)$. Then (5.40) gives

$$\|Tu_n - Tu_m\|_{L^2(\partial\Omega)} \le C\|u_n - u_m\|_{H^1(\Omega)},$$

and so $\{Tu_n\}$ is a Cauchy sequence in $L^2(\partial\Omega)$. It follows that Tu_n converges to some $v \in L^2(\partial\Omega)$, and we define $Tu = v$. Clearly, T is bounded.

If $u \in H^1(\Omega) \cap C^0(\overline{\Omega})$ then the sequence u_n, provided by Theorem 5.21, in fact converges uniformly on $\overline{\Omega}$ (see note after the proof of that theorem), and so for such u we have $Tu = u|_{\partial\Omega}$. ∎

The following theorem gives an important characterisation of the space $H^1_0(\Omega)$.

Theorem 5.36. $u \in H^1_0(\Omega)$ iff $u \in H^1(\Omega)$ and $Tu = 0$.

Proof. Clearly if $u \in H^1_0(\Omega)$ then $u \in H^1(\Omega)$, and there exists a sequence of functions $u_n \in C^\infty_c(\Omega)$ that converges to u in the H^1 norm. Since $Tu_n = 0$ for every n, it follows from the continuity of $T : H^1(\Omega) \to L^2(\partial\Omega)$, (5.38), that $Tu = 0$.

To prove the converse, we use a partition of unity and straighten the boundary, which reduces the problem to the case of a function $u \in H^1(\mathbb{R}^m_+)$ with compact support in $\overline{\mathbb{R}^m_+}$ and $Tu = 0$ on $\partial\mathbb{R}^m_+$. We need to show that $u \in H^1_0(\mathbb{R}^m_+)$.

Choose a function $\phi \in C^\infty_c(\mathbb{R})$ with $\|\phi\|_\infty \le 1$ and such that

$$\phi(r) = \begin{cases} 1, & r \ge 2, \\ 0, & r \notin [0, 2]. \end{cases}$$

We consider the sequence $u_n(x) = u(x)\phi(nx_m)$ and show that this converges to u in $H^1(\mathbb{R}^m_+)$.

First, convergence in $L^2(\mathbb{R}^m_+)$ follows, since

$$\int_{\mathbb{R}^m_+} |u_n(x) - u(x)|^2 \, dx = \int_{\mathbb{R}^m_+} |u(x)|^2 |1 - \phi(nx_m)|^2 \, dx$$

$$= \int_0^{2/n} |1 - \phi(nx_m)|^2 \int_{\mathbb{R}^{m-1}} |u(x)|^2 \, dx' \, dx_m$$

$$\le \int_0^{2/n} \int_{\mathbb{R}^{m-1}} |u(x)|^2 \, dx' \, dx_m,$$

which tends to zero as $n \to \infty$.

To show convergence of the derivative terms, note that

$$D_j u_n = \begin{cases} (D_j u)\phi(nx_m), & 1 \le j \le m-1, \\ (D_m u)\phi_n(x_m) + nu\phi'(nx_m), & j = m. \end{cases}$$

Convergence of $D_j u_n$ to $D_j u$ in $L^2(\mathbb{R}^m)_+$ follows exactly as above unless $j = m$. In this case we have

$$D_m u - D_m u_n = (D_m u)(1 - \phi(nx_m)) + nu\phi'(nx_m),$$

where we can deal easily with the first term but have to consider more carefully

$$I_n = n^2 \int_{\mathbb{R}^m_+} |u(x)|^2 |\phi'(nx_m)|^2 \, dx.$$

If $u \in H^1(\mathbb{R}^m_+)$ with compact support in $\overline{\mathbb{R}^m_+}$ and $Tu = 0$ on $\partial\mathbb{R}^m_+$, we have the following inequality:

$$\int_{\mathbb{R}^{m-1}} |u(x', x_m)|^2 \, dx' \le C x_m \int_0^{x_m} \int_{\mathbb{R}^{m-1}} |Du(x', y_m)|^2 \, dx' \, dy_m. \tag{5.41}$$

To prove this, consider a sequence v_n of functions in $C^1(\overline{\mathbb{R}^m_+})$ such that $v_n \to u$ and $v_n|_{\partial\mathbb{R}^m_+} = T v_n \to 0$ in $L^2(\mathbb{R}^{m-1})$. Then for each $x_m > 0$ we have

$$|v_n(x', x_m)| \le |v_n(x', 0)| + \int_0^{x_m} |D_m v_n(x', y_m)| \, dy_m,$$

from which (5.41) follows for u by taking limits [since $T v_n \to 0$ in $L^2(\mathbb{R}^{m-1})$].

We therefore have

$$I_n = n^2 \int_0^{2/n} \int_{\mathbb{R}^{m-1}} |u(x)|^2 |\phi'(nx_m)|^2 \, dx' \, dx_m$$

$$\le C n^2 \int_0^{2/n} \int_{\mathbb{R}^{m-1}} |u(x)|^2 \, dx' \, dx_m$$

$$\le C n^2 \int_0^{2/n} \left(x_m \int_0^{x_m} \int_{\mathbb{R}^{m-1}} |Du(x', y_m)|^2 \, dx' \, dy_m \right) dx_m$$

$$\le C n^2 \left(\int_0^{2/n} x_m \, dx_m \right) \left(\int_0^{2/n} \int_{\mathbb{R}^{m-1}} |Du(x', y_m)|^2 \, dx' \, dy_m \right)$$

$$\le C \int_0^{2/n} \int_{\mathbb{R}^{m-1}} |Du(x', y_m)|^2 \, dx' \, dy_m,$$

which tends to zero as $n \to \infty$.

So $u_n \to u$ in $H^1(\mathbb{R}^m_+)$. Now, given $\epsilon > 0$ we choose n large enough that

$$\|u_n - u\|_{H^1(\mathbb{R}^m_+)} < \epsilon/2,$$

and then we mollify u_n, choosing h small enough that

$$\|(u_n)_h - u\|_{H^1(\mathbb{R}^m_+)} < \epsilon$$

and that the support of $(u_n)_h$ is a compact subset of $\mathbb{R}^m_+$; we can do this since u_n has compact support in $\mathbb{R}^m_+$.

We have obtained a sequence of functions $\{(u_n)_h\}$ in $C_c^\infty(\mathbb{R}^m_+)$ that tends to u in the $H^1(\mathbb{R}^m_+)$ norm, and hence $u \in H_0^1(\mathbb{R}^m_+)$ as claimed. ∎

5.10 Sobolev Spaces of Periodic Functions

In this final section we introduce Sobolev spaces of periodic functions. Such functions can be represented as Fourier series, which makes their analysis significantly more straightforward than that for the spaces on bounded domains we have been considering until now. Furthermore there are no boundary issues to deal with. Because of these simplifications, we will use periodic boundary conditions when we discuss the Navier–Stokes equations in later chapters.

We will treat Sobolev spaces $H_p^s(Q)$ of L-periodic functions on the m-dimensional domain $Q = [0, L]^m$. We will denote by $C_p^\infty(Q)$ the space of restrictions to Q of infinitely differentiable functions that are L-periodic in each direction,

$$u(x + Le_j) = u(x), \qquad j = 1, \ldots, m,$$

where e_j is a unit vector in the jth component.

Definition 5.37. *The Sobolev space $H_p^s(Q)$ is the completion of $C_p^\infty(Q)$ with respect to the H^s norm*

$$\|u\|_{H^s} = \left(\sum_{0 \le |\alpha| \le s} \|D^\alpha u\|^2_{L^2(Q)} \right)^{1/2}.$$

Now, by definition $C_p^\infty(Q)$ is dense in $H_p^s(Q)$ – so we have already obtained the analogue of Theorem 5.21 in the periodic setting with minimal effort.

We now give another characterisation of the spaces $H_p^s(Q)$ by using Fourier series. We enforce the periodicity of the function u by writing it as a formal

Fourier series,

$$u = \sum_{k \in \mathbb{Z}^m} c_k e^{2\pi i k \cdot x / L} \qquad \text{with} \qquad c_{-k} = \bar{c}_k.$$

For any function $u \in C_p^\infty(Q)$, this Fourier series is uniformly convergent, and the derivatives of u are given by

$$D^\alpha u(x) = \left(\frac{2\pi i}{L}\right)^{|\alpha|} \sum_{k \in \mathbb{Z}^m} c_k k^\alpha e^{2\pi i k \cdot x / L}.$$

It follows from Parseval's identity that

$$|D^\alpha u|^2 = L^m \left(\frac{2\pi}{L}\right)^{2|\alpha|} \sum_{k \in \mathbb{Z}^m} |c_k|^2 |k^{2\alpha}|. \tag{5.42}$$

This suggests that the H_p^s norm will be related to some expression involving powers of $|k|$ and the Fourier coefficients. In fact, we can show that the H_p^s norm is equivalent to the "H_f^s norm,"

$$\|u\|_{H_f^s} = \left(\sum_{k \in \mathbb{Z}^m} (1 + |k|^{2s}) |c_k|^2\right)^{1/2}.$$

Proposition 5.38. *The H_p^s norm and the H_f^s norm are equivalent:*

$$C_s' \|u\|_{H_f^s} \leq \|u\|_{H_p^s} \leq C_s \|u\|_{H_f^s}.$$

In the proof C_s will be used to denote any constant that depends on only s and m.

Proof. Write

$$\|u\|_{H_p^s}^2 = \sum_{0 \leq |\alpha| \leq s} |D^\alpha u|^2$$

$$= L^m \sum_{0 \leq |\alpha| \leq s} (2\pi/L)^{2|\alpha|} \left(\sum_{k \in \mathbb{Z}^m} k^{2\alpha} |c_k|^2\right)$$

$$\leq C_s \sum_{k \in \mathbb{Z}^m} \left(\sum_{0 \leq |\alpha| \leq s} |k|^{2|\alpha|}\right) |c_k|^2$$

$$\leq C_s \left(\sum_{k \in \mathbb{Z}^m} (1 + |k|^{2s}) |c_k|^2\right) = C_s \|u\|_{H_f^s}^2.$$

The other inequality follows from (5.42):

$$\|u\|_{H_p^s}^2 \geq C \sum_{k \in \mathbb{Z}^m} \left(1 + \sum_{j=1}^m |k_j|^{2s}\right) |c_k|^2. \qquad \blacksquare$$

The completion of $C_p^\infty(Q)$ in the H_p^s norm is therefore the same as its completion in the H_f^s norm, and we can identify $H_p^s(Q)$ with the collection of all formal Fourier series such that the norm H_f^s is finite.

Proposition 5.39. *The Sobolev space of periodic functions $H_p^s(Q)$ is the same as*

$$\left\{ u : u = \sum_{k \in \mathbb{Z}^m} c_k e^{2\pi i k \cdot x/L}, \ \bar{c}_k = c_{-k}, \ \sum_{k \in \mathbb{Z}^m} |k|^{2s} |c_k|^2 < \infty \right\}. \qquad (5.43)$$

Note that we ensure finiteness of the H_f^s norm by requiring only

$$\sum_{k \in \mathbb{Z}^m} |k|^{2s} |c_k|^2 < \infty.$$

In general, however, these two expressions do not give rise to equivalent norms, since if $c_0 \neq 0$ the contribution from the constant term is lost in the second expression. However, for the subspace $\dot{H}_p^s(Q)$ of functions with $c_0 = 0$, i.e. those with zero mean over Q,

$$\int_Q u(x) \, dx = 0,$$

the two norms are equivalent. This is because we have a Poincaré inequality for these spaces, just as we did for $H_0^1(\Omega)$ in Proposition 5.8.

Lemma 5.40 (Poincaré's inequality). *If $u \in \dot{H}_p^1(Q)$ then*

$$|u| \leq \left(\frac{L}{2\pi}\right) |Du|. \qquad (5.44)$$

Proof. See Exercise 5.11. $\blacksquare$

In this situation the norm $|Du|$ is equivalent to the standard H_p^1 norm on $\dot{H}_p^1$ (cf. (5.13)), and in general we will use

$$\left(\sum_{k \in \mathbb{Z}^m} |k|^{2s} |c_k|^2\right)^{1/2} \qquad (5.45)$$

for the norm on $\dot{H}_p^s$ (cf. (5.16)). The corresponding "$\dot{H}_p^s$" inner product is

$$((u, v))_{\dot{H}_p^s} = \sum_{k \in \mathbb{Z}^m} |k|^{2s} c_k d_k,$$

when

$$u = \sum_{k \in \mathbb{Z}^m} c_k e^{2\pi i k \cdot x / L} \qquad \text{and} \qquad v = \sum_{k \in \mathbb{Z}^m} d_k e^{2\pi i k \cdot x / L}.$$

We define $H_p^{-s}(Q)$ as the dual space of $\dot{H}_p^s(Q)$. This can also be characterised as in Proposition 5.39 (cf. Proposition 5.11).

The Sobolev embedding Theorem 5.31 and the compactness result of Theorem 5.32 are also valid in this setting. The proofs, which are very simple, are given in Appendix A.

Exercises

5.1   Show that if $u \in \mathcal{D}'(\Omega)$ and $\phi_n \to \phi$ in $\mathcal{D}(\Omega)$ then

$$\langle D^\alpha u, \phi_n \rangle \to \langle D^\alpha u, \phi \rangle,$$

and so $D^\alpha u$ is a distribution.

5.2   For $\psi \in C_c^\infty(\Omega)$ and $u \in \mathcal{D}'(\Omega)$ we can define the distribution ψu by

$$\langle \psi u, \phi \rangle = \langle u, \psi \phi \rangle \qquad \text{for all} \qquad \phi \in C_c^\infty(\Omega).$$

Show that we do indeed have $\psi u \in \mathcal{D}'(\Omega)$ and that

$$D(u\psi) = u D\psi + \psi D u.$$

5.3   Show that if $\{f_n\}$ is a bounded sequence in $L^2(\Omega)$ such that $f_n \to f$ in $\mathcal{D}'(\Omega)$ then $f_n \rightharpoonup f$ in $L^2(\Omega)$.

5.4   Prove that, under the conditions of Proposition 5.8, there exists a constant $C(k)$ such that

$$\|u\|_{H^k}^2 \leq C \sum_{|\alpha|=k} |D^\alpha u|^2 \qquad \text{for all} \qquad u \in H_0^k(\Omega). \tag{5.46}$$

(Hint: Use induction along with the argument of Proposition 5.8.)

5.5   Show that if $\psi \in C_c^\infty(\Omega)$ and $u \in H^k(\Omega)$ then $\psi u \in H^k(\Omega)$ and

$$\|\psi u\|_{H^k(\Omega)} \leq C(\psi) \|u\|_{H^k(\Omega)}.$$

5.6   Show that the unbounded function

$$\log\log\left(1 + \frac{1}{|x|}\right)$$

is still an element of $H^1(B(0, 1))$, where $B(0, 1)$ is the unit ball in $\mathbb{R}^2$.

5.7   Show that if $\Omega \subset \mathbb{R}^3$ is a bounded C^1 domain then, for $u \in H^1(\Omega)$,

$$\|u\|_{L^3} \leq C|u|^{1/2}\|u\|_{H^1}^{1/2}.$$

[Hint: Use (5.33) to write

$$|u(x)|^3 \leq \left(6\int_{-\infty}^{\infty} |uD_1u|\, dy_1 \int_{-\infty}^{\infty} |uD_2u|\, dy_2 \int_{-\infty}^{\infty} |uD_3u|\, dy_3\right)^{1/2},$$

$$(5.47)$$

and then follow an argument similar to that of Proposition 5.23.]

5.8   Prove that if $\Omega \subset \mathbb{R}^m$ is a bounded C^k domain and $k > (m/2) + j$ then each $u \in H^k(\Omega)$ is an element of $C^j(\overline{\Omega})$, with

$$\|u\|_{C^j} \leq C\|u\|_{H^k}.$$

5.9   Let V be the subspace of $H^1(\Omega)$ consisting of functions with zero integral over Ω:

$$V = \left\{u \in H^1(\Omega) : \int_{\Omega} u(x)\, dx = 0\right\}.$$

Arguing by contradiction, use the compactness result of Theorem 5.32 to show that there exists a constant C such that we have a Poincaré inequality

$$|u| \leq C|\nabla u| \qquad \text{for all} \qquad u \in V. \qquad (5.48)$$

(You may assume that if $Du = 0$ then u is almost everywhere constant.)

5.10   Show that $L^2(\Omega)$ is compactly embedded in $H^{-1}(\Omega)$.

5.11   Prove that if $u \in \dot{H}^1_p(Q)$ then $|u| \leq (L/2\pi)|Du|$.

Notes

Adams' book *Sobolev Spaces* (1975) is the main reference for the material in this chapter, although because it treats the subject in maximum generality most of his proofs are not

directly suitable to this kind of simplified presentation. Nonetheless, anybody wishing
to go into this subject in more depth should certainly consult his book.

Distribution theory is treated didactically but in much greater detail than the rather
cursory treatment here by Friedlander & Joshi (1999). Rudin (1991) and Yosida (1980)
offer more technical treatments.

A slightly different class of distributions from those used here provides a natural
setting in which to treat the Fourier transform, defined as

$$\mathcal{F}[u](\xi) = \hat{u}(\xi) = (2\pi)^{-m/2} \int_{\mathbb{R}^m} e^{-i\xi \cdot x} u(x)\, dx,$$

initially for $u \in L^1(\mathbb{R}^m)$. The definition can be extended to all distributions by defining

$$\langle \hat{u}, \phi \rangle = \langle u, \hat{\phi} \rangle \qquad \text{for all} \qquad \phi \in \mathcal{S}$$

($\mathcal{S}$ is a larger class of test functions that $\mathcal{D}$). Since one can show that

$$\mathcal{F}[D^\alpha u] = (-i\xi)^\alpha \hat{u}$$

and that

$$|\hat{u}|_{L^2} = |u|_{L^2},$$

it follows that $u \in H^k(\mathbb{R}^m)$ iff

$$(1 + |\xi|^2)^{k/2} \mathcal{F}[u] \in L^2(\mathbb{R}^m). \tag{5.49}$$

This characterisation forms the main basis of the treatment in Renardy & Rogers
(1992). Particularly important is that it provides a way of defining $H^k(\mathbb{R}^m)$ when k
is not an integer, simply by imposing (5.49) for the appropriate value of k. These fractional index Sobolev spaces can be useful, but we will not need them in what follows.

Evans (1998) avoids introducing the distribution derivative, which somewhat limits
his discussion of the spaces $H^{-k}(\Omega)$, but otherwise his treatment is very similar to that
here. Indeed, the development in this chapter owes much to him and also to Gilbarg &
Trudinger (1983), who manage to present a lot of material very clearly and concisely;
the proof of Theorem 5.29 is taken from there.

There are many other results concerning Sobolev spaces that are not included here.
One notable result (proofs can be found in Evans, Gilbarg & Trudinger, and Renardy
& Rogers) is that if $u \in H^k(\Omega)$ and $m/2 < k < (m/2) + 1$ then in fact u is Hölder
continuous with exponent $k - (m/2)$. Coupled with the Arzelà–Ascoli theorem this
shows that $H^k(\Omega)$ is compactly embedded in $C^0(\overline{\Omega})$ if $k > m/2$. Many of these

results, and various inequalities, are usefully summarised in Chapter 2 of Temam (1988).

Sobolev spaces of periodic functions are treated briefly in Temam (1995) and Constantin & Foias (1988). They also form the basis of the analysis in Doering & Gibbon (1995), although this is not expressed using the Sobolev space formalism.

The proof of Ladyzhenskaya's inequality in Lemma 5.27 is taken from Ladyzhenskaya (1963). It will be useful in our treatment of the Navier–Stokes equations in Chapter 9.

Part II
Existence and Uniqueness Theory

6

The Laplacian

Although we are ultimately interested in treating time-dependent nonlinear equations, this chapter lays some foundations by considering a time-independent linear example, Poisson's equation. This equation is

$$-\Delta u = f(x), \tag{6.1}$$

where Δu denotes the Laplacian,

$$\Delta u = \sum_{j=1}^{m} \frac{\partial^2 u}{\partial x_j^2}.$$

[Note that we write Poisson's equation with a minus sign; this is because $-\Delta$ is a positive operator; see (3.23).]

We will consider (6.1) with two choices of boundary condition. The simpler case is that of periodic boundary conditions,

$$u(x + Le_j) = u(x) \qquad \text{for all} \qquad j = 1, \ldots, m,$$

which we can analyse very easily using the Fourier expansion introduced in the last section of the previous chapter. However, most of our efforts will go towards understanding the case of Dirichlet boundary conditions, where u is specified on the boundary of a smooth bounded domain Ω,

$$u(x) = 0 \qquad \text{for all} \qquad x \in \partial\Omega.$$

We abbreviate these boundary conditions to $u|_{\partial\Omega} = 0$.

An understanding of linear operators such as this example is necessary for the nonlinear evolution equations we will treat later. We will study existence and uniqueness of solutions of Poisson's equation in a variety of settings, from "weak" solutions to fully classical solutions.

6.1 Classical, Strong, and Weak Solutions

There are three distinct senses in which we can understand the notion of a "solution" to Equation (6.1).

The first is classically, where the problem can be stated formally as

$$\text{given} \quad f \in C^0(\Omega),$$

find $\quad u \in C^2(\Omega) \cap C^0(\overline{\Omega}) \quad$ with $\quad u = 0 \quad$ on $\quad \partial \Omega,$ $\quad$ such that

$$-\Delta u(x) = f(x), \quad \text{classically, for each } x \in \Omega.$$

We wish to generalise the class of possible "solutions," in the hope that the equation will become easier to solve in a less restrictive form. This essentially means that, rather than considering each term to be a continuous function, we will take each term to be in some less regular space. A natural way to do this is to allow each term to be in L^2. This will immediately involve Sobolev spaces in the description of the problem. A *strong solution* is a solution of the problem

$$\text{given} \quad f \in L^2(\Omega),$$

find $\quad u \in H^2(\Omega) \cap H_0^1(\Omega), \quad$ such that

$-\Delta u = f$ in $L^2,$ $\quad$ with the derivatives understood in the weak sense.

We have not only weakened the regularity of u, we have also weakened the acceptable regularity for f. Furthermore, we have incorporated the boundary conditions into the space in which the solution u is required to lie (cf. Theorem 5.36). Since the equality here is equality of functions in L^2, the equation $-\Delta u = f$ is only required to hold almost everywhere in Ω. Finally, it is easy to check that a classical solution is also a strong solution.

In this chapter we will concentrate first on *weak solutions*, where we relax even further the smoothness required for u (and also f). We will see that in this case each term is viewed as an element of $H^{-1}(\Omega)$.

To show that a weak (or perhaps strong) solution is in fact classical we try to show that the solution is bounded in higher order Sobolev spaces. We can then use the embedding Theorem 5.31 to show that $u(t)$ is a classically differentiable function. We will do this for periodic boundary conditions in Section 6.3 and for Dirichlet boundary conditions in Sections 6.4 and 6.5.

6.2 Weak Solutions of Poisson's Equation

We will start by discussing weak solutions of Poisson's equation (6.1), which is a straightforward introduction to the framework that we will use for fully nonlinear, time-varying equations. We will consider first the case of Dirichlet boundary conditions.

Suppose that we want to weaken further the differentiability required for our solution u. What we can do is multiply (6.1) by a smooth function v (say a function in $C_c^1(\Omega)$) and integrate:

$$- \int_\Omega [\Delta u(x)] v(x)\, dx = \int_\Omega f(x) v(x)\, dx.$$

If we integrate the first term by parts and use the fact that $v = 0$ on $\partial\Omega$, we get

$$\sum_{i=1}^m \int_\Omega D_i u(x) D_i v(x)\, dx = \int_\Omega \nabla u \cdot \nabla v\, dx = \int_\Omega f(x) v(x)\, dx, \qquad (6.2)$$

an equality we require to hold for all $v \in C_c^1(\Omega)$. If $u \in C^2(\overline{\Omega})$ and $f \in C^0(\overline{\Omega})$ we can show that (6.2) implies that u is a classical solution of (6.1) (see Exercise 6.1).

However, (6.2) requires u to be only a C^1 function. If in addition we use the density of $C_c^1(\Omega)$ in $H_0^1(\Omega)$ we can deduce that (6.2) must hold for all $v \in H_0^1(\Omega)$. In fact, (6.2) makes sense whenever $u \in H_0^1(\Omega)$ and $f \in H^{-1}(\Omega)$, provided that we understand the derivatives in a weak sense. If we write the left-hand side in the form $a(u, v)$, so that

$$a(u, v) = \int_\Omega \nabla u \cdot \nabla v\, dx, \qquad (6.3)$$

we can re-pose the original problem in its *weak form*:

$$\text{given} \quad f \in H^{-1}(\Omega),$$

$$\text{find} \quad u \in H_0^1(\Omega), \quad \text{such that} \qquad (6.4)$$

$$a(u, v) = \langle f, v \rangle \quad \text{for all} \quad v \in H_0^1(\Omega),$$

where $\langle f, v \rangle$ denotes the pairing between $f \in H^{-1}$ and $v \in H_0^1$.

Note that $a(u, v)$ is a bilinear form on $H_0^1 \times H_0^1$, which (using the Cauchy–Schwarz inequality) satisfies

$$|a(u, v)| \le \|u\|_{H_0^1} \|v\|_{H_0^1}. \qquad (6.5)$$

In fact, in this case the bilinear form $a(u, v)$ is exactly (5.13), the reduced inner product on $H_0^1(\Omega)$ obtained using Poincaré's inequality. The problem is therefore equivalent to finding a $u \in H_0^1$ such that

$$((u, v))_{H_0^1} = \langle f, v \rangle \quad \text{for all} \quad v \in H_0^1. \qquad (6.6)$$

Now, the Riesz representation theorem (Theorem 4.9) immediately tells us that there exists a unique $u \in H_0^1$ such that (6.6) holds and that the map $f \mapsto u$ is continuous.

Theorem 6.1. *For $f \in H^{-1}(\Omega)$, the weak form of Poisson's equation (6.4) has a unique solution $u \in H_0^1(\Omega)$, and*

$$\|u\|_{H_0^1(\Omega)} = \|f\|_{H^{-1}(\Omega)}.$$

There is another way to consider Equation (6.6), which will become more useful later. Observe that for each fixed $u \in H_0^1$ the map

$$v \mapsto a(u, v)$$

is a linear functional on H_0^1. It follows that we can define a linear operator $A : H_0^1 \to H^{-1}$ by

$$\langle Au, v \rangle = a(u, v) \qquad \text{for all} \qquad v \in H_0^1(\Omega). \qquad (6.7)$$

Observe that (6.5) shows that A is a bounded operator. We can now write the equation $a(u, v) = \langle f, v \rangle$ as

$$Au = f,$$

an equation in H^{-1}.

We will now show that in fact A is a symmetric operator with compact inverse, and we will apply Corollary 3.26 to deduce that there is a basis of $L^2(\Omega)$ consisting of eigenfunctions of A.

We have already shown in Theorem 6.1 that A is an invertible mapping. To show that A is compact, observe that we can identify $L^2(\Omega)$ with a subset of $H^{-1}(\Omega)$, since an element of $L^2(\Omega)$ gives rise to a linear functional on $H_0^1(\Omega)$ via the L^2 inner product,

$$\langle u, v \rangle = (u, v) \qquad \text{for all} \qquad v \in H_0^1(\Omega).$$

Since

$$|\langle u, v \rangle| \le |u||v| \le C|u|\|v\|_{H_0^1}$$

(using Poincaré's inequality),

$$\|u\|_{H^{-1}} \le C|u|.$$

It follows from this that

$$\|A^{-1}f\|_{H_0^1} \le C|f| \qquad \text{for all} \qquad f \in L^2, \qquad (6.8)$$

so that A^{-1} is a bounded map from L^2 into H_0^1. Since a bounded subset of H_0^1 is compact in L^2 (Theorem 5.32), A^{-1} is in fact a compact map from L^2 into itself. Finally, A is clearly symmetric, since

$$\langle Au, v \rangle = a(u, v) = a(v, u) = \langle Av, u \rangle.$$

It follows via Corollary 3.26 that there is a basis of L^2 consisting of eigenfunctions $\{w_j\}$ of A. Since $Aw_j = \lambda_j w_j$ and $w_j \in L^2(\Omega)$, it follows that $w_j \in H_0^1(\Omega)$. We collect these facts in the following theorem.

Theorem 6.2. *There is a basis of $L^2(\Omega)$ consisting of eigenfunctions of the operator $A = -\Delta$ with Dirichlet boundary conditions. These eigenfunctions are elements of $H_0^1(\Omega)$.*

It is now readily apparent that translating an equation into its weak form can make it extremely straightforward to treat. In this case, the Laplacian led to a bilinear form that was symmetric, and this allowed us to use the Riesz representation theorem.

In a more general situation than (6.1) the weak formulation will give rise to a bilinear form that is not symmetric (see Exercise 6.2), and we need to use the Lax–Milgram lemma to deal with this case. When $a(u, v)$ is a bilinear form on some general Hilbert space V (with norm $\| \cdot \|$ and dual V^*), the symmetry of $a(u, v)$ has to be replaced with the *coercivity* condition,

$$a(v, v) \ge \alpha \|v\|^2 \qquad \text{for all} \qquad v \in V$$

for some $\alpha > 0$. A precise statement of the Lax–Milgram lemma is

Lemma 6.3. *If a is bilinear and coercive on V, then A is an isomorphism from V onto V^*: i.e. if $f \in V^*$ then the equation $Au = f$ has a unique solution.*

We will not treat this more complicated situation in any detail; instead we will concentrate on higher regularity for Poisson's equation. For some discussion of the more general case, see Exercises 6.3 and 6.4. Exercise 6.5 treats the problem with Neumann boundary conditions ($\nabla u \cdot n$ specified on $\partial\Omega$).

6.3 Higher Regularity for the Laplacian I:
Periodic Boundary Conditions

What we have found is that if $f \in H^{-1}(\Omega)$ then the solution of $-\Delta u = f$ ("$Au = f$") is an element of $H_0^1(\Omega)$. However, suppose that we know that f is a smoother function, for example, $f \in L^2(\Omega)$. For such f, what is the smoothness of u? From the discussion in Chapter 3, this amounts to finding $D(A)$, the domain of A in $L^2(\Omega)$.

In this section we will treat Poisson's equation on a periodic box $Q = [0, L]^m$ (as discussed in the final section of the Sobolev space chapter, Chapter 5), making use of the Fourier expansion to simplify the analysis.

However, without a little care we cannot expect unique solutions in this case, since we could add a constant c to any solution u, and $u + c$ would still be a solution. For this reason we will restrict ourselves to finding a solution whose average over Q is zero,

$$u \in L^2 \quad \text{such that} \quad \int_Q u = 0. \tag{6.9}$$

The space of all such functions was denoted by $\dot{L}^2(Q)$ in Chapter 5. Furthermore, note that unless f also satisfies such a condition we cannot hope to find a solution at all, because

$$\int_Q \Delta u \, dx = 0$$

after an integration by parts (cf. Exercise 6.6).

We now cast the problem into its weak form. This process, along with the use of the Riesz representation theorem to prove the existence of a solution, is almost exactly as above in the Dirichlet case: we take the inner product with a suitable smooth function v, in this case $v \in \dot{C}_p^1(Q)$, to get

$$a(u, v) = (f, v) \tag{6.10}$$

for all $v \in \dot{C}_p^1(Q)$, where we have defined the bilinear form as before (6.3):

$$a(u, v) = \int_Q \nabla u \cdot \nabla v \, dx.$$

The density of $\dot{C}_p^1(Q)$ in $\dot{H}_p^1(Q)$ means that we also have (6.10) for all $v \in \dot{H}_p^1(Q)$. It follows from using the Riesz representation theorem that for all $f \in H_p^{-1}(Q)$ there is a unique solution of (6.10) that lies in $\dot{H}_p^1(Q)$.

To investigate the regularity of this solution more closely we will use the Fourier series expansions of f and u introduced in Section 5.10 and show that if $f \in \dot{L}^2(Q)$ then in fact $u \in \dot{H}^2_p(Q)$. We write f as a Fourier series,

$$f = \sum_{k \in \mathbb{Z}^m} e^{2\pi i k \cdot x / L} f_k,$$

where $f_0 = 0$ and $\sum |f_k|^2 < \infty$ (since $f \in \dot{L}^2(Q)$). Now, if we expand u similarly,

$$u = \sum_{k \in \mathbb{Z}^m} e^{2\pi i k \cdot x / L} u_k,$$

we have

$$-\Delta u = \left(\frac{4\pi^2}{L^2} \right) \sum_{k \in \mathbb{Z}^m} e^{2\pi i k \cdot x / L} |k|^2 u_k,$$

and comparing coefficients in $-\Delta u = f$ we get

$$\frac{4\pi^2 |k|^2}{L^2} u_k = f_k, \qquad (6.11)$$

so that for $k \neq 0$ we have

$$u_k = \frac{L^2}{4\pi^2} \frac{f_k}{|k|^2}. \qquad (6.12)$$

To make sure that u satisfies (6.9) we set $u_0 = 0$ [note that this gives equality in (6.11), since $f_0 = 0$].

Theorem 6.4. *If $f \in \dot{L}^2(Q)$ and $u \in \dot{H}^1_p(Q)$ satisfies (6.10), then in fact $u \in \dot{H}^2_p(Q)$, and there exists a constant C such that*

$$\|u\|_{\dot{H}^2_p} \le C|f|.$$

Proof. Writing C for any constant that does not depend on the choice of f, we can use the Fourier expansion to write

$$\|u\|^2_{\dot{H}^2_p} = \sum_{k \in \mathbb{Z}^m} |k|^4 |u_k|^2 \le C \sum_{k \in \mathbb{Z}^m} |f_k|^2 = C|f|^2. \qquad \blacksquare$$

Viewing (6.10) as the equation $Au = f$ and observing that Equation (6.12) defines the map A^{-1} from $\dot{L}^2(Q)$ into $D(A)$, we also have precisely

$$D(A) = \dot{H}^2_p(Q).$$

In this case it is very easy to characterise $D(A)$, and we can see that A^{-1} is a compact operator, since a bounded set in $\dot{L}^2(Q)$ becomes a bounded set in

$\dot{H}^2_p(Q)$, i.e. a compact subset of $\dot{L}^2(Q)$. Since A^{-1} is a compact symmetric operator, it has a set of eigenfunctions that form a basis for $\dot{L}^2(Q)$. These are the real and imaginary parts of the exponential functions we have been using in our Fourier expansion:

$$w^{(s)}_k = \sqrt{2/L^m}\sin 2\pi k\cdot x/L \quad \text{and} \quad w^{(c)}_k = \sqrt{2/L^m}\cos 2\pi k\cdot x/L,$$
$$\text{with} \quad Aw^{(\cdot)}_k = (4\pi^2/L^2)|k|^2 w^{(\cdot)}_k. \tag{6.13}$$

Note that k in (6.13) runs over all multi-indices with $k > 0$ (we usually index the eigenfunctions by a single integer n), and we have included a normalising factor of $\sqrt{2/L^m}$ (so that $|w^{(\cdot)}_k| = 1$).

Use of the Fourier expansion again makes it easy to deduce higher regularity results, showing that if $f \in \dot{H}^s_p$ then $u \in \dot{H}^{s+2}_p$.

Theorem 6.5. *If $f \in \dot{H}^s_p(Q)$ and $u \in \dot{H}^1_p(Q)$ satisfies (6.10), then in fact $u \in \dot{H}^{s+2}_p(Q)$ and*

$$\|u\|_{\dot{H}^{s+2}_p} \leq C\|f\|_{\dot{H}^s_p}.$$

Proof. If $f \in \dot{H}^s_p(Q)$ then

$$\|f\|^2_{\dot{H}^s_p} = \sum_{k\in\mathbb{Z}^m} |k|^{2s}|f_k|^2 < \infty,$$

and it then follows from (6.12) that

$$\|u\|^2_{\dot{H}^{s+2}_p(Q)} = \sum_{k\in\mathbb{Z}^m}|k|^{2(s+2)}|u_k|^2 = \left(\frac{L^2}{4\pi^2}\right)^2\sum_{k\in\mathbb{Z}^m}|k|^{2s}|f_k|^2 = C\|f\|^2_{\dot{H}^s_p(Q)}.$$

Note that C does not depend on s. ∎

We can use this result along with the Sobolev embedding in Theorem 5.31 to show that we obtain smooth solutions when f is smooth. This corollary is particularly significant, since it shows that we can use the Sobolev space formulation, with the generalised derivatives that it involves, and end up with results about smooth classical solutions.

Corollary 6.6. *If $f \in \dot{C}^\infty_p(Q)$ and $u \in \dot{H}^1_p(Q)$ is a solution of (6.10), then in fact $u \in \dot{C}^\infty_p(Q)$.*

Proof. $f \in \dot{H}^s_p(Q)$ for all $s \geq 0$, and so $u \in \dot{H}^s_p(Q)$ for all $s \geq 0$, using Theorem 6.5. It follows that $u \in \dot{C}^r_p(Q)$ for all $r \geq 0$, using the Sobolev embedding $H^k \subset C^j$ if $k > j + (m/2)$ from Theorem 5.31. ∎

As a second corollary, we obtain the important result that the eigenfunctions of $-\Delta$ are elements of $\dot{C}_{\mathrm{p}}^{\infty}(Q)$. Of course, since in this case we know that the eigenfunctions are given explicitly by the trigonometric functions in (6.13) this is not a surprise. However, we will also obtain a similar result in the Dirichlet boundary condition case, when we cannot (in general) find an explicit form for the eigenfunctions.

Corollary 6.7. *In the case of periodic boundary conditions, the eigenfunctions of the Laplacian are all elements of $\dot{C}_{\mathrm{p}}^{\infty}(Q)$.*

Proof. We know that each eigenfunction w is an element of $\dot{L}^{2}(Q)$. Since w solves the equation

$$Aw = \lambda w,$$

it follows from Theorem 6.5 that in fact $w \in \dot{H}_{\mathrm{p}}^{2}(Q)$. We then obtain similarly $w \in \dot{H}_{\mathrm{p}}^{4}(Q)$, and by induction that $w \in \dot{H}_{\mathrm{p}}^{s}(Q)$ for any $s \geq 0$. It follows as in the previous corollary that $w \in \dot{C}_{\mathrm{p}}^{\infty}(Q)$. ∎

Finally, we note that we can use this regularity theory to characterise the Sobolev spaces $\dot{H}_{\mathrm{p}}^{s}(Q)$ in terms of the fractional powers of the operator $A = -\Delta$ (see Section 3.10). Indeed, from Proposition 5.39 we have

$$\dot{H}_{\mathrm{p}}^{s}(Q) = \left\{ u : u = \sum_{k \neq 0} c_k e^{2\pi i k \cdot x / L}, \ \bar{c}_k = c_{-k}, \ \sum_{k \neq 0} |k|^{2s} |c_k|^2 < \infty \right\},$$

whereas, using the definition of fractional powers of A from (3.25), we have

$$D(A^{s/2}) = \left\{ u : u = \sum_{k>0, \, t=s,c} a_k^{(t)} w_k^{(t)}, \ \sum_{k>0, \, t=s,c} \lambda_k^s |a_k^{(t)}|^2 < \infty \right\}.$$

Since

$$c_k e^{2\pi i k \cdot x / L} + c_{-k} e^{-2\pi i k \cdot x / L} = \mathrm{Re}(c_k) \cos \frac{2\pi k \cdot x}{L} - \mathrm{Im}(c_k) \sin \frac{2\pi k \cdot x}{L}$$

$$= \sqrt{\frac{L^m}{2}} \left(\mathrm{Re}(c_k) w_k^{(c)} - \mathrm{Im}(c_k) w_k^{(s)} \right),$$

we have

$$\sum_{k \neq 0} c_k e^{2\pi i k \cdot x / L} = \sqrt{\frac{L^m}{2}} \left(\sum_{k>0} \mathrm{Re}(c_k) w_k^{(c)} - \mathrm{Im}(c_k) w_k^{(s)} \right),$$

and, using $\lambda_k = (4\pi^2/L^2)|k|^2$, we have

$$\|u\|_{\dot{H}_p^s}^2 = \sum_{k>0} |k|^{2s}|c_k|^2$$

$$= \frac{L^m}{2}\left(\frac{L^2/4}{\pi^2}\right)^s \sum_{k>0,\ t=s,c} \lambda_k^s |a_k^{(t)}|^2$$

$$= C|A^{s/2}u|^2.$$

Therefore

$$|A^{s/2}u| = C\|u\|_{\dot{H}_p^s(Q)}, \tag{6.14}$$

and so $D(A^{s/2}) = \dot{H}_p^s(Q)$.

The very simple arguments of this section become much more involved in the case of Dirichlet boundary conditions on a bounded domain $\Omega \subset \mathbb{R}^m$, which we treat in the remainder of this chapter.

6.4 Higher Regularity for the Laplacian II: Dirichlet Boundary Conditions

We now want to reproduce the results of the previous section in the case of Dirichlet boundary conditions,

$$-\Delta u = f, \qquad u|_{\partial\Omega} = 0,$$

where Ω is a bounded C^2 domain. We know that if $f \in H^{-1}(\Omega)$ then there is a solution that lies in $H_0^1(\Omega)$; we have to investigate what happens when we take $f \in L^2(\Omega)$.

6.4.1 A Heuristic Estimate

We will first give a heuristic argument indicating that we can still expect a result along the lines of Theorem 6.4. We will try to show initially that if $\Omega' \subset\subset \Omega$ then

$$\|u\|_{H^2(\Omega')} \le C\|f\|_{L^2(\Omega)}.$$

This result, which shows that the second derivatives of u are square integrable on any compact subdomain of Ω, is known as an interior regularity result. To extend this result all the way up to the boundary of Ω will require some care and messy manipulations.

For the remainder of this chapter we will use the abbreviations

$$u_{,i} = D_i u, \qquad u_{,ij} = D_i D_j u, \qquad u_{,ijj} = D_i D_j^2 u,$$

etc. This will cut down heavily on the notation, since many derivatives will be involved.

To obtain our heuristic estimate, we treat all the terms in the equation as if they were smooth. Since we want to avoid any effect of the boundaries, we introduce a cutoff function $\zeta \in C_c^\infty(\Omega)$ such that $0 \leq \zeta \leq 1$ and $\zeta = 1$ on Ω'. Now we can write

$$\int_\Omega \zeta^2 |\Delta u|^2 \, dx = \sum_{i,j=1}^m \int_\Omega u_{,ii} u_{,jj} \zeta^2 \, dx$$

$$= -\sum_{i,j=1}^m \int_\Omega u_{,i} u_{,jji} \zeta^2 \, dx - \sum_{i,j=1}^m \int_\Omega u_{,i} u_{,jj} 2\zeta \zeta_i \, dx$$

$$= \sum_{i,j=1}^m \int_\Omega u_{,ij} u_{,ij} \zeta^2 \, dx - \sum_{i,j=1}^m \int_\Omega u_{,i} u_{,jj} 2\zeta \zeta_{,i} - u_{,i} u_{,ij} 2\zeta \zeta_{,j} \, dx.$$

Note that we get no boundary terms in our integrations by parts, since ζ and all its derivatives are zero on $\partial\Omega$.

By using Young's inequality "with ϵ" (1.15) to estimate the terms involving derivatives of ζ, for example

$$|u_{,i} u_{,jj} 2\zeta \zeta_{,i}| \leq \tfrac{1}{4} u_{,jj}^2 \zeta^2 + 4 u_{,i}^2 \zeta_{,i}^2,$$

we obtain

$$\int_\Omega |\nabla^2 u|^2 \zeta^2 \, dx \leq \tfrac{1}{2} \int_\Omega |\nabla^2 u|^2 \zeta^2 \, dx + C \int_\Omega |\nabla u|^2 |\nabla \zeta|^2 \, dx + \int_\Omega \zeta^2 |\Delta u|^2 \, dx.$$

Note that here and in what follows we write

$$|\nabla^2 u(x)| = \sum_{i,j=1}^m |u_{,ij}(x)|^2,$$

so that $|\nabla^2 u|$ gives an L^2 estimate on all the second derivatives of u. (This is very different from the engineering usage $\nabla^2 u = \Delta u$.)

We therefore have

$$\int_{\Omega'} |\nabla^2 u|^2 \, dx \leq \int_\Omega |\nabla^2 u|^2 \zeta^2 \, dx$$

$$\leq C \left[\int_\Omega |\Delta u|^2 \, dx + \int_\Omega |\nabla u|^2 \, dx \right]$$

$$\leq C \left[|f|^2 + \|u\|_{H_0^1(\Omega)}^2 \right].$$

Since we already know from (6.8) that

$$\|u\|_{H_0^1(\Omega)} \le C|f|,$$

we in fact have

$$\int_{\Omega'} |\nabla^2 u|^2 \, dx \le C|f|^2,$$

yielding the estimate

$$\|u\|_{H^2(\Omega')} \le C_{\Omega'}|f|.$$

We now have to make these arguments rigorous. To do this we will have to discuss difference quotients,

$$D_i^h u(x) = \frac{u(x + h e_i) - u(x)}{h},$$

which will enable us to treat "derivatives" of functions that we do not know a priori to be differentiable.

6.4.2 Difference Quotients

The usefulness of difference quotients is ensured by the following two results. The first is unsurprising and shows that if $u \in H^1(\Omega)$ then there is a uniform bound on the L^2 norm of its difference quotients.

Lemma 6.8. If $u \in H^1(\Omega)$ then for any $\Omega' \subset\subset \Omega$ we have

$$\left\| D_i^h u \right\|_{L^2(\Omega')} \le \|D_i u\|_{L^2(\Omega)},$$

for every $h < \mathrm{dist}(\Omega', \partial\Omega)$.

Proof. First consider the case $u \in C^1(\overline{\Omega})$. Then for any $x \in \Omega'$ we have, using the fundamental theorem of calculus,

$$D_i^h u(x) = \frac{u(x + h e_i) - u(x)}{h} = \frac{1}{h} \int_0^h u_{,i}(x + \xi e_i) \, d\xi.$$

It follows that

$$\int_{\Omega'} \left| D_i^h u(x) \right|^2 dx = \int_{\Omega'} \frac{1}{h^2} \left(\int_0^h D_i u(x + \xi e_i) \, d\xi \right)^2 dx,$$

and so, using the Cauchy–Schwarz inequality and changing the order of integration, we get

$$\int_{\Omega'} |D_i^h u(x)|^2 \, dx \leq \int_{\Omega'} \frac{1}{h} \int_0^h |D_i u(x + \xi e_i)|^2 \, d\xi \, dx$$

$$\leq \frac{1}{h} \int_0^h \int_{\Omega} |D_i u(x)|^2 \, dx \, d\xi$$

$$\leq |D_i u|^2.$$

The result now follows for a general $u \in H^1(\Omega)$ by taking limits, since $C^1(\overline{\Omega})$ is dense in $H^1(\Omega)$ (Theorem 5.21). ∎

More useful is the following converse, which shows that if the difference quotients are uniformly bounded in L^2 then we can deduce that the weak derivative exists and is also an element of L^2.

Lemma 6.9. *Suppose that $\Omega' \subset\subset \Omega$ and that for every $h < \operatorname{dist}(\Omega', \partial\Omega)$ we have*

$$\|D_i^h u\|_{L^2(\Omega')} \leq C.$$

Then $D_i u \in L^2(\Omega')$ with

$$\|D_i u\|_{L^2(\Omega')} \leq C. \tag{6.15}$$

Proof. Since $D_i^h u$ is bounded in $L^2(\Omega')$, there is a sequence $h_n \to 0$ such that $D_i^{h_n} u$ converges weakly to some $v \in L^2(\Omega')$ as $n \to \infty$. It follows from Lemma 4.14 that $\|v\|_{L^2(\Omega')} \leq C$.

We now need to show that in fact $v = D_i u$. For any $\phi \in C_c^\infty(\Omega')$ we have

$$\int_{\Omega'} u D_i \phi \, dx = \int_{\Omega} u D_i \phi \, dx$$

$$= \lim_{n \to \infty} \int_{\Omega} u D_i^{h_n} \phi \, dx$$

$$= -\lim_{n \to \infty} \int_{\Omega'} \left(D_i^{-h_n} u\right) \phi \, dx$$

$$= -\int_{\Omega'} v \phi \, dx$$

$$= -\int_{\Omega} v \phi \, dx,$$

and so $v = D_i u$ in the weak sense, and (6.15) follows. ∎

We will also need the following three properties of the difference quotient. The proof is left as an exercise (Exercise 6.7).

Lemma 6.10. *If* $\Omega' \subset\subset \Omega$ *and* $h < \mathrm{dist}(\Omega', \partial\Omega)$ *then for any* $u \in L^2(\Omega)$ *and any* $v \in C^0(\overline{\Omega})$ *with* supp $v \subseteq \Omega'$ *we have the product rule*

$$D_i^h(uv)(x) = u(x)D_i^h(x) + D_i^h u(x)v(x + he_i) \qquad (6.16)$$

and the integration by parts formula

$$\int_\Omega D_i^h u(x)v(x)\, dx = -\int_\Omega u(x)D_i^{-h} v(x)\, dx. \qquad (6.17)$$

Finally, if $u \in H^1(\Omega)$ *then*

$$D_i D_j^h u = D_j^h D_i u$$

(differencing and derivatives commute).

6.4.3 Interior Regularity Result

We now use the difference quotients of the previous section to perform a rigorous analysis along the lines of that in Section 6.4.1. We replace derivatives with differences and then use Lemma 6.9 to deduce that $u \in H^2$. Otherwise, the only real trick is in choosing an appropriate "test function" v, which we do in (6.21).

Theorem 6.11. *Suppose that* $f \in L^2(\Omega)$, *and that* $u \in H_0^1(\Omega)$ *satisfies*

$$((u, v))_{H_0^1} = (f, v) \qquad for\ all \qquad v \in H_0^1(\Omega). \qquad (6.18)$$

Then in fact $u \in H^2(\Omega')$ *for any* $\Omega' \subset\subset \Omega$, *with*

$$\|u\|_{H^2(\Omega')} \le C_{\Omega'}|f|. \qquad (6.19)$$

Proof. We first write (6.18) explicitly as

$$\sum_{i=1}^m \int_\Omega u_{,i} v_{,i}\, dx = \int_\Omega f v\, dx \qquad (6.20)$$

for all $v \in H_0^1(\Omega)$. For each $\Omega' \subset\subset \Omega$ we choose a cutoff function $\zeta \in C_c^\infty(\Omega)$ with $\zeta \equiv 1$ on Ω', and

$$|\nabla\zeta| < \frac{2}{\mathrm{dist}(\Omega', \partial\Omega)}.$$

We now consider, for each $k = 1, \ldots, m$ the test function*

$$v = D_k^{-h}\left[\zeta^2 D_k^h u\right] \qquad (6.21)$$

and substitute this in (6.20) (clearly $v \in H_0^1(\Omega)$) to obtain

$$\sum_{i=1}^m \int_\Omega u_{,i}\left(D_k^{-h}\left[\zeta^2 D_k^h u\right]\right)_{,i} dx = \int_\Omega f D_k^{-h}\left[\zeta^2 D_k^h u\right] dx.$$

Use of Lemma 6.8 allows us to bound the right-hand side by

$$
\begin{aligned}
|f|\left|\nabla\left(\zeta^2 D_k^h u\right)\right| &\leq |f|\left(\left|2\zeta(\nabla\zeta)D_k^h u\right| + \left|\zeta^2\nabla D_k^h u\right|\right) \\
&\leq |f|\left(C\left|D_k^h u\right| + \left|\zeta\nabla D_k^h u\right|\right) \\
&\leq C\left(|f|^2 + \|u\|_{H^1}^2\right) + \tfrac{1}{4}\int_\Omega \left|\zeta\nabla D_k^h u\right|^2 dx,
\end{aligned}
$$

where Young's inequality "with ϵ" (1.15) was used in the last line.

We now treat the left-hand side, first "integrating by parts" using (6.17), and then applying the derivative to the terms in the square bracket to obtain

$$-\sum_{i=1}^m \int_\Omega \left(D_k^h u_{,i}\right)\left[\zeta^2 D_k^h u_{,i} + 2\zeta \zeta_{,i} D_k^h u\right] dx.$$

The first term here is equal to

$$\sum_{i=1}^m \int_\Omega \left|\zeta D_k^h u_{,i}\right|^2 dx = \int_\Omega \left|\zeta D_k^h \nabla u\right|^2 dx,$$

and we can estimate the second, using Young's inequality with ϵ, by

$$
\begin{aligned}
C\int_\Omega \zeta\left|D_k^h\nabla u\right|\left|D_k^h u\right| dx &\leq \tfrac{1}{4}\int_\Omega \zeta^2\left|D_k^h\nabla u\right|^2 dx + C\int_\Omega \left|D_k^h u\right|^2 dx \\
&\leq \tfrac{1}{4}\int_\Omega \left|\zeta\nabla D_k^h u\right|^2 dx + C\|u\|_{H^1}^2,
\end{aligned}
$$

using Lemma 6.8 again and the fact that ∇ and D_k^h commute (Lemma 6.10).

We can now combine these estimates to obtain

$$\int_\Omega \left|\zeta D_k^h \nabla u\right|^2 dx \leq C\left(|f|^2 + \|u\|_{H^1(\Omega)}^2\right).$$

---

* This is not a test function in the technical sense (i.e. not an element of $C_c^\infty(\Omega)$), but rather a function against which we "test" u in (6.20).

It follows that

$$\left|D_k^h \nabla u\right|^2_{L^2(\Omega')} \le C\big(|f|^2 + \|u\|^2_{H^1(\Omega)}\big),$$

and so, using the result of Lemma 6.9, we have

$$\left|\nabla^2 u\right|^2_{L^2(\Omega')} \le C\big(|f|^2 + \|u\|^2_{H^1(\Omega)}\big).$$

Using the result of (6.8) ($\|u\|_{H^1} \le C|f|$), we obtain (6.19). ∎

We now improve this result, using induction, to show that the solutions are in $H^{s+2}_{\text{loc}}(\Omega)$ when $f \in H^s(\Omega)$.

Theorem 6.12. *Suppose that $f \in H^s(\Omega)$ and that $u \in H_0^1(\Omega)$ satisfies (6.18). Then $u \in H^{s+2}_{\text{loc}}(\Omega)$: for each $\Omega' \subset\subset \Omega$ we have the estimate*

$$\|u\|_{H^{s+2}(\Omega')} \le C_{\Omega'} \|f\|_{H^s(\Omega)}.$$

Note that this interior regularity result does not depend on the smoothness of the boundary.

Proof. Supposing that the statement is true for $s \le \sigma$, we show that it is then true for $s \le \sigma + 1$. Take $f \in H^{\sigma+1}(\Omega)$; then we know from the induction hypothesis that $u \in H^{\sigma+2}_{\text{loc}}(\Omega)$, with

$$\|u\|_{H^{\sigma+2}(\Omega')} \le C_{\Omega'} \|f\|_{H^\sigma(\Omega)}.$$

We take a multi-index α with $|\alpha| = \sigma + 1$, and for any $w \in C_c^\infty(\Omega')$ use the "test function"

$$v = (-1)^{|\alpha|} D^\alpha w$$

in (6.20). Integrating by parts $|\alpha|$ times we obtain

$$((D^\alpha u, w))_{H_0^1} = (D^\alpha f, w) \qquad \text{for all} \qquad w \in C_c^\infty(\Omega'),$$

and so $D^\alpha u$ is a weak solution of the equation with f replaced with $D^\alpha f$. Since $f \in H^{\sigma+1}(\Omega)$, $D^\alpha f \in L^2(\Omega)$, and so we must have

$$|D^\alpha u|_{H^2(\Omega')} \le C|D^\alpha f|_{L^2(\Omega)}.$$

Since this holds for every multi-index α with $|\alpha| = \sigma + 1$, it follows that

$$\|u\|_{H^{\sigma+3}(\Omega')} \le C\|f\|_{H^{\sigma+1}(\Omega)},$$

and the theorem follows by induction. ∎

We have the following important corollary, which shows that if the right-hand side is smooth (in $C^\infty(\Omega)$) then the solution is also smooth.

Corollary 6.13. *If $f \in C^\infty(\Omega)$ then the solution of*

$$-\Delta u = f \qquad in \ \Omega$$

satisfies $u \in C^\infty(\Omega)$.

Proof. Theorem 6.12 shows that $u \in H^s_{\text{loc}}(\Omega)$ for every s. The Sobolev embedding $H^k \subset C^j$ if $k > j + (m/2)$ from Corollary 5.30 shows then that $u \in C^k(\Omega)$ for every k, and so $u \in C^\infty(\Omega)$. ∎

In particular, the eigenfunctions of the Laplacian are elements of $C^\infty(\Omega)$. This will be extremely important in what follows, when we use these eigenfunctions as a basis in which to expand trial solutions of parabolic PDEs.

Corollary 6.14. *The eigenfunctions of the Laplacian with Dirichlet boundary conditions are elements of $C^\infty(\Omega) \cap H^1_0(\Omega)$.*

Proof. That an eigenfunction w is an element of $H^1_0(\Omega)$ follows from Theorem 6.2. The smoothness of w on the interior of Ω follows using Theorem 6.12 repeatedly on the equation $Aw = \lambda w$, starting with $w \in L^2(\Omega)$, to deduce that $u \in H^s_{\text{loc}}(\Omega)$ for every $s \geq 0$. ∎

6.5 Boundary Regularity for the Laplacian

To improve the results of the previous section so that we obtain regularity up to the boundary requires a more careful treatment. The idea is to treat first a simple problem with a flat boundary, and then to use transformation ideas (as discussed in Chapter 5) to treat the case of a general boundary. Once the flat boundary case is dealt with the general boundary is relatively straightforward.

6.5.1 Regularity up to a Flat Boundary

We will have to treat a more general second-order elliptic equation near a flat boundary, since the coordinate transformations that straighten a general boundary will change the form of the equation from $-\Delta u = f$. Although the equation is more complicated, the simple geometry enables us to perform the analysis.

The plan is to obtain bounds on all the second-order derivatives $u_{,ij}$ apart from $u_{,mm}$ by using methods similar to those of Theorem 6.11, and then to use the equation itself to deduce bounds on $u_{,mm}$.

For some $0 < \lambda < 1$ we take

$$D^+ = B(0, R) \cap \mathbb{R}^m_+$$

and consider the equation

$$L(x)u = g \qquad \text{with} \qquad g \in L^2(D^+)$$

on D^+, where

$$L(x)u = - \sum_{i,j=1}^m \frac{\partial}{\partial x_i} \left(a_{ij}(x) \frac{\partial u}{\partial x_j} \right),$$

and the coefficients $a_{ij} \in W^{1,\infty}(D^+)$ are chosen such that

$$\sum_{i,j=1}^m a_{ij}(x)\xi_i\xi_j \geq \theta |\xi|^2, \tag{6.22}$$

for some $\theta > 0$ (we say that L is "uniformly elliptic").

A weak solution of this equation is expressed in terms of the associated bilinear form

$$B(u, v) = \sum_{i,j=1}^m a_{ij} \frac{\partial u}{\partial x_i} \frac{\partial u}{\partial x_j}.$$

We also call B uniformly elliptic when (6.22) holds.

Proposition 6.15. *Suppose that $u \in H^1(D^+)$ satisfies*

$$B(u, v) = (g, v) \qquad \text{for all} \qquad v \in H^1_0(D^+) \tag{6.23}$$

with $u = 0$ on $\partial \mathbb{R}^m_+$ in the sense of trace. Then for any $0 < \lambda < 1$ we have

$$\|u\|_{H^2(\lambda D^+)} \leq C \|g\|_{L^2(D^+)}.$$

Proof. Choose an index $k \neq m$, pick $h \in (0, \frac{1}{2}R(1 - \lambda))$, and choose a cutoff function $\zeta \in C_b^\infty(D^+)$ such that $0 \leq \zeta \leq 1$, $\zeta \equiv 1$ on λD^+ and $\operatorname{supp} \zeta \subset B(0, \frac{1}{2}(1+\lambda)R)$. For v in (6.23) we use the same "test function" as in the proof of Theorem 6.11:

$$v = D_k^{-h}\left[\zeta^2 D_k^h u\right].$$

Now, v is in fact given by

$$v(x) = D_k^{-h}\left[\zeta^2(x)\frac{u(x + he_k) - u(x)}{h}\right]$$

$$= \frac{1}{h}\left[\zeta^2(x - he_k)\frac{u(x) - u(x - he_k)}{h} - \zeta^2(x)\frac{u(x + he_k) - u(x)}{h}\right].$$

This complicated looking expression is useful, since it shows that $v \in H_0^1(D^+)$, because $\zeta = 0$ near the curved section of the boundary and $u = 0$ on $\{x_m = 0\}$.

Writing $w = \zeta D_k^h u$, so that $v = D_k^{-h}[\zeta w]$, we have from (6.23)

$$\int_{D^+} g D_k^{-h}(\zeta w)\, dx = \sum_{i,j=1}^m \int_{D^+} a_{ij} u_{,i} v_{,j}\, dx$$

$$= \sum_{i,j=1}^m \int_{D^+} a_{ij} u_{,i} D_k^{-h}(\zeta w)_{,j}\, dx$$

$$= \sum_{i,j=1}^m \int_{D^+} D_k^h(a_{ij} u_{,i})(\zeta w)_{,j}\, dx.$$

Using the product differencing formula (6.16) on this we get

$$\sum_{i,j=1}^m \int_{D^+} a_{ij}(x + he_k)(D_k^h u_{,i})(\zeta w)_{,j}\, dx + \sum_{i,j=1}^m \int_{D^+} (D_k^h a_{ij}) u_{,i}(\zeta w)_{,j}\, dx,$$

and so we have

$$\int_{D^+} g D_k^{-h}(\zeta w)\, dx = \sum_{i,j=1}^m \int_{D^+} a_{ij}(x + he_k) w_{,i} w_{,j}\, dx$$

$$- \sum_{i,j=1}^m \int_{D^+} a_{ij}(x + he_k)\zeta_{,i}(D_k^h u) w_{,j}\, dx$$

$$+ \sum_{i,j=1}^m \int_{D^+} a_{ij}(x + he_k)(D_k^h u_{,i})\zeta_{,j} w\, dx$$

$$+ \sum_{i,j=1}^m \int_{D^+} (D_k^h a_{ij}) u_{,i}(\zeta w)_{,j}\, dx. \tag{6.24}$$

Since

$$\int_{\lambda D^+} \left| D_k^h \nabla u \right|^2 dx = \int_{\lambda D^+} |\nabla w|^2 \, dx \leq \int_{D^+} |\nabla w|^2 \, dx$$

and the uniformly ellipticity of L implies that

$$\int_{D^+} a_{ij}(x + he_k) w_{,i} w_{,j} \, dx \geq \theta \int_{D^+} |\nabla w|^2 \, dx,$$

we can bound

$$\int_{\lambda D^+} \left| D_k^h \nabla u \right|^2 dx$$

by estimating

$$\int_{D^+} a_{ij}(x + he_k) w_{,i} w_{,j} \, dx.$$

To do this, we bound all the other integrals from (6.24). First, we have

$$\left| \int_{D^+} g D_k^{-h}(\zeta w) \, dx \right| \leq |g| \left| D_k^{-h}(\zeta w) \right|$$

$$\leq |g|(|D_k \zeta| |w| + |\zeta| |D_k w|)$$

$$\leq C |g| \|w\|_{H^1},$$

using Lemma 6.8 to bound the difference terms. Next, we have

$$\left| \int_{D^+} a_{ij}(x + he_k) \zeta_{,i}(D_k^h u) w_{,j} \, dx \right| \leq \|a_{ij}\|_\infty \|\nabla \zeta\|_\infty \left| D_k^h u \right| |w_{,j}|$$

$$\leq C \|u\|_{H^1} \|w\|_{H^1},$$

$$\left| \int_{D^+} a_{ij}(x + he_k)(D_k^h u_{,i}) \zeta_{,j} w \, dx \right| = \left| \int_{D^+} D_k^{-h}(a_{ij}(x + he_k) \zeta_{,j} w) u_{,i} \, dx \right|$$

$$\leq |a_{ij}|_{W^{1,\infty}} |\nabla \zeta|_{W^{1,\infty}} \|w\|_{H^1} |u_{,i}|$$

$$\leq C \|w\|_{H^1} \|u\|_{H^1},$$

and finally

$$\left| \int_{D^+} (D_k^h a_{ij}) u_{,i}(\zeta w)_{,j} \, dx \right| \leq \|a_{ij}\|_{W^{1,\infty}} |u_{,i}| \|\zeta\|_{C^1} \|w\|_{H^1}$$

$$\leq C \|u\|_{H^1} \|w\|_{H^1}.$$

Assembling these estimates we have

$$\int_{\lambda D^+} \left| D_k^h \nabla u \right|^2 dx \le C |g| \|u\|_{H^1} + C \|w\|_{H^1} \|u\|_{H^1}.$$

Since

$$\|w\|_{H^1} = \left\| \zeta D_k^h u \right\|_{H^1} \le C \left\| D_k^h u \right\|_{H^1} \le C \left(\|u\|_{H^1} + \left| D_k^h \nabla u \right| \right),$$

we can use Young's inequality as before to obtain

$$\int_{\lambda D^+} \left| D_k^h \nabla u \right|^2 dx \le C \left(|g|^2 + \|u\|_{H^1}^2 \right) \le C |g|^2.$$

Using Lemma 6.9 we deduce a bound on all the second-order derivatives apart from $u_{,mm}$:

$$\sum_{i+j<2m} \int_{D_\lambda^+} |u_{,ij}|^2 dx \le C|g|^2. \tag{6.25}$$

To estimate the derivative $u_{,mm}$ we use the previous interior regularity result and the equation itself. We know that, since $u \in H^2(\Omega')$ for any $\Omega' \subset\subset \Omega$, within Ω the equation is satisfied as an equality in L^2 and so it is also satisfied as an equality almost everywhere. Rewrite the equation as

$$a_{mm} u_{,mm} + a_{mm,m} u_{,m} + \sum_{i+j<2m} \frac{\partial}{\partial x_i} a_{ij} \frac{\partial u}{\partial x_j} = g.$$

Now, the uniform ellipticity condition (6.22) implies that $a_{mm} \ge \theta > 0$ (just set $\xi = (0, \ldots, 0, 1)$), and so we can rearrange the equation to give

$$u_{,mm} = \frac{1}{a_{mm}} \left(g - a_{mm,m} u_{,m} - \sum_{i+j<2m} \frac{\partial}{\partial x_i} \left[a_{ij} \frac{\partial u}{\partial x_j} \right] \right).$$

We already know that the right-hand side is in $L^2(D_\lambda^+)$, and it follows from (6.25) that

$$\int_{D_\lambda^+} |u_{,mm}|^2 dx \le C \left(|g|^2 + \|u\|_{H^1}^2 \right). \tag{6.26}$$

Combining (6.25) and (6.26) we obtain

$$\sum_{i,j=1}^m \int_{D_\lambda^+} |u_{,ij}|^2 dx \le C \left(|g|^2 + \|u\|_{H^1}^2 \right),$$

and an appeal to (6.8) yields the result. ∎

6.5.2 Regularity Up to a C^2 Boundary

Of course, we now need to improve this result so that it holds not only for simple geometries but for general domains. We will assume that Ω is a bounded C^2 domain as in Definition 5.19. Recall that, in this case, near any point $x_0 \in \partial\Omega$ we can find a neighbourhood U and a C^2 transformation Φ such that Φ takes $\Omega \cap U$ into a set $\tilde{\Omega} \subset \mathbb{R}^m$ (with coordinates "y"; see Figure 5.1), so that $\partial\Omega \cap U$ is mapped onto $y_m = 0$ and x_0 is mapped onto the origin. The inverse of this transformation we will denote by Ψ. Furthermore, we can choose Φ such that $\det \nabla\Psi = 1$ (see Exercise 6.8).

We now have to write down the equation satisfied by the transformed function $\tilde{u}(y) = u(\Psi(y))$. Since $u(x) = \tilde{u}(\Phi(x))$, the equation

$$((u, v))_{H_0^1} = (f, v)$$

becomes

$$\int_\Omega \sum_{j=1}^m \sum_{k,l=1}^m \frac{\partial \tilde{u}}{\partial y_k}(\Phi(x))\frac{\partial \Phi_k}{\partial x_j}(x)\frac{\partial \tilde{v}}{\partial y_l}(\Phi(x))\frac{\partial \Phi_l}{\partial x_j}(x)\,dx$$

$$= \int_\Omega \tilde{f}(\Phi(x))\tilde{v}(\Phi(x))\,dx.$$

Changing variables in the integral expression on both sides by using (5.18) and $\det \nabla\Psi = 1$ gives

$$\int_{\tilde{\Omega}} \sum_{k,l=1}^m \tilde{a}_{kl}(y)\tilde{u}_{,k}\tilde{v}_{,l}\,dy = \int_{\tilde{\Omega}} F(y)\tilde{v}(y)\,dy,$$

where

$$\tilde{a}_{kl}(y) = \sum_{j=1}^m \Phi_{k,j}(\Psi(y))\Phi_{l,j}(\Psi(y)),$$

and $F(y) = \tilde{f}(y)$. Using Theorem 5.18, $F \in L^2(\Omega')$ with

$$\|F\|_{L^2(\Omega')} \le C\|f\|_{L^2(\Omega)}.$$

To check that the new equation is uniformly elliptic, we write

$$\sum_{k,l=1}^m \tilde{a}_{kl}(y)\xi_k\xi_l = \sum_{j=1}^m \sum_{k,l=1}^m \Phi_{k,j}\xi_k\Phi_{l,j}\xi_l$$

$$= \sum_{i,j=1}^m \eta_j^2 = |\eta|^2,$$

where $\eta(y) = (\nabla\Phi)^T \xi$, that is,

$$\eta_i = \sum_{k=1}^{m} \Phi_{k,i}(\Psi(y))\xi_k.$$

Since $(\nabla\Phi)^T$ is invertible with inverse uniformly bounded over the domain Ω', it follows that we have $|\xi| \leq C|\eta|$, for some C, and so

$$\sum_{k,l} \tilde{a}_{kl}(y)\xi_k\xi_l \geq \tilde{\theta}|\xi|^2,$$

ensuring that

$$\tilde{B}(\tilde{u}, \tilde{v}) = \int_{\tilde{\Omega}} \sum_{k,l=1}^{m} \tilde{a}_{kl}(y)\tilde{u}_{,k}\tilde{v}_{,l}\, dy$$

is uniformly elliptic on $\tilde{\Omega}$.

It follows from Proposition 6.15 that $\tilde{u} \in H^2(\tilde{\Omega})$, with

$$\|\tilde{u}\|_{H^2(\tilde{\Omega})} \leq C\|\tilde{f}\|_{L^2(\tilde{\Omega})}.$$

Since we have already shown in Theorem 5.18 that a C^2 coordinate transformation induces a bounded map from $H^2(\Omega \cap U)$ into $H^2(\tilde{\Omega})$, and from $L^2(\Omega \cap U)$ into $L^2(\tilde{\Omega})$, we have

$$\|u\|_{H^2(\Omega \cap U)} \leq C\|f\|_{L^2(\Omega)}.$$

Combining this with the interior regularity result of Theorem 6.11, we have the following global regularity result.

Theorem 6.16. *Suppose that Ω is a bounded C^2 domain and that $u \in H_0^1(\Omega)$ is a weak solution of Poisson's equation*

$$-\Delta u = f \qquad and \qquad u|_{\partial\Omega} = 0,$$

where $f \in L^2(\Omega)$. Then $u \in H^2(\Omega)$ with

$$\|u\|_{H^2(\Omega)} \leq C\|f\|_{L^2(\Omega)}.$$

Proof. For each point $x_0 \in \partial\Omega$, the above result provides an estimate in $H^2(\Omega \cap U(x_0))$ for some neighbourhood $U(x_0)$ of x_0,

$$\|u\|_{H^2(\Omega \cap U(x_0))} \leq C(x_0)\|f\|_{L^2(\Omega)}.$$

Since $\partial\Omega$ is compact, a finite number (N, say) of these regions cover $\partial\Omega$. We add to these one additional set $\Omega' \subset\subset \Omega$, such that

$$\Omega' \cup \bigcup_{j=1}^{N}\left(\Omega \cap U\left(x_0^j\right)\right) \supset \Omega.$$

We know that u is in $H^2(\Omega')$ by using the interior regularity result of Theorem 6.11. Combining these we get

$$\|u\|_{H^2(\Omega)} \leq \left(C_{\Omega'} + \sum_{j=1}^{N} C\left(x_0^j\right)\right)|f|,$$

and so

$$\|u\|_{H^2(\Omega)} \leq C|f|_{L^2(\Omega)}$$

as required. ∎

Note that this also shows that we can use $|Au| = |\Delta u|$ to bound the H^2 norm; in fact $|Au|$ and $\|u\|_{H^2}$ are equivalent norms:

$$|Au| \leq \|u\|_{H^2(\Omega)} \leq c|Au|. \tag{6.27}$$

There is a higher regularity result, corresponding to the interior regularity given in Theorem 6.12, that we will only state. A proof can be found in Evans (1998). Note that to obtain regularity up to the boundary we have to assume that the boundary is smooth, in contrast with the interior regularity result of Theorem 6.12.

Theorem 6.17. *Suppose that $f \in H^s(\Omega)$ and that $u \in H_0^1(\Omega)$ is a weak solution of Poisson's equation*

$$-\Delta u = f \qquad with \qquad u|_{\partial\Omega} = 0.$$

Then, provided that Ω is a bounded C^{s+2} domain, we have $u \in H^{s+2}(\Omega)$, with

$$\|u\|_{H^{s+2}(\Omega)} \leq C\|f\|_{H^s(\Omega)}.$$

If Ω is a bounded C^∞ domain (i.e. a C^k domain for each $k \geq 0$) then we have two immediate corollaries: (i) If $f \in C^\infty(\overline{\Omega})$ then $u \in C^\infty(\overline{\Omega})$ (cf. Corollary 6.13) and (ii) the eigenvalues of the Laplacian are elements of $C^\infty(\overline{\Omega})$ (cf. Corollary 6.14).

6.5.3 $H^{2k}(\Omega)$ and Domains of A^k

Finally we show that we can once again use the fractional powers of A (the negative Laplacian) to provide norms on $D(A^{k/2})$ equivalent to the $H^k(\Omega)$ norms, so that

$$|A^{k/2}u| \leq \|u\|_{H^k(\Omega)} \leq C_k|A^{k/2}u|,$$

along the lines of (6.14). Of course, this is exactly the consequence of the regularity theorem given in Equation (6.27) above for $k = 2$, and it follows from the definition of the bilinear form that

$$|A^{1/2}u|^2 = (A^{1/2}u, A^{1/2}u) = (Au, u) = a(u, u) = \|u\|^2_{H^1_0(\Omega)} \qquad (6.28)$$

for $k = 1$. For general k we simply combine these two equalities.

Proposition 6.18. *For $k \in \mathbb{Z}^+$, if $u \in D(A^{k/2})$ and Ω is of class C^k, then*

$$|A^{k/2}u| \leq \|u\|_{H^k(\Omega)} \leq C_k|A^{k/2}u|. \qquad (6.29)$$

Proof. First we show (6.29) for $k = 2n$, with n an integer. The left-hand part of the inequality is clear, since

$$|A^n u|^2 \leq C \sum_{|\alpha_j|=2,\, 1\leq j\leq n} |D^{\alpha_1+\cdots+\alpha_n}u|^2 \leq C\|u\|^2_{H^{2n}}.$$

We show the right-hand inequality by induction on n. We have the result for $n = 1$ in (6.27). If it is true for $n = 1, \ldots, N$, then for $n = N + 1$ we have

$$\|u\|^2_{H^{2(N+1)}} \leq \sum_{|\alpha|\leq 2N,|\beta|\leq 2} |D^{\alpha+\beta}u|^2$$
$$\leq C \sum_{|\alpha|\leq 2N} |D^\alpha u|^2_{H^2}$$
$$\leq C \sum_{|\alpha|\leq 2N} |AD^\alpha u|^2$$
$$= C \sum_{|\alpha|\leq 2N} |D^\alpha Au|^2$$
$$\leq C|A^{N+1}u|^2,$$

as required (note that A and D^α commute because A just consists of second derivatives).

We now use (6.28) to prove the result when k is odd. In this case

$$\left|A^{k+\frac{1}{2}}u\right|^2 = \|A^k u\|_{H_0^1}^2$$

$$= \sum_{|\alpha|=1} |D^\alpha A^k u|^2 = C \sum_{|\alpha|=1} |A^k D^\alpha u|^2.$$

This expression is bounded above by

$$C\|u\|_{H^{2k+1}}^2$$

and bounded below by

$$C\left(\|u\|_{H^{2k+1}}^2 - |u|_{L^2}^2\right).$$

Since $u \in H_0^1(\Omega)$ if $u \in D(A^{k/2})$, we can use the Poincaré inequality to deduce that the left-hand inequality holds. ∎

Although this proposition is potentially useful, we need to take $u \in D(A^{k/2})$ in order to apply it. Note that (6.29) shows that $D(A^{k/2}) \subset H^k(\Omega)$; the next result gives a class of functions that form a subset of $D(A^{k/2})$.

Proposition 6.19. *If k is even then*

$$H^k(\Omega) \cap H_0^{k-1}(\Omega) \subset D(A^{k/2}),$$

and if k is odd then

$$H_0^k(\Omega) \subset D(A^{k/2}).$$

Proof. For $k = 2s$ take $u \in H^k(\Omega) \cap H_0^{k-1}(\Omega)$. If w_j is an eigenfunction of A then we have

$$\lambda_j^{k/2}(u, w_j) = (u, A^s w_j) = (u, (-\Delta)^s w_j) = ((-\Delta)^s u, w_j),$$

where we can perform all the integrations by parts since $(-\Delta)^j u \in H_0^{k-1-2j}$ for all $j = 0, 1, \ldots, s - 1$. It follows that

$$|A^s u|^2 = \sum_{j=1}^\infty \lambda_j^{k/2}(u, w_j) \le |(-\Delta)^s u|^2 \le \|u\|_{H^k}^2.$$

If $k = 2s + 1$, then take $u \in H_0^k(\Omega)$ and consider

$$\lambda_j^{k/2}(u, w_j) = (u, A^s A^{1/2} w_j) = ((-\Delta)^s u, A^{1/2} w_j) = (A^{1/2}(-\Delta)^s u, w_j)$$

(the integrations by parts are valid because $(-\Delta)^j u \in H_0^{k-2j}(\Omega)$ for all $j = 0, 1, \ldots, s$). Now, since $(-\Delta)^s u \in H_0^1(\Omega)$, we have

$$|A^{1/2}(-\Delta)^s u| = \|(-\Delta u)^s u\|_{H_0^1} \le \|u\|_{H^k},$$

and the proof is concluded as in the even case. ∎

These two results enable us to deduce interesting facts about the Sobolev spaces $H^k(\Omega)$; see Exercise 6.9, for example.

Exercises

6.1   Prove that if $f \in C^0(\Omega)$, and $u \in C^2(\Omega) \cap C^0(\overline{\Omega})$ satisfies $u \in H_0^1(\Omega)$ and

$$\sum_{i=1}^{m} \int_{\Omega} D_i u(x) D_i v(x)\, dx = \int_{\Omega} f(x) v(x)\, dx$$

for all $v \in C_c^1(\Omega)$, then u is a classical solution of

$$-\Delta u = f \qquad \text{with} \qquad u|_{\partial\Omega} = 0.$$

6.2   Find the weak form of the equation

$$Lu = f \qquad \text{with} \qquad u|_{\partial\Omega} = 0,$$

where

$$Lu = -\sum_{i,j=1}^{m} \frac{\partial}{\partial x_i}\left(a_{ij}(x)\frac{\partial u}{\partial x_j}\right) + \sum_{i=1}^{m} b_i(x)\frac{\partial u}{\partial x_i} + c(x)u. \qquad (6.30)$$

6.3   Suppose that the operator in (6.30) is *uniformly elliptic* [i.e. that Equation (6.22) holds] and assume that a_{ij} (for $i, j = 1, \ldots, m$), b_k (for $k = 1, \ldots, m$), and c are in $L^\infty(\Omega)$. Show that there are constants λ and C such that

$$a(u, u) + \lambda|u|^2 \ge C\|u\|_{H_0^1}^2 \qquad \text{for all} \qquad u \in H_0^1(\Omega).$$

(This is a particular case of a more general result known as Gårding's inequality.)

6.4   Use the result of the previous exercise and the Lax–Milgram lemma (Lemma 6.3) to prove that for any $\alpha \ge \lambda$ and any $f \in H^{-1}(\Omega)$, the equation

$$Lu + \alpha u = f, \qquad u|_{\partial\Omega} = 0$$

has a unique weak solution in $H_0^1(\Omega)$.

6.5   If n is the outwards normal to $\partial\Omega$, Neumann boundary conditions are
       $\nabla u \cdot n = 0$ on $\partial\Omega$. The weak form of Poisson's equation with these
       boundary conditions is

$$\text{given} \quad f \in H^{-1}(\Omega),$$

$$\text{find} \quad u \in H^1(\Omega), \quad \text{such that} \quad \quad (6.31)$$

$$a(u, v) = \langle f, v \rangle \quad \text{for all} \quad v \in H^1(\Omega),$$

where $a(u, v)$ is as in (6.3). Show that this problem has a solution iff

$$\int_\Omega f(x)\,dx = 0.$$

(Hint: You will need to use the result of Exercise 5.9.)

6.6   Give an example to show that the assumption of coercivity in the Lax–
       Milgram lemma (Lemma 6.3) cannot be relaxed.

6.7   Prove Lemma 6.10.

6.8   Let Ω be a C^2 domain and $z \in \partial\Omega$. Then after a suitable renumbering
       and reordering of the coordinates there is a neighbourhood U of z for
       which we can write

$$\partial\Omega \cap U = \{x \in U : x_m = \psi(x_1, \ldots, x_{(m-1)})\}$$

and

$$\Omega \cap U = \{x \in U : x_m > \psi(x_1, \ldots, x_{(m-1)})\}.$$

Now, the C^2 map $x \mapsto \Phi(x) = y$ defined by

$$y_i = \begin{cases} x_i - z_i, & 1 \le i \le m - 1 \\ x_m - \psi(x_1, \ldots, x_{m-1}), & i = m \end{cases}$$

maps $\Omega \cap U$ into a set with a flat boundary $y_m = 0$. Write down $\Psi = \Phi^{-1}$, and hence deduce that Φ is a C^2-diffeomorphism with $\det \nabla\Psi = 1$.

6.9   Suppose that Ω is a bounded C^k domain. Use the equivalence of norms
       given in Proposition 6.18 along with the interpolation inequality for
       fractional powers of A (the result of Lemma 3.27) to deduce the Sobolev
       interpolation inequality for $u \in H^k(\Omega)$,

$$\|u\|_{H^s} \le C \|u\|_{H^l}^{(k-s)/(k-l)} \|u\|_{H^k}^{(s-l)/(k-l)}, \quad \quad (6.32)$$

for $0 \le l < s < k$.

(Hint: Prove first for $u \in H_0^k(\Omega')$, and then use Theorem 5.20 to prove
for $u \in H^k(\Omega)$.)

Notes

The theory of regularity of elliptic operators is an important subject with a large literature. Various approaches are covered in Gilbarg & Trudinger (1983), in addition to the Sobolev approach we use here.

Treating a general elliptic equation involves only slightly more algebra than for Poisson's equation, and details of this case, along with more on the Lax–Milgram lemma and higher-order elliptic equations, can be found in both Evans (1998) and Renardy & Rogers (1992). The arguments given here (and there) are originally due to Douglis & Nirenberg (1955) and Agmon *et al.* (1959, 1964). The monograph by Agmon (1965) also covers these topics in detail.

The proof of Proposition 6.19 is adapted from Constantin & Foias (1988).

7

Weak Solutions of Linear Parabolic Equations

In this chapter we will consider how to solve the equation

$$\frac{\partial u}{\partial t} - \Delta u = f(x, t), \qquad \text{with} \qquad u|_{\partial \Omega} = 0$$

$$\text{for} \qquad x \in \Omega, \quad t \geq 0, \tag{7.1}$$

where $f : \Omega \to \mathbb{R}$ is a time-dependent function and Ω is a bounded C^2 domain. As discussed in the Introduction, from a dynamical systems point of view, the natural way to think of a solution $u(x, t)$ is as a trajectory in some infinite-dimensional phase space. In other words, we regard $u(x, t)$ as a sequence of functions $u(t)$ (a convenient abuse of notation), each defined on Ω, so that

$$[u(t)](x) = u(x, t) \qquad \text{with} \qquad u(t) : \Omega \to \mathbb{R}.$$

Since we have already seen that the natural solution spaces for linear equations are Sobolev spaces, it is likely that we will want to consider spaces of functions that map a time interval into some Sobolev space.

7.1 Banach-Space Valued Function Spaces

We can define these function spaces in a natural way. Suppose that I is an interval (usually $(0, T)$ or $[0, T]$) and X is a Banach space – say $L^2(\Omega)$, in the above example. Then the space of continuous functions from I into X, $C^0(I, X)$, consists of those $u(t) : I \to X$ such that $u(t) \to u(t_0)$ in X as $t \to t_0$. Note that "$u \in C^0(I, X)$" says nothing about the smoothness of $u(\cdot, t)$ as a function of X.

The Lebesgue spaces $L^p(I, X)$ consist of all those functions $u(t)$ that take values in X for almost every $t \in I$, such that the L^p norm of $\|u(t)\|_X$ is finite. Since $\|u(t)\|_X$ is (almost everywhere) a real-valued function, this removes the need for any direct definition of the integral of a Banach-space valued function. However, we will need this briefly in Proposition 7.1 and in Chapter 11, so we treat the theory in a cursory fashion in Exercises 7.1 and 7.2. We could also define $L^p(I, X)$ (cf. Theorem 1.13) as the completion of $C^0(I, X)$ with respect to the norm

$$\|u\|_{L^p(I,X)} = \left(\int_I \|u(t)\|_X^p \, dt \right)^{1/p}.$$

In some cases we can write these spaces in another way; for example, the norm on $L^p(I, L^p(\Omega))$ is

$$\left(\int_I \|u(t)\|_{L^p(\Omega)}^p \, dt \right)^{1/p} = \left(\int_I \int_\Omega |u(x,t)|^p \, dx \, dt \right)^{1/p}$$
$$= \left(\int_{I \times \Omega} |u(x,t)|^p \, dx \, dt \right)^{1/p},$$

which is precisely the norm on $L^p(I \times \Omega)$, and so these two spaces consist of the same functions.

We will often write $L^p(0, T; X)$ for $L^p((0, T); X)$ to avoid too clumsy a notation.

If we need to consider the dual of such a space, then some care is needed. For example, the dual of $L^p(0, T; L^r(\Omega))$ is $L^q(0, T; L^s(\Omega))$, with (p, q) and (r, s) conjugate indices – we need to take the dual space of "both parts," since the pairing between u and v is naturally

$$\int_0^T \int_\Omega uv \, dx \, dt.$$

Some particular examples we will need are $L^2(0, T; H_0^1)$ in duality with $L^2(0, T; H^{-1})$ (in this chapter) and $L^2(0, T; L^p(\Omega))$ in duality with $L^2(0, T; L^q(\Omega))$ (in the next chapter).

We can also define Sobolev spaces in this setting; for example, we say that $u \in H^1(0, T; L^2(\Omega))$ if u and Du are both in $L^2(0, T; L^2(\Omega))$. The embedding, density, and compactness results we obtained in Chapter 5 can be proved in the same way as they were for scalar-valued functions, and we will occasionally use these in what follows.

Just as we found it helpful to weaken the requirement that the spatial derivatives existed in a classical sense to obtain weak solutions of elliptic equations in the last chapter, so in this chapter we will find it useful to weaken the temporal regularity of solutions and relax the condition that du/dt exists classically. To do this, we will also interpret the equation in a weak sense in the time variable.

We define the weak derivative du/dt by analogy with (5.2), so that

$$\int_0^T du/dt(t)\phi(t)\,dt = -\int_0^T u(t)(d\phi/dt)(t)\,dt$$

for all $\phi \in C_c^\infty(0, T)$. The Sobolev embedding Theorem 5.31 was the key to deducing the existence of classical solutions in the purely spatial case in the last chapter, and to obtain temporal regularity we will need similar results involving the Banach-space valued function spaces we have just introduced.

Proposition 7.1. *Suppose that $u \in W^{1,p}(0, T; X)$, $1 \le p \le \infty$. Then*

$$u(t) = u(s) + \int_s^t \frac{du}{dt}(\tau)\,d\tau \qquad \text{for every} \qquad 0 \le s \le t \le T,$$

and $u \in C^0([0, T]; X)$ (with the usual caveat). Furthermore, we have the estimate

$$\sup_{0 \le t \le T} \|u(t)\|_X \le C \|u\|_{W^{1,p}(0, T; X)}.$$

Proof. If $p = \infty$, replace it with some $p < \infty$. Extending u to be zero outside $[0, T]$, we mollify u (in t) to give a function u_h. Then $(du_h/dt) = (du/dt)_h$, as in Lemma 5.12, and so, using Proposition 1.18, we have

$$u_h \to u \quad \text{in} \quad L^p(0, T; X)$$

and

$$du_h/dt \to u \quad \text{in} \quad L^p(0, T; X),$$

as $h \to 0$. For $h > 0$ we can write

$$u_h(t) = u_h(s) + \int_s^t \frac{du_h}{dt}(\tau)\,d\tau,$$

and then, taking the limit as $h \to 0$, we obtain

$$u(t) = u(s) + \int_s^t \frac{du}{dt}(\tau)\,d\tau$$

for almost every $0 \le s \le t \le T$. Since the map

$$t \mapsto \int_0^t \frac{du}{dt}(\tau)\,d\tau$$

is continuous, the first part of the proposition follows.

To obtain the estimate, note that we have

$$\|u(0)\|_X \le \|u(t)\|_X + \int_0^t \|u'(s)\| \, ds,$$

and so, integrating between 0 and T, we obtain

$$T\|u(0)\|_X \le \int_0^T \|u(t)\|_X dt + \int_0^T \int_0^t \|u'(t)\|_X \, ds \, dt$$

$$\le \|u\|_{L^1(0,T;X)} + T\|u'\|_{L^1(0,T;X)}$$

$$\le T^{1/q} \|u\|_{L^p(0,T;X)} + T^{(1+q)/q} \|u'\|_{L^p(0,T;X)}$$

(where p and q are conjugate). Therefore, we have

$$\|u(t)\|_X \le \|u(0)\|_X + T^{1/q} \|u'\|_{L^p(0,T;X)}$$

$$\le T^{(1-q)/1} \|u\|_{L^p(0,T;X)} + 2T^{1/q} \|u'\|_{L^p(0,T;X)}$$

$$\le C\|u\|_{W^{1,p}(0,T;X)},$$

as required. ∎

In Section 7.4 we will be able to prove the existence of a solution $u(t)$ with $u \in L^2(0, T; H^1(\Omega))$ and with time derivative $du/dt \in L^2(0, T; H^{-1}(\Omega))$. Since $H^1 \subset H^{-1}$, it follows that $u \in H^1(0, T; H^{-1}(\Omega))$, so that $u \in C^0([0, T]; H^{-1}(\Omega))$ by the above result. As the next theorem shows, it is in fact possible to do better than this and to conclude that u is continuous (as a function of time) into $L^2(\Omega)$.

Theorem 7.2. *Suppose that*

$$u \in L^2(0, T; H^1(\Omega)) \qquad and \qquad du/dt \in L^2(0, T; H^{-1}(\Omega)).$$

Then

(i) u is continuous* from $[0, T]$ into $L^2(\Omega)$, with

$$\sup_{t \in [0,T]} |u(t)| \le C\big(\|u\|_{L^2(0,T;H^1)} + \|du/dt\|_{L^2(0,T;H^{-1})}\big), \tag{7.2}$$

and
(ii)

$$\frac{d}{dt}|u|^2 = 2\langle du/dt, u\rangle \tag{7.3}$$

---

* With the caveat that it may have to be adjusted on a set of measure zero, cf. the discussion before Proposition 5.22.

for almost every $t \in [0, T]$, *that is,*

$$|u(t)|^2 = |u_0|^2 + 2 \int_0^t \langle du/dt(s), u(s) \rangle \, ds.$$

We can generalise this result to the case when V and H are Hilbert spaces (where V^* is the dual of V) with

$$V \subset\subset H = H^* \subset V^*,$$

replacing H^1 with V, H^{-1} by V^*, and L^2 by H. [For two Banach spaces A and B we write $A \subset\subset B$ if A is compactly embedded in B.] We will need this in Chapter 17 with $V = H_0^2(0, L)$ and $H = L^2(0, L)$.

Proof. We will use a nonstandard pairing $\langle v^*, u \rangle$ between an element $v^* \in H^{-1}$ and an element $u \in H^1$. Note that H^{-1} is the dual space of H_0^1, which is a proper subset of H^1. However, we know (by the Riesz representation theorem) that there is an element $v \in H_0^1$ such that

$$((v, u))_{H^1} = \langle v^*, u \rangle$$

for all $u \in H_0^1$. Here we are using the standard H^1 norm and inner product for H_0^1, rather than the equivalent norm involving only first derivatives. Not only does the Riesz theorem guarantee the existence of such a u, it also asserts that the norm of v in H^1 is the same as that of v^* in H^{-1}.

We now define an extension of v^* to a linear functional on all of H^1 by

$$\langle v^*, u \rangle = ((v, u))_{H^1}, \qquad u \in H^1.$$

Since we have used the standard H^1 norm on H_0^1, it follows that

$$|((v, u))_{H^1}| \le \|v\|_{H^1} \|u\|_{H^1} = \|v^*\|_{H^{-1}} \|u\|_{H^1},$$

and so this extension has the same norm as v^* does in H^{-1}.

(i) By extending u outside $[0, T]$ by zero and setting $u_n = (u)_{1/n}$, a mollified version of u (with respect to the variable t), we can approximate u by a sequence $u_n \in C^1([0, T]; H^1)$, which converges to u in the sense that

$$u_n \to u \qquad \text{in} \qquad L^2(0, T; H^1)$$

and

$$du_n/dt \to du/dt \qquad \text{in} \qquad L^2(0, T; H^{-1}).$$

Then, for any t^*,

$$|u_n(t)|^2 = |u_n(t^*)|^2 + 2\int_{t^*}^{t} \langle \dot{u}_n(s), u_n(s) \rangle \, ds. \qquad (7.4)$$

Choose t^* such that

$$|u_n(t^*)|^2 = \frac{1}{T}\int_0^T |u_n(t)|^2 \, dt,$$

estimate the integrand in (7.4) by

$$|\langle \dot{u}_n(s), u_n(s) \rangle| \leq \|\dot{u}_n\|_{H^{-1}} \|u_n\|_{H^1},$$

and so obtain

$$|u_n(t)|^2 \leq \frac{1}{T}\int_0^T |u_n(t)|^2 \, dt + 2\int_0^T \|\dot{u}_n\|_{H^{-1}} \|u_n\|_{H^1} \, dt.$$

From this it follows that

$$\sup_{t\in[0,T]} |u_n(t)|^2 \leq \frac{1}{T}\|u_n\|_{L^2(0,T;L^2)}^2 + 2\|\dot{u}_n\|_{L^2(0,T;H^{-1})} \|u_n\|_{L^2(0,T;H^1)},$$

and so

$$\sup_{t\in[0,T]} |u_n(t)| \leq C\left(\|u_n\|_{L^2(0,T;H^1)} + \|du_n/dt\|_{L^2(0,T;H^{-1})}\right). \qquad (7.5)$$

Since u_n is a Cauchy sequence in $L^2(0, T; H^1)$ and du_n/dt is a Cauchy sequence in $L^2(0, T; H^{-1})$, it follows that u_n is a Cauchy sequence in $C([0, T];$ $L^2(\Omega))$, and so $u \in C([0, T]; L^2(\Omega))$. Taking limits in (7.5) yields the estimate (7.2), and taking limits in (7.4) yields part (ii) of the theorem. ∎

To obtain greater regularity of solutions we will also need the following similar result. We obtain the equivalent of (7.3) in terms of fractional powers of the negative Laplacian, $A = -\Delta$, since we saw in (6.18) (periodic boundary conditions) and Proposition 6.18 (Dirichlet boundary conditions) that

$$|A^{k/2}u| \leq \|u\|_{H^k} \leq C|A^{k/2}u|, \qquad u \in D(A^{k/2}). \qquad (7.6)$$

Corollary 7.3. *For some $k \geq 0$, suppose that*

$$u \in L^2(0, T; H^{k+1}(\Omega)) \qquad and \qquad du/dt \in L^2(0, T; H^{k-1}(\Omega)).$$

Then u is continuous from $[0, T]$ *into* $H^k(\Omega)$. *Furthermore, if*

$$u \in L^2\big(0, T; D\big(A^{(k+1)/2}\big)\big) \qquad and \qquad du/dt \in L^2\big(0, T; D\big(A^{(k-1)/2}\big)\big),$$

then we have

$$\frac{d}{dt}|A^{k/2}u|^2 = \left\langle A^{k/2}\frac{du}{dt}, A^{k/2}u \right\rangle. \tag{7.7}$$

Proof. For $k = 0$ the result is just Theorem 7.2. For $k > 0$, consider $v = D^\alpha u$, with $|\alpha| \le k + 1$. Then

$$v \in L^2(0, T; H^1), \qquad dv/dt \in L^2(0, T; H^{-1}),$$

so that v is continuous from $[0, T]$ into $L^2(\Omega)$. Since this holds for each α, summing over all $|\alpha| \le k + 1$, we obtain the result.

To obtain (7.7), set $v = A^{k/2}u$, and observe, using (7.6), that

$$v \in L^2(0, T; D(A^{1/2})) \qquad and \qquad dv/dt \in L^2(0, T; D(A^{-1/2})).$$

It follows that $v \in L^2(0, T; H^1)$ and that $dv/dt \in L^2(0, T; H^{-1})$, and so

$$\frac{d}{dt}|v|^2 = 2\langle dv/dt, v \rangle.$$

Equation (7.7) follows, since $v = A^{k/2}u$. ∎

Note that although the two conditions in the corollary are not identical, they are similar. We will see below (Section 7.5) that we usually obtain estimates that allow us to apply both parts of this result.

7.2 Weak Solutions of Parabolic Equations

We will investigate solutions of

$$\frac{\partial u}{\partial t} - \Delta u = f(x, t), \qquad u|_{\partial\Omega} = 0, \qquad u(x, 0) = u_0(x)$$

that evolve continuously in $L^2(\Omega)$. Since we are thinking of $u(x, t)$ as "$[u(t)](x)$," we will be rewriting the governing equation as

$$du/dt + Au = f(t), \qquad u(0) = u_0, \tag{7.8}$$

where $A = -\Delta$ with Dirichlet boundary conditions and $f(t) = [f(t)](x)$ is a function in L^2 (say) for each t (see later for details). For example, we will take

$f(t) \in L^{\infty}_{\text{loc}}(0, \infty, L^2(\Omega))$ in Theorem 7.6. This form of the equation serves to emphasise that the spatial dependence is treated on a different basis from the time dependence. Note that we have incorporated the boundary condition in the definition of the operator A (strictly in the definition of its domain). We will show the existence and uniqueness of solutions, along with continuous dependence on initial conditions, in Theorem 7.6.

To simplify the notation we will denote $L^2(\Omega)$ by H and, as usual, the L^2 norm by $|\cdot|$. Since we will also use the space $H^1_0(\Omega)$ throughout the argument, we will denote this by V.

Since Ω is bounded the Poincaré inequality

$$|u| \leq C|\nabla u|$$

is valid (see Proposition 5.8). For the norm on V we will therefore use the H^1_0 norm containing only derivative terms (5.12), and we denote this by $\|u\|$:

$$\|u\|^2 = \sum_{|\alpha|=1} |D^{\alpha} u|^2.$$

We will denote the norm on $V^* = H^{-1}(\Omega)$ as $\|\cdot\|_*$.

Recall that in the previous chapter we defined the bilinear form

$$a(u, v) = \langle Au, v \rangle = \sum_{j=1}^{m} (D_j u, D_j v),$$

which is the same as the H^1_0 inner product on V (5.13). We denote this inner product by $((u, v))$, so that

$$((u, v)) = a(u, v) = \langle Au, v \rangle, \qquad u, v \in V. \tag{7.9}$$

We then obtained a weak solution of the linear equation (Poisson's equation)

$$Au = f, \tag{7.10}$$

which we could view in two alternative ways: either as an equality in H^{-1} or, alternatively (but equivalently), in the weak form

$$a(u, v) = \langle f, v \rangle \qquad \text{for all} \qquad v \in H^1_0. \tag{7.11}$$

For the time-dependent problem, we will similarly have two views. We find a solution to the equation (cf. (7.10))

$$\frac{du}{dt} + Au = f(t),$$

where the equality holds in $L^2(0, T; V^*)$. This is equivalent to the statement that

$$\frac{d}{dt}(u, v) + a(u, v) = \langle f, v \rangle \qquad \text{for all} \qquad v \in L^2(0, T; V), \qquad (7.12)$$

for almost every $t \in [0, T]$ (cf. (7.11)). This follows (with $g = du/dt + Au - f$) from the following lemma, where we prove a slightly more general result.

Lemma 7.4. *Let V be a Hilbert space and V^* its dual. If $g \in L^p(0, T; V^*)$ $(1 \le p < \infty)$ then the following two statements are equivalent:*

(i) $g = 0$ in $L^p(0, T; V^*)$.
(ii) For every $v \in V$,

$$\langle g(t), v \rangle = 0$$

for almost every $t \in [0, T]$.

Proof. [(i) implies (ii)] That $g(t) = 0$ in $L^p(0, T; V^*)$ means that for each $\psi \in L^q(0, T; V)[(p, q)$ conjugate]

$$\int_0^T \langle g(t), \psi \rangle \, dt = 0.$$

Now, fix $v \in V$. Then for any $\alpha(t) \in L^q(0, T)$, $\psi = v\alpha(t)$ is an element of $L^q(0, T; V)$. Therefore

$$\int_0^T \langle g(t), v \rangle \alpha(t) \, dt = 0$$

for all $\alpha(t) \in L^q(0, T)$. Since $g(t) \in L^p(0, T; V^*)$, it follows that

$$\langle g(t), v \rangle \in L^p(0, T).$$

If we write $G(t) = \langle g(t), v \rangle$, we know that

$$\int_0^T G(t)\alpha(t) \, dt = 0 \qquad \text{for all} \qquad \alpha(t) \in L^q(0, T).$$

It follows that $\langle g(t), v \rangle = G(t) = 0$ in $L^p(0, T)$ and hence is zero for almost all t.

[(ii) implies (i)] Conversely, elements of the form

$$\psi = \sum_{j=1}^n v_j \alpha_j(t), \qquad \text{some} \qquad n < \infty, \qquad (7.13)$$

with $v_j \in V$ and $\alpha_j \in L^q(0, T)$ are dense in $L^q(0, T; V)$ (see Exercise 7.3). Since

$$\langle g(t), \psi(t) \rangle = \sum_{j=1}^{n} \langle g(t), v_j \rangle \alpha_j(t),$$

and for each j, $\langle g(t), v_j \rangle = 0$ for almost every $t \in [0, T]$, it follows that $\langle g(t), \psi(t) \rangle = 0$ for a.e. $t \in [0, T]$. Therefore

$$\int_0^T \langle g(t), \psi(t) \rangle \, dt = 0 \qquad (7.14)$$

for all ψ of the form (7.13). Since such ψ are dense in $L^q(0, T; V)$ it follows that (7.14) holds for all $\psi \in L^q(0, T; V)$, and so $g = 0$ in $L^p(0, T; V^*)$. ∎

7.3 The Galerkin Method: Truncated Eigenfunction Expansions

We are trying to find a solution of the PDE

$$du/dt + Au = f(t), \qquad u(0) = u_0. \qquad (7.15)$$

Rather than treating the PDE directly, we will approximate it by finite systems of ODEs, which we can treat easily using the theory of Chapter 2.

What we will do closely parallels the method in Theorem 2.6 of showing existence of a solution for the equation

$$\dot{x} = f(x), \qquad x(0) = x_0,$$

where f is continuous but not Lipschitz. Recall that the approach we adopted was to approximate $f(x)$ by a sequence of Lipschitz functions $f_n(x)$, so that we could apply standard results to the equations

$$\dot{x}_n = f(x_n), \qquad x_n(0) = x_0$$

to give solutions $x_n(t)$. By finding a convergent subsequence of these, $\{x_{n_j}\}$ with

$$x_{n_j} \to x \qquad \text{uniformly on} \qquad [0, T],$$

we obtained a solution of the original equation.

Not just for this example, but also for the reaction–diffusion equations of the next chapter and the Navier–Stokes equations in Chapter 9, we will expand a solution of (7.15) using the eigenfunctions of A (here $-\Delta$) and then truncate the series to approximate the PDE by ever larger systems of ODEs. This is known as the Galerkin method.

Existence and uniqueness for these approximations will follow from standard ODE results. Some uniform bounds on the solutions (i.e. bounds that do not depend on the truncation) will be used to find a subsequence that converges (weakly) using the Alaoglu compactness theorem (Theorem 4.18). This weak limit will give us our solution.

Since we will need some properties of these truncated eigenfunction expansions repeatedly in the next few chapters, we will gather them here. We treat the case when A is the linear operator associated with the bilinear form $a(u, v)$ and the eigenfunction decomposition is ensured by the Hilbert–Schmidt theorem. For example, for the Laplacian on a general domain we do not know (nor do we need to know) the explicit form of the eigenfunctions – but we do know that we have a set $\{w_i\}$ with eigenvalues λ_i,

$$Aw_i = \lambda_i w_i.$$

The regularity results of the previous chapter ensure that each w_i is an element of $V = H_0^1(\Omega)$ and of $C^\infty(\Omega)$ (Corollary 6.14).

The Hilbert–Schmidt theorem tells us that these eigenfunctions form an orthonormal basis for H,

$$(w_j, w_k) = \delta_{jk}.$$

They also form an orthogonal (though not orthonormal) basis for V, since

$$((w_j, w_k)) = (Aw_j, w_k) = (\lambda_j w_j, w_k) = \lambda_j \delta_{jk}.$$

If we take an element of H and project it onto the space spanned by the first n eigenfunctions, we get

$$P_n u = \sum_{j=1}^{n} (u, w_j) w_j.$$

This definition clearly makes sense for elements of $V \subset H$. Also, if $f \in V^*$ we attach meaning to $P_n f$ by

$$\langle P_n f, v \rangle = \langle f, P_n v \rangle \qquad \text{for all} \qquad v \in V.$$

We also define the projection orthogonal to P_n, $Q_n = I - P_n$,

$$Q_n u = \sum_{j=n+1}^{\infty} (u, w_j) w_j.$$

Lemma 7.5. *If $X = H$, V, or V^*, then*

$$\|P_n u\|_X \leq \|u\|_X \qquad and \qquad P_n u \to u \quad in \quad X.$$

Proof. If we write $u \in V$ as

$$u = \sum_{j=1}^{\infty} (u, w_j) w_j,$$

we have

$$\|u\|^2 = \sum_{j=1}^{\infty} \lambda_j |(u, w_j)|^2, \qquad \|P_n u\|^2 = \sum_{j=1}^{n} \lambda_j |(u, w_j)|^2,$$

and so

$$\|P_n u\| \leq \|u\|.$$

Clearly, $P_n u \to u$ in V, since

$$\|u - P_n u\|^2 = \sum_{j=n+1}^{\infty} \lambda_j |(u, w_j)|^2,$$

and

$$\sum_{j=1}^{\infty} \lambda_j |(u, w_j)|^2 = \|u\|^2 < \infty.$$

The argument for the convergence and norm bounds in H is identical. In V^*, the norm bound follows from

$$\|P_n f\|_{V^*} = \sup_{\|v\| \leq 1} |\langle P_n f, v \rangle| = \sup_{\|v\| \leq 1} |\langle f, P_n v \rangle| \leq \sup_{\|w\| \leq 1} |\langle f, w \rangle| = \|f\|_{V^*}.$$

For convergence, observe that each linear functional $f \in V^*$ corresponds to an element $\phi \in V$ using the Riesz representation theorem (Theorem 4.9), so that

$$\langle f, v \rangle = ((\phi, v)) \qquad \text{for all} \qquad v \in V.$$

Now,

$$
\begin{aligned}
\langle P_n f, v \rangle - \langle f, v \rangle &= \langle f, P_n v \rangle - \langle f, v \rangle \\
&= \langle f, Q_n v \rangle \\
&= ((\phi, Q_n v)) \\
&= ((Q_n \phi, v)),
\end{aligned}
$$

so that

$$|\langle P_n f, v \rangle - \langle f, v \rangle| \le \|Q_n \phi\| \|v\|.$$

Therefore

$$\sup_{\{v:\ \|v\|=1\}} |\langle P_n f - f, v \rangle| \le \|Q_n \phi\|$$

and so tends to zero as $n \to \infty$ by the previous convergence result for $\phi \in V$. ■

7.4 Weak Solutions

We now apply these ideas to prove existence and uniqueness of weak solutions on $H = L^2(\Omega)$.

Theorem 7.6 (Weak Solutions). *If $f \in L^2_{\text{loc}}(0, \infty; V^*)$ then the equation*

$$du/dt + Au = f(t), \qquad u(0) = u_0 \in H \qquad (7.16)$$

has a unique weak solution $u(t)$ with, for any $T > 0$,

$$u \in L^2(0, T; V), \qquad du/dt \in L^2(0, T; V^*), \qquad and \qquad u \in C([0, T]; H).$$

Furthermore $u_0 \mapsto u(t)$ is continuous on H.

Equation (7.16) holds as an equality in $L^2(0, T; V^*)$ [i.e. in the sense of (7.12)]. The assumption on f is motivated by this.

Proof. We use the Galerkin method. First we look for an approximate solution $u_n(t)$ lying in the finite-dimensional space spanned by the first n eigenfunctions of A and solving an ODE. We know that this ODE has a solution by using the techniques from Chapter 2.

We then show that the solutions $u_n(t)$ of these ODEs in fact exist for all time. Use of Lemma 2.4 involves showing that the solutions cannot "blow up" in finite time. In fact we will obtain bounds on the solutions $u_n(t)$ in various spaces; the important thing is that these bounds are independent of n. For example, we will show that for each $T > 0$ there is a constant $K = K(T)$ such that

$$\int_0^T \|u_n(t)\|^2 \, dt \le K \qquad (7.17)$$

for all n. In other words, $\{u_n\}$ is a bounded sequence in $L^2(0, T; V)$. We can then use the Alaoglu compactness theorem (Corollary 4.19) to find a subsequence

u_{n_j} that converges weakly in $L^2(0, T; V)$ to some u. The function u is our candidate solution for the infinite-dimensional equation, and with a little work we can show that u does indeed satisfy (7.16) in a weak sense (as in Lemma 7.4).

7.4.1 The Galerkin Approximations

We look for an approximate solution $u_n(t)$ that lies in the finite-dimensional space spanned by the first n eigenfunctions,

$$u_n(t) = \sum_{j=1}^{n} u_{nj}(t) w_j, \tag{7.18}$$

and solves

$$(du_n/dt, w_j) + (Au_n, w_j) = \langle f, w_j \rangle, \qquad 1 \leq j \leq n,$$

with $(u_n(0), w_j) = (u_0, w_j)$. The final term makes sense since $w_j \in V$.

From (7.18) we have $u_{nj} = (u_n, w_j)$; also

$$(du_n/dt, w_j) = du_{nj}/dt \qquad \text{and} \qquad (Au_n, w_j) = \lambda_j u_{nj}.$$

So we have a set of n ODEs for the components u_{nj}:

$$du_{nj}/dt + \lambda_j u_{nj} = \langle f(t), w_j \rangle. \tag{7.19}$$

We could also write this as

$$du_n/dt + Au_n = P_n f, \tag{7.20}$$

where u_n is the composite function given in terms of the components by (7.18). In fact, this second form of the equation is more useful in deriving the estimates that follow.

Existence and uniqueness results for ODEs imply that we have a unique solution of (7.19) for the u_{nj}, at least on some short time interval $[0, T_n]$. We can extend this time interval to infinity if we know that the u_{nj} are bounded. Note that if we consider u_n as an n-component vector (i.e. an element of $\mathbb{R}^n$) we need to show that

$$\sum_{j=1}^{n} u_{nj}^2 \tag{7.21}$$

remains bounded.

7.4.2 Uniform Bounds on u_n in Various Spaces

Now, using (7.18), observe that (7.21) is proportional to the square of the L^2 norm of u_n. We can therefore try to obtain an L^2 bound on u_n using (7.20). Take the inner product of (7.20) with u_n to get

$$(du_n/dt, u_n) + (Au_n, u_n) = \langle P_n f, u_n \rangle.$$

Now, $(Au_n, u_n) = a(u_n, u_n) = \|u_n\|^2$ (7.9), and we can bound the right-hand side by

$$\langle P_n f, u_n \rangle \le \|P_n f\|_* \|u_n\|,$$

giving

$$\frac{1}{2} \frac{d}{dt} |u_n|^2 + \|u_n\|^2 \le \|P_n f\|_* \|u_n\|.$$

Using Young's inequality ($ab \le \frac{1}{2}(a^2 + b^2)$) and rearranging yields

$$\frac{d}{dt} |u_n|^2 + \|u_n\|^2 \le \|P_n f\|_*^2. \tag{7.22}$$

Note that all these manipulations are rigorous (for example, we have taken second derivatives in the Au_n term), because we know that the truncated Fourier series (u_n) is C^∞, since the eigenfunctions $\{w_j\}$ are.

We now integrate (7.22) between 0 and T to get

$$|u_n(T)|^2 + \int_0^T \|u_n(t)\|^2 \, dt \le |u_n(0)|^2 + \int_0^T \|P_n f(t)\|_*^2 \, dt.$$

Since we know that $f \in L^2_{\text{loc}}(0, \infty; V^*)$ [i.e. $f \in L^2(0, T; V^*)$ for all $T < \infty$] and furthermore (by Lemma 7.5) that

$$|u_n(0)| = |P_n u_0| \le |u_0| \qquad \text{and} \qquad \|P_n f\|_* \le \|f\|_*,$$

we have

$$|u_n(T)|^2 + \int_0^T \|u_n(t)\|^2 \, dt \le |u_0|^2 + \int_0^T \|f(t)\|_*^2 \, dt, \tag{7.23}$$

a bound that is uniform in n. We can reexpress the two bounds contained in (7.23) by saying that u_n is uniformly bounded (in n) in the two spaces

$$L^\infty(0, T; H) \qquad \text{and} \qquad L^2(0, T; V)$$

for all $T > 0$ (cf. (7.17)). [We could say that $\{u_n\}$ is a bounded sequence in $L^\infty(0, T; H)$, but the above terminology serves to emphasise that the bound has to be uniform in n.]

To show that u is continuous into H, using Theorem 7.2, we will need to obtain a bound on du/dt in $L^2(0, T; V^*)$. We therefore work with du_n/dt and show that we can find a uniform bound for this sequence. In fact this is almost immediate, since

$$du_n/dt = -Au_n + P_n f.$$

We know that $P_n f \in L^2(0, T; V^*)$, and $Au_n \in L^2(0, T; V^*)$ too, since $u_n \in L^2(0, T; V)$ and A is a bounded linear map from V into V^* (see (6.7)). Thus

$$du_n/dt \quad \text{is uniformly bounded in} \quad L^2(0, T; V^*).$$

7.4.3 Extraction of an Appropriate Subsequence

We now use the Alaoglu compactness theorem (Theorem 4.18 and its corollary) repeatedly to extract a subsequence (which we will repeatedly and wickedly relabel u_n) that converges in various different senses. First, u_n is uniformly bounded in $L^2(0, T; V)$, so by extracting a subsequence we can ensure that

$$u_n \rightharpoonup u \quad \text{in} \quad L^2(0, T; V).$$

Also, we have a uniform bound on du_n/dt in $L^2(0, T; V^*)$, so by extracting a second subsequence we can also guarantee that

$$du_n/dt \overset{*}{\rightharpoonup} \dot{u} \quad \text{in} \quad L^2(0, T; V^*).$$

We have written $\dot{u}$ because it is not immediately obvious that in fact $\dot{u} = du/dt$ (the weak time derivative). However, if we use the definition of weak-* convergence of du_n/dt to $\dot{u}$ we have

$$\int_0^T \frac{du_n}{dt}(t)\psi(t)\,dt \to \int_0^T \dot{u}(t)\psi(t)\,dt \quad \text{for all} \quad \psi \in L^2(0, T; V).$$

Now, if $\psi(t) \in C_c^\infty(0, T; V)$ then we can integrate the left-hand side by parts to give

$$\int_0^T \frac{du_n}{dt}(t)\psi(t)\,dt = -\int_0^T u_n(t)\frac{d\psi}{dt}(t)\,dt$$
$$\to -\int_0^T u(t)\frac{d\psi}{dt}(t)\,dt,$$

using the weak convergence of u_n to u, since $d\psi/dt \in C_c^\infty(0, T; V) \subset L^2(0, T; V)$. We therefore have

$$\int_0^T \dot{u}(t)\psi(t)\, dt = -\int_0^T u(t)\frac{d\psi}{dt}(t)\, dt \qquad \text{for all} \qquad \psi \in C_c^\infty(0, T; V),$$

and so $\dot{u} = du/dt$ as required.

So, we know that

$$du_n/dt \overset{*}{\rightharpoonup} du/dt \qquad \text{in} \qquad L^2(0, T; V^*).$$

To obtain the same convergence of Au_n, we use the fact that A is a bounded linear operator from V into V^*, so that the weak convergence of u_n to u in $L^2(0, T; V)$ implies weak-* convergence of Au_n to Au in $L^2(0, T; V^*)$:

$$\int_0^T \langle Au_n, \psi \rangle \, dt = \int_0^T \langle u_n, A\psi \rangle \, dt \to \int_0^T \langle u, A\psi \rangle \, dt$$

$$= \int_0^T \langle Au, \psi \rangle \, dt, \qquad \psi \in L^2(0, T; V).$$

Finally, to show that $P_n f \overset{*}{\rightharpoonup} f$ in $L^2(0, T; V^*)$, observe that functions of the form

$$\psi = \sum_{j=1}^k \psi_j \alpha_j(t), \qquad \psi_j \in V, \ \alpha_j \in L^2(0, T; \mathbb{R})$$

are dense in $L^2(0, T; V)$. (We used this, the result of Exercise 7.3, in the proof of Lemma 7.4.) For such ψ

$$\int_0^T \langle P_n f(t), \psi(t) \rangle \, dt = \int_0^T \langle f(t), P_n \psi(t) \rangle \, dt$$

$$= \int_0^T \sum_{j=1}^k \langle f(t), P_n \psi_j \rangle \alpha_j(t) \, dt.$$

Now, since $P_n \psi_j \to \psi_j$ in V for each j (see Lemma 7.5), this converges to

$$\int_0^T \sum_{j=1}^k \langle f(t), \psi_j \rangle \alpha_j(t) \, dt = \int_0^T \langle f(t), \psi \rangle \, dt.$$

This gives weak-* convergence of $P_n f$ to f in $L^2(0, T; V^*)$, and so we have weak-* convergence in $L^2(0, T; V^*)$ of all the terms in (7.20).

7.4.4 Properties of the Weak Solution

We therefore have

$$du/dt + Au = f$$

as an equality in $L^2(0, T; V^*)$, and Lemma 7.4 shows that this is equivalent to

$$\langle du/dt, v \rangle + a(u, v) = \langle f, v \rangle \qquad (7.24)$$

for every $v \in V$ and for almost every $t \in [0, T]$.

Note that since the limit value u has

$$u \in L^2(0, T; V) \qquad \text{and} \qquad du/dt \in L^2(0, T; V^*),$$

it follows from Theorem 7.2 that u is actually continuous into H:

$$u \in C([0, T]; H).$$

[Note that this is much better than the $u \in L^\infty(0, T; H)$ we could have obtained by taking advantage of the uniform bound on u_n in this space and extracting another subsequence – in fact, we did not use the L^∞ bound above.]

Before checking uniqueness, we still have to show that $u(0) = u_0$. Note that although $u \in C([0, T]; H)$ and $u = \lim_{n \to \infty} u_n$ (in various senses) with $u_n(0) = P_n u_0$, this does not follow automatically: For example, suppose that

$$x_n(t) = \begin{cases} 1, & t = 0, \\ t, & 0 < t \le 1; \end{cases}$$

then $x_n(t) \to x(t)$ in L^2, where $x(t) = t$ and $x(0) = 0$.

Choosing a function $\phi \in C^1([0, T]; V)$ with $\phi(T) = 0$, in the "limit equation" (7.24) we can integrate from 0 to T and then do the first integral by parts to obtain

$$\int_0^T -\langle u, \phi' \rangle + a(u, \phi) \, ds = \int_0^T (f, \phi) \, ds + (u(0), \phi(0)). \qquad (7.25)$$

If we do the same in the Galerkin equation (7.20) we get

$$\int_0^T -\langle u_n, \phi' \rangle + a(u_n, \phi) \, ds = \int_0^T (P_n f, \phi) \, ds + (u_n(0), \phi(0)).$$

We have seen above that we can take limits in all these terms, so this becomes

$$\int_0^T -\langle u, \phi' \rangle + a(u, \phi)\, ds = \int_0^T (f, \phi)\, ds + (u_0, \phi(0)), \qquad (7.26)$$

since $u_n(0) \to u_0$. Since $\phi(0)$ is arbitrary, a comparison of (7.25) and (7.26) shows that indeed $u(0) = u_0$, as required.

7.4.5 Uniqueness and Continuous Dependence on Initial Conditions

To show that the solution is unique, suppose that we have two solutions u and v of

$$du/dt + Au = f(t).$$

Since the equation is linear, the difference of these two solutions, $w(t) = u(t) - v(t)$, satisfies

$$dw/dt + Aw = 0.$$

Using Theorem 7.2 we can take the inner product of this equation with w to give

$$\tfrac{1}{2} \frac{d}{dt} |w|^2 + \|w\|^2 = 0,$$

whence

$$\frac{d}{dt} |w|^2 \le 0.$$

It follows that

$$|w(t)| \le |w(0)|,$$

which gives continuous dependence on initial conditions and, in the case when $w(0) = 0$, uniqueness. ∎

7.5 Strong Solutions

We now show how increasing the regularity of f and u_0 results in more regular solutions. In particular, we will see that if $u_0 \in V$ and $f \in L^2_{\mathrm{loc}}(0, \infty; H)$ then the solutions are continuous into V and are elements of H for almost every t. We call such solutions *strong solutions* (cf. discussion in Section 6.1).

Theorem 7.7 (Strong Solutions). *Suppose that* $f \in L^2_{\text{loc}}(0, \infty; H)$ *and that* $u_0 \in V$. *Then there exists a unique solution of*

$$du/dt + Au = f(t) \qquad (7.27)$$

satisfying

$$u \in L^2(0, T; D(A)), \qquad du/dt \in L^2(0, T; H), \qquad and \qquad u \in C^0([0, T]; V).$$

Furthermore, the map $u_0 \mapsto u(t)$ *is continuous in* V. *The sense of Equation (7.27) is now an equality in* $L^2(0, T; H)$; *in particular, it holds in* $H = L^2(\Omega)$ *for almost every* $T \in [0, T]$.

In the statement of the theorem, $D(A)$ is the Hilbert space with the inner product

$$((u, v))_{D(A)} = (Au, Av);$$

see (3.22). In particular, using the regularity result

$$\|u\|_{H^2(\Omega)} \leq C|Au|$$

of Theorem 6.16, we see that a bound in $L^2(0, T; D(A))$ is equivalent to a bound in $L^2(0, T; H^2(\Omega))$.

Proof. The idea is essentially the same as the proof of the previous theorem, except that we have to ensure that we have some better estimates on the u_n and their derivatives. So, rather than take the inner product with u_n, we take the inner product of

$$du_n/dt + Au_n = P_n f$$

with Au_n,

$$(du_n/dt, Au_n) + |Au_n|^2 = (P_n f, Au_n).$$

Now, since u_n is smooth we have $Au_n = -\Delta u_n$, and so, using the boundary condition $u_n = 0$ on $\partial\Omega$, we get

$$-\int_\Omega \Delta u_n \frac{\partial u_n}{\partial t} dx = \sum_{i=1}^m \int_\Omega \frac{\partial u_n}{\partial x_i} \frac{\partial^2 u_n}{\partial t \partial x_i} dx = \frac{1}{2}\frac{d}{dt}\|u_n\|^2. \qquad (7.28)$$

Therefore

$$\frac{1}{2}\frac{d}{dt}\|u_n\|^2 + |Au_n|^2 \le |P_n f||Au_n|,$$

so that, using Young's inequality on the right-hand side, we have

$$\frac{d}{dt}\|u_n\|^2 + |Au_n|^2 \le |P_n f|^2 \le |f|^2,$$

and, integrating between 0 and T, we get

$$\|u_n(T)\|^2 + \int_0^T |Au_n(t)|^2\, dt \le \|u_0\|^2 + \int_0^T |f(t)|^2\, dt.$$

Now we have uniform bounds for u_n in

$$L^\infty(0, T; V) \qquad \text{and} \qquad L^2(0, T; D(A)).$$

Furthermore, since

$$du_n/dt = -Au_n + P_n f$$

it follows that du_n/dt is uniformly bounded in $L^2(0, T; H)$. Arguing as before, we take subsequences by using Corollary 4.19 such that

$$u_n \rightharpoonup u \qquad \text{in} \qquad L^2(0, T; D(A))$$

and

$$du_n/dt \rightharpoonup du/dt \qquad \text{in} \qquad L^2(0, T; H).$$

We can show that all terms in the equation converge weakly in $L^2(0, T; H)$ – applying Lemma 7.4 with $V = H$ shows that Equation (7.27) holds in H for almost every $t \in [0, T]$. Also, Corollary 7.3 shows that $u \in C^0([0, T]; V)$.

Finally, uniqueness follows from the previous argument, since any two strong solutions would both be weak solutions. We can obtain continuity in V with respect to the initial condition u_0 by taking the inner product of

$$\frac{dw}{dt} + Aw = 0$$

with Aw (using Corollary 7.3) and dropping the $|Aw|^2$ term,

$$\frac{d}{dt}\|w\|^2 \le 0.$$

Therefore

$$\|w(t)\| \leq \|w(0)\|,$$

yielding uniqueness and continuous dependence on initial conditions as before.

∎

7.6 Higher Regularity: Spatial and Temporal

If we assume ever greater regularity for $f(t)$ and u_0, then the spatial regularity of solutions increases.

Theorem 7.8 (More regular solutions). *If $k \geq 0$, Ω is a bounded C^k domain, $f \in L^2_{\text{loc}}(0, \infty; D(A^{(k-1)/2}))$, and $u_0 \in D(A^{k/2})$, then there exists a unique solution $u(t)$ of*

$$du/dt + Au = f(t)$$

with

$$u \in L^2(0, T; H^{k+1}(\Omega)) \qquad and \qquad u \in C^0([0, T]; H^k(\Omega)).$$

We will not give a proof here, which relies on methods similar to the proof of Corollary 7.3 and is relatively straightforward (see Exercise 7.4).

To obtain temporal regularity so that the time derivative du/dt becomes a classical one, we need some further assumption on the derivatives of f. We give a simple result, which gives differentiability into V^*.

Theorem 7.9 (Temporal regularity). *Suppose that $f \in L^2(0, T; V)$ and that in addition $df/dt \in L^2(0, T; V^*)$. If $u_0 \in D(A)$ then*

$$u \in C^1([0, T]; V^*) \qquad and \qquad u \in C([0, T]; D(A)).$$

Proof. Note first that Theorem 7.2 implies that $f \in C^0([0, T]; H)$. Now consider the formal equation for $v = \dot{u} = du/dt$, which is

$$dv/dt + Av = \dot{f}(t), \qquad v(0) = -Au_0 + f(0).$$

This is an equation for v in the form of (7.8), since

$$\dot{f} \in L^2(0, T; V^*) \qquad and \qquad v(0) \in H.$$

It follows from Theorem 7.6 that $v \in L^2(0, T; V) \cap H^1(0, T; V^*)$.

Since solutions of Equation (7.8) are unique, it follows that v must actually be equal to $\dot{u}$, so that $\dot{u} \in L^2(0, T; V) \cap H^1(0, T; V^*)$. Therefore u itself is an element of $H^2(0, T; V^*) \cap H^1(0, T; V)$, and so it follows from the Sobolev embedding theorem (Corollary 5.30) that $u \in C^1([0, T]; V^*)$. Noting that we have

$$Au(t) = f(t) - du/dt,$$

and the right-hand side is in $L^2(0, T; V)$, we can use the regularity theory for the Laplacian (Theorem 6.17) to deduce that $u \in L^2(0, T; H^3)$. Corollary 7.3 now guarantees that $u \in C^0([0, T]; D(A))$. ∎

It is possible to continue in this vein and to show that if f and u_0 are smooth then so is $u(x, t)$. Since we will be mainly interested in the dynamical systems generated on L^2 and H^1 by the solutions of PDEs, we do not pursue these higher-order estimates here. More details can be found in Evans (1998).

Exercises

7.1 For a Banach space X we will now define the integral of function $f : [0, T] \to X$. We say that f is integrable if the function $\langle L, f(t) \rangle : [0, T] \to \mathbb{R}$ is integrable for every $L \in X^*$, and there exists an element $y \in X$ such that

$$\langle L, y \rangle = \int_0^T \langle L, f(t) \rangle \, dt \qquad \text{for all} \qquad L \in X^*. \qquad (7.29)$$

In this case we define

$$\int_0^T f(t) \, dt = y. \qquad (7.30)$$

Show that if X is reflexive and

$$\int_0^T \|f(t)\|_X \, dt < \infty \qquad (7.31)$$

then such a y exists and that Definition (7.30) is then unambiguous. (Hint: Use Lemma 4.4 for the second part.)

7.2 Show that if $f : I \to X$ is integrable (as in Exercise 7.1) then

$$\left\| \int_0^T f(t)\, dt \right\|_X \leq \int_0^T \| f(t) \|_X \, dt.$$

(Hint: Use Corollary 4.5.)

7.3 For $1 \leq p < \infty$ show that elements of the form

$$\sum_{j=1}^n v_j \alpha_j(t), \tag{7.32}$$

with $v_j \in V$ and $\alpha_j \in L^p(0, T)$, are dense in $L^p(0, T; V)$. Show that the same result is true if we take $\alpha_j \in C^1([0, T])$ or $v_j \in C_c^\infty(\Omega)$.

7.4 Suppose that Ω is a bounded C^{k+1} domain, $u_0 \in D(A^{k/2})$, and

$$f \in L^2_{\text{loc}}\left(0, \infty; D\left(A^{(k-1)/2}\right)\right).$$

By taking the inner product of the Galerkin approximation

$$du_n/dt + Au_n = P_n f(t) \tag{7.33}$$

with $A^k u_n$, show that u_n is uniformly bounded in

$$L^\infty(0, T; D(A^{k/2})) \cap L^2\left(0, T; D\left(A^{(k+1)/2}\right)\right) \tag{7.34}$$

and that

$$du_n/dt \in L^2\left(0, T; D\left(A^{(k-1)/2}\right)\right). \tag{7.35}$$

Deduce the result of Theorem 7.8 by using Corollary 7.3.

7.5 Let $\{w_j\}_{j=1}^\infty$ be an orthonormal basis for $L^2(\Omega)$, orthogonal in $H_0^1(\Omega)$. Use the Galerkin method to prove the existence of a solution for the problem

$$-\Delta u = f, \qquad u|_{\partial\Omega} = 0 \tag{7.36}$$

for $f \in L^2(\Omega)$ by considering the solution $u_n = \sum_{k=1}^n u_{nk} w_k$ of

$$((u_n, w_k)) = (f, w_k) \qquad \text{for all} \qquad k = 1, \ldots, n.$$

Show that u_n converges in $H_0^1(\Omega)$ to some u, and deduce that this is a solution of (7.36).

Notes

The presentation in this chapter is designed to be consistent with that in the next two chapters dealing with nonlinear equations. Temam (1984) covers the material in Section 7.1 that gives sense to weak solutions, and concise proofs of the results in this chapter are given in the introduction to his 1988 book.

Evans (1998) covers much of this material in similar detail, including the higher regularity results mentioned in the final section. Renardy & Rogers (1992) give a very brief account of these methods; Theorem 7.9 is taken from their treatment. These two fine books do not overlap with this volume from now on.

The approach to Banach-space valued integration in Exercises 7.1 and 7.2 is adapted from pages 77–81 in Rudin (1991). A more general theory can be found in Yosida (1980), Chapter V, Sections 4 and 5.

8

Nonlinear Reaction–Diffusion Equations

In this chapter we will study reaction–diffusion equations, a simple class of nonlinear parabolic equations. As the name suggests, they model chemical reactions in extended spatial domains where diffusion is important. If a chemical concentration evolves locally according to the ordinary differential equation (ODE)

$$du/dt = f(u)$$

then with the addition of diffusion we obtain the scalar reaction–diffusion equation

$$\frac{\partial u}{\partial t} - \Delta u = f(u). \tag{8.1}$$

For questions concerning the existence and uniqueness of weak (and strong) solutions, we will treat this equation on a bounded domain $\Omega \subset \mathbb{R}^m$, imposing Dirichlet boundary conditions $u|_{\partial \Omega} = 0$ and "sufficient smoothness" of Ω, which we will specify below. We will also require that $f(s)$ be a C^1 function that satisfies the bounds

$$-k - \alpha_1|s|^p \le f(s)s \le k - \alpha_2|s|^p, \qquad p > 2, \tag{8.2}$$

and

$$f'(s) \le l,$$

both for all $s \in \mathbb{R}$. One example of such a function $f(s)$ is an odd degree polynomial,

$$f(s) = \sum_{j=1}^{2n-1} c_j s^j,$$

213

where $\alpha_j < 0$. This gives $p = 2n$ above. We will consider the particular example $f(s) = s - s^3$ in more detail in Chapter 11.

It is possible to treat systems of reaction–diffusion equations, and of course these are better models of real chemical reactions. Some indications of possible generalisations are given in the notes, but we stick to the scalar case for the sake of simplicity.

8.1 Results to Deal with the Nonlinear Term

Although the linear terms are well behaved (we have already dealt with them, more or less, in the previous chapter), the nonlinear term is more problematic. In particular, although we will easily be able to ensure that there exists a χ such that $f(u_n) \to \chi$ in some appropriate sense [where u_n is our truncated version of the solution u of (8.1)], it is much less straightforward to show that $\chi = f(u)$, i.e. that the limiting function actually solves the equation.

The problem is essentially that the extraction of a subsequence usually gives a weakly convergent sequence, and this is not enough to show that $\chi = f(u)$ directly. We will need in addition some strong convergence of u_n (given by the compactness result in Section 8.1.1) and a weak version of the dominated convergence theorem (given in Section 8.1.2).

8.1.1 A Compactness Theorem

In the last chapter we saw that weak solutions of the linear parabolic equation

$$\frac{\partial u}{\partial t} - \Delta u = f(t)$$

had $u \in L^2(0, T; H_0^1)$ and $du/dt \in L^2(0, T; H^{-1})$. The derivative was L^2 into H^{-1} rather than into H_0^1. Nevertheless this enabled us to conclude something about u as a function into L^2, namely (Theorem 7.2) that $u \in C^0([0, T]; L^2)$. The same conclusion would follow, of course, if we knew that $u \in H^1(0, T; L^2)$. In this brief section we will state another result along similar lines. This compactness result would hold from the standard theory of Sobolev spaces if $\{u_n\}$ were a bounded sequence in $H^1(0, T; L^2)$, but it is not so straightforward if u_n is bounded in $L^2(0, T; H_0^1)$ and du_n/dt is bounded in $L^2(0, T; H^{-1})$. Nonetheless, the conclusion is the same – that there is a subsequence of $\{u_n\}$ that converges strongly in $L^2(0, T; L^2)$.

Theorem 8.1. *Let $X \subset\subset H \subset Y$ be Banach spaces, with X reflexive. Suppose that u_n is a sequence that is uniformly bounded in $L^2(0, T; X)$, and du_n/dt is* *uniformly bounded in $L^p(0, T; Y)$, for some $p > 1$. Then there is a subsequence that converges strongly in $L^2(0, T; H)$.*

Above we focussed on the case $X = H_0^1$, $Y = H^{-1}$, and $p = 2$. Note the possibility of weakening the assumption on du_n/dt discussed above [bounded in $L^2(0, T; H^{-1})$] to boundedness in $L^p(0, T; H^{-1})$ for any $p > 1$. In the proof, which can be skipped on first reading, we will need the following result, known as Ehrling's lemma.

Lemma 8.2 (Ehrling's Lemma). *Suppose that X, H, and Y are three Banach spaces with $X \subset\subset H \subset Y$. Then for each $\eta > 0$ there exists a constant c_η such that*

$$\|u\|_H \leq \eta \|u\|_X + c_\eta \|u\|_Y \qquad \text{for all} \qquad u \in X.$$

Proof. Suppose not. Then there exists an $\eta > 0$ such that for each $n \in \mathbb{Z}^+$ there is a u_n with

$$\|u_n\|_H \geq \eta \|u_n\|_X + n \|u_n\|_Y.$$

Now consider the normalised sequence

$$v_n = u_n / \|u_n\|_X,$$

which satisfies

$$\|v_n\|_H \geq \eta + n \|v_n\|_Y. \tag{8.3}$$

The sequence v_n is bounded in X (since $\|v_n\|_X = 1$), and so it is bounded in H (as $X \subset\subset H$). It follows from (8.3) that $\|v_n\|_Y \to 0$ as $n \to \infty$. However, since $X \subset\subset H$, $v_n \to v$ in H; clearly this limit should be zero. However, this contradicts

$$\|v_n\|_H \geq \eta,$$

which follows from (8.3). ∎

We can now prove Theorem 8.1.

Proof. Since X is reflexive, so is $L^2(0, T; X)$, and so, using the corollary of the Alaoglu theorem (Corollary 4.19), there is a subsequence u_n with

$$u_n \rightharpoonup u \qquad \text{in} \qquad L^2(0, T; X).$$

Now, we need to show that $v_n = u_n - u \to 0$ in $L^2(0, T; H)$. We start off by showing that

$$v_n \to 0 \quad \text{in} \quad L^2(0, T; Y) \qquad (8.4)$$

will be sufficient to guarantee that $v_n \to 0$ in $L^2(0, T; H)$.

Indeed, we know that since X is compactly embedded in H, we can use Ehrling's lemma to guarantee that for each $\eta > 0$ there exists a c_η such that

$$\|u\|_H^2 \le \eta \|u\|_X^2 + c_\eta \|u\|_Y^2.$$

Therefore

$$\|v_n\|_{L^2(0,T;H)}^2 \le \eta \|v_n\|_{L^2(0,T;X)}^2 + c_\eta \|v_n\|_{L^2(0,T;Y)}^2,$$

and since v_n is bounded in $L^2(0, T; X)$,

$$\|v_n\|_{L^2(0,T;H)}^2 \le C\eta + c_\eta \|v_n\|_{L^2(0,T;Y)}^2.$$

If $v_n \to 0$ in $L^2(0, T; Y)$ then we have

$$\limsup_{n\to\infty} \|v_n\|_{L^2(0,T;H)}^2 \le C\eta,$$

valid for any $\eta > 0$, and so this would give

$$\lim_{n\to\infty} \|v_n\|_{L^2(0,T;H)} = 0,$$

as required.

So we now have to try to show (8.4). It follows from Proposition 7.1 that $v_n \in C^0([0, T]; Y)$ and

$$|v_n(t)|_Y \le C\big(\|v_n\|_{L^2(0,T;X)} + \|dv_n/dt\|_{L^p(0,T;Y)}\big) \le M$$
$$\text{for all} \quad t \in [0, T] \qquad (8.5)$$

uniformly in n. We now show that for almost every $t \in [0, T]$, $v_n(t) \to 0$ strongly in Y, and then the result follows using Lebesgue's dominated convergence theorem (Theorem 1.7).

We have (writing $\dot{}$ for d/dt)

$$v_n(t) = v_n(u) - \int_t^u \dot{v}_n(\tau)\, d\tau,$$

and integrating this with respect to u from t to $t + s$ gives

$$v_n(t) = \frac{1}{s}\left[\int_t^{t+s} v_n(u)\, du - \int_t^{t+s}\int_t^u \dot{v}_n(\tau)\, d\tau\, du\right]$$
$$= a_n + b_n,$$

where

$$a_n = \frac{1}{s} \int_t^{t+s} v_n(u)\, du \qquad \text{and} \qquad b_n = \frac{1}{s} \int_t^{t+s} (t + s - u)\dot{v}_n(u)\, du.$$

Now take $\epsilon > 0$ and estimate

$$\|b_n\|_Y \le \int_t^{t+s} \|dv_n/dt\|_Y\, du \le \left(\int_t^{t+s} \|dv_n/dt\|_Y^p\, du \right)^{1/p} s^{1/q}$$

$$\le s^{1/q} \|dv_n/dt\|_{L^p(0,T;Y)}.$$

Choose s so that

$$\|b_n\|_Y \le \epsilon/2.$$

For this value of s, observe that

$$\int_t^{t+s} v_n(u)\, du \rightharpoonup 0 \qquad \text{in} \qquad X. \tag{8.6}$$

Indeed, if χ is the indicator function of $[t, t + s]$, and $\phi \in X^*$, then $\chi\phi$ is an element of $L^2(0, T; X^*)$, and

$$\int_0^T \langle v_n(t), \chi\phi \rangle\, dt = \int_t^{t+s} \langle v_n(t), \phi \rangle\, dt = \left\langle \int_t^{t+s} v_n(t)\, dt, \phi \right\rangle,$$

where we used (7.29) from the definition of such Banach-space valued integrals (Exercise 7.1). Since $v_n \rightharpoonup 0$ in $L^2(0, T; X)$, (8.6) follows, and hence a_n converges weakly to zero in X. Since X is compactly embedded in Y, $a_n \to 0$ strongly in Y. It follows that for n large enough we have

$$\|a_n\|_Y \le \epsilon/2,$$

and so

$$\|v_n(t)\|_Y \le \epsilon.$$

Since $v_n(t) \to 0$ in Y and $v_n(t)$ is bounded in Y over $t \in [0, T]$ (8.5), an application of Lebesgue's dominated convergence theorem (Theorem 1.7) completes the proof. ■

8.1.2 A Weak Version of the Dominated Convergence Theorem

Lebesgue's dominated convergence theorem (Theorem 1.7) guarantees that if a sequence $g_j \in L^p$ is bounded by an L^p function and converges to g almost everywhere then $g_j \to g$ in L^p. In the next result we weaken the condition, requiring only that the sequence $\{g_j\}$ is bounded in L^p, and obtain $g_j \rightharpoonup g$ in L^p.

Lemma 8.3. *Let $\mathcal{O}$ be a bounded open set in $\mathbb{R}^m$, and let g_j be a sequence of functions in $L^p(\mathcal{O})$ with*

$$\|g_j\|_{L^p(\mathcal{O})} \leq C \qquad \text{for all} \qquad j \in \mathbb{Z}^+.$$

If $g \in L^p(\mathcal{O})$ and $g_j \to g$ almost everywhere then $g_j \rightharpoonup g$ in $L^p(\mathcal{O})$.

Proof. For each n, set

$$E_n = \{x \mid x \in \mathcal{O}, \, |g_j(x) - g(x)| \leq 1 \text{ for all } j \geq n\}.$$

These sets E_n are increasing with n, and the measure of E_n increases to the measure of $\mathcal{O}$ as $n \to \infty$ (since $g_j \to g$ a.e.).

Now let Φ_n be the set of functions in $L^q(\mathcal{O})$ ($p^{-1} + q^{-1} = 1$) with support in E_n, and let

$$\Phi = \bigcup_{n=1}^{\infty} \Phi_n.$$

Φ is dense in $L^q(\mathcal{O})$; for $\phi \in L^q(\mathcal{O})$ take

$$\phi_n = \chi[E_n]\phi,$$

where $\chi[E]$ is the characteristic function of the set E. Then $\phi_n \to \phi$ a.e., and since

$$|\phi_n(x)| \leq |\phi(x)|,$$

Lebesgue's dominated convergence theorem gives $\phi_n \to \phi$ in $L^q(\mathcal{O})$.

Now, if we take $\phi \in \Phi$, then we can use Lebesgue's dominated convergence theorem again to show that

$$\int_{\mathcal{O}} \phi(g_j - g)\, dx \to 0 \qquad \text{as} \qquad j \to \infty.$$

Indeed, $\phi \in \Phi_{n_0}$ for some n_0, and then for $j \geq n_0$, we have

$$|\phi(x)(g_j(x) - g(x))| \leq |\phi(x)|,$$

dominating $\phi(g_j - g)$, and $\phi(g_j - g)$ tends to zero a.e.

Finally we use the density of Φ in $L^q(\mathcal{O})$; take $u \in L^q(\mathcal{O})$, and, given $\epsilon > 0$, choose $\phi \in \Phi$ such that

$$\|u - \phi\|_{L^q(\mathcal{O})} < \epsilon/4C$$

and M such that

$$\int_{\mathcal{O}} \phi(g_j - g)\, dx < \epsilon/2 \qquad \text{for all} \qquad j \geq M.$$

Then

$$\int_{\mathcal{O}} (g_j - g)(u - \phi + \phi)\, dx < 2C(\epsilon/4C) + \epsilon/2 = \epsilon,$$

which shows that $g_j \rightharpoonup g$ in $L^p(\mathcal{O})$, as claimed. ∎

8.2 The Basis for the Galerkin Expansion

As in the previous chapter we will write $H = L^2(\Omega)$ and $V = H_0^1(\Omega)$, and we will use $\|u\|^2$ to denote the "reduced" norm on $H_0^1(\Omega)$, which is the same as the bilinear form $a(u, u)$,

$$\|u\|^2 = a(u, u) = |\nabla u|^2.$$

We again use $\|\cdot\|_*$ for the norm on $V^* = H^{-1}(\Omega)$, and we will write Ω_T for $\Omega \times (0, T)$. Note that

$$L^r(0, T; L^r(\Omega)) = L^r(\Omega_T)$$

for any r.

For linear parabolic equations (in Chapter 7) we performed our Galerkin truncation by using the eigenfunctions of the Laplacian. We will do the same here, but now we want to ensure that the eigenfunctions are all elements of $L^p(\Omega)$, where p comes from the condition on f given in (8.2). Using the Sobolev embedding theorem (Theorem 5.31) there is an $s \in \mathbb{Z}^+$ such that

$$H^s(\Omega) \subset L^p(\Omega). \tag{8.7}$$

We need

$$s \geq m(p-2)/2p; \tag{8.8}$$

note that this depends on m. In particular, if $m = 1$ or $m = 2$ we can take $s = 1$, no matter what the value of p. This means that in one or two dimensions, $V \subset L^p(\Omega)$ for any p (we will come back to this later).

We now assume that Ω is a bounded C^s domain, with s chosen as in (8.8), and then Theorem 6.17 ensures that the eigenfunctions of the Laplacian are elements of $D(A^{s/2}) \subset H^s(\Omega)$, and so of $L^p(\Omega)$ by (8.7). This is the "sufficient smoothness" of Ω referred to at the beginning of the chapter.

As in the last chapter we will be projecting the equation onto the span of the first n eigenfunctions, so we define P_n as the orthogonal projection in L^2 onto the span of $\{w_1, \ldots, w_n\}$:

$$P_n u = \sum_{j=1}^{n} (u, w_j) w_j.$$

Lemma 7.5 shows that if $u \in L^2(\Omega)$ then we have $P_n u \to u$ in $L^2(\Omega)$. We need to strengthen this to show that if $u \in L^p(\Omega)$ is sufficiently smooth then $P_n u \to u$ in $L^p(\Omega)$. We can do this by showing that the eigenfunctions form a basis for $D(A^{s/2})$.

To prove that $\{w_j\}$ form a basis for $D(A^{s/2})$ it suffices to show that if

$$((u, w_j))_{D(A^{s/2})} = 0 \qquad \text{for all} \qquad j \tag{8.9}$$

then $u = 0$ (since then only 0 is orthogonal to the linear span of the $\{w_j\}$; cf. proof of Proposition 1.23). If (8.9) holds then

$$((u, w_j))_{D(A^{s/2})} = (A^{s/2}u, A^{s/2}w_j) = (u, A^s w_j) = \lambda_j^s (u, w_j),$$

and so $(u, w_j) = 0$ for all j. Since the $\{w_j\}$ form a basis for $L^2(\Omega)$ it follows that $u = 0$. In particular, the expansion

$$u = \sum_{j=1}^{\infty} (u, w_j) w_j \tag{8.10}$$

converges to u in $D(A^{s/2})$ and so in $H^s(\Omega)$ (using $\|u\|_{H^s} \leq C|A^{s/2}u|$ from Proposition 6.18). Using (8.7), we find that the expansion also converges in L^p. In particular, if $Q_n = I - P_n$, then

$$Q_n u \to 0 \qquad \text{in} \qquad L^p(\Omega), \quad u \in D(A^{s/2}). \tag{8.11}$$

Note also that, having chosen s so that $v \in H_0^s(\Omega)$ implies that $v \in L^p(\Omega)$, it follows that, if $u \in L^q(\Omega)$, the dual of $L^p(\Omega)$, then $u \in H^{-s}(\Omega)$, the dual of $H_0^s(\Omega)$, so that

$$L^q(\Omega) \subset H^{-s}(\Omega). \tag{8.12}$$

This will be important when we come to apply the compactness theorem

(Theorem 8.1) in this situation. (Again, in one and two dimensions this in fact shows that $L^q \subset V^*$ for any q.)

8.3 Weak Solutions

Apart from technical problems caused by the nonlinear term (we will need both results from Section 8.1 and will have to make careful estimates in some L^p spaces with $p \neq 2$) we proceed much as in the previous chapter. We will obtain a solution by using the basis $\{w_j\}$ to approximate the equation by ever larger systems of ODEs. Existence and uniqueness for these approximations will follow from the ODE theorem (Theorem 2.3). We will then take a (weak) limit by using the Alaoglu compactness theorem. Use of Theorem 8.1 and Lemma 8.3 will guarantee that we actually have a solution.

As in the previous chapter, we rewrite (8.1) as

$$du/dt + Au = f(u), \tag{8.13}$$

where $A = -\Delta$ with Dirichlet boundary conditions.

Theorem 8.4 (Weak solutions). *Equation (8.13), with f a C^1 function satisfying*

$$-k - \alpha_1|s|^p \leq f(s)s \leq k - \alpha_2|s|^p, \qquad p > 2, \tag{8.14}$$

and

$$f'(s) \leq l, \tag{8.15}$$

has a unique weak solution: for any $T > 0$ given $u_0 \in L^2(\Omega)$ there exists a solution u with

$$u \in L^2\big(0, T; H_0^1(\Omega)\big) \cap L^p(\Omega_T), \qquad u \in C^0([0, T]; L^2(\Omega)),$$

and $u_0 \mapsto u(t)$ is continuous on $L^2(\Omega)$. Equation (8.13) holds as an equality in $L^q(0, T; H^{-s}(\Omega))$, where q is conjugate to p from (8.14).

Since the equation holds as an equality in $L^q(0, T; H^{-s}(\Omega))$, this means (using Lemma 7.4) that for any $v \in L^p(0, T; H^s(\Omega))$, we have

$$\langle du/dt, v \rangle + a(u, v) = \langle f(u), v \rangle \tag{8.16}$$

for almost every $t \in [0, T]$. For $m = 1, 2$ we can take $s = 1$, and so the equation holds in the more familiar space $L^q(0, T; V^*)$, and (8.16) holds for all $v \in V$, as in the previous chapter.

Proof. As before, we consider the ordinary differential system

$$du_n/dt + Au_n = P_n f(u_n), \qquad u_n(0) = P_n u_0 \qquad (8.17)$$

for the n-dimensional approximation

$$u_n = \sum_{j=1}^{n} u_{nj} w_j.$$

Since the nonlinearity in (8.17) is locally Lipschitz, Theorem 2.3 shows that the finite-dimensional system has a unique solution on some finite time interval.

We have to show that the solutions are bounded in time and uniformly bounded in n. To obtain an L^2-type bound, multiply (8.17) by u_n and integrate. Noting that

$$(P_n f(u_n), u_n) = (f(u_n), P_n u_n) = (f(u_n), u_n),$$

we get

$$\tfrac{1}{2}\frac{d}{dt}|u_n|^2 + \|u_n\|^2 = \int_\Omega f(u_n) u_n \, dx.$$

Now we can use (8.14) to write

$$\tfrac{1}{2}\frac{d}{dt}|u_n|^2 + \|u_n\|^2 \le \int_\Omega k - \alpha_2 |u_n|^p \, dx. \qquad (8.18)$$

Integrating both sides between 0 and T gives

$$\tfrac{1}{2}|u_n(T)|^2 + \int_0^T \|u_n(t)\|^2 \, dt + \alpha_2 \int_0^T \int_\Omega |u_n(x,t)|^p \, dx \, dt \le \tfrac{1}{2}|u_0|^2 + kT|\Omega|,$$

where $|\Omega|$ is the measure of Ω, $|\Omega| = \int_\Omega dx$. It follows from the last inequality that

$$\sup_{t\in[0,T]} |u_n(t)|^2 \le 2K,$$

$$\int_0^T \|u(t)_n\|^2 \, dt \le K,$$

and

$$\int_0^T \int_\Omega |u_n|^p \, dx \, dt \le K/\alpha_2,$$

where $K = \tfrac{1}{2}|u_0|^2 + kT|\Omega|$, bounded for bounded sets of initial conditions (in H) and bounded time intervals. One can write these as

$$u_n \qquad \text{is uniformly bounded in} \qquad L^\infty(0, T; H),$$
$$u_n \qquad \text{is uniformly bounded in} \qquad L^2(0, T; V), \qquad (8.19)$$
$$u_n \qquad \text{is uniformly bounded in} \qquad L^p(\Omega_T).$$

We now use the bound on u_n in $L^p(\Omega_T)$ to obtain bounds on the nonlinear term $f(u_n)$. Since

$$|f(s)| \le \beta(|s|^{p-1} + 1),$$

from (8.14), the bound on u_n in $L^p(\Omega_T)$ gives a bound on $f(u_n)$ in $L^q(\Omega_T)$, where (p, q) are conjugate:

$$\|f(u_n)\|_{L^q(\Omega_T)}^q = \int_0^T \left(\int_\Omega |f(u_n)|^q \, dx \right) dt$$
$$\le \beta \int_0^T \left(\int_\Omega (|u_n|^{p-1} + 1)^q \, dx \right) dt$$
$$\le C \int_0^T \left(\int_\Omega |u_n|^{q(p-1)} + 1 \, dx \right) dt,$$

and since $p^{-1} + q^{-1} = 1$ it follows that $q(p-1) = p$ and so we have

$$f(u_n) \qquad \text{is uniformly bounded in} \qquad L^q(\Omega_T), \qquad (8.20)$$

as claimed.

Finally, we need a uniform bound on the derivative. Note first that both $L^2(0, T; V^*)$ and $L^q(0, T; L^q(\Omega))$ are continuously included in $L^q(0, T; H^{-s}(\Omega))$ (we have $q < 2$ since $p > 2$, and we have used (8.12)). It follows that since

$$du_n/dt = -Au_n + P_n f(u_n),$$

we have

$$du_n/dt \qquad \text{is uniformly bounded in} \qquad L^q(0, T; H^{-s}(\Omega)). \qquad (8.21)$$

We now use Corollary 4.19 to extract a weakly convergent subsequence, u_n, with

$$u_n \rightharpoonup u \qquad \text{in} \qquad L^2(0, T; V),$$
$$u_n \rightharpoonup u \qquad \text{in} \qquad L^p(\Omega_T),$$
$$f(u_n) \rightharpoonup \chi \qquad \text{in} \qquad L^q(\Omega_T).$$

With an application of the compactness theorem (Theorem 8.1) we can extract a further subsequence such that additionally

$$u_n \to u \quad \text{in} \quad L^2(0, T; H). \tag{8.22}$$

Indeed, we have that u_n is uniformly bounded in $L^2(0, T; H_0^1(\Omega))$ (in (8.19)) and that du_n/dt is uniformly bounded in $L^q(0, T; H^{-s}(\Omega))$ (in (8.21)). Since $H_0^1(\Omega) \subset\subset L^2(\Omega) \subset H^{-s}(\Omega)$ and $H_0^1(\Omega)$ is reflexive (it is a Hilbert space), we can apply the compactness theorem to deduce (8.22).

We have a little more to do, since we actually want $P_n f(u_n) \rightharpoonup \chi$ in $L^q(\Omega_T)$. However, we can write

$$\int_{\Omega_T} (P_n f(u_n) - \chi) \phi \, dx \, dt = \int_{\Omega_T} (f(u_n) - \chi) \phi \, dx \, dt$$

$$- \int_{\Omega_T} Q_n f(u_n) \phi \, dx \, dt$$

for all $\phi \in L^p(\Omega_T)$. We already know that the first term on the right-hand side here tends to zero. For the second term, we note that functions of the form

$$\phi = \sum_{j=1}^{n} \alpha_j(t) \phi_j$$

with $\alpha_j \in L^p(0, T)$ and $\phi_j \in C_c^\infty(\Omega)$ are dense in $L^p(\Omega_T)$ (see Exercise 7.3), and for such functions

$$\int_{\Omega_T} Q_n f(u_n) \left(\sum_{j=1}^{n} \alpha_j(t) \phi_j \right) dx \, dt = \int_{\Omega_T} f(u_n) \left(\sum_{j=1}^{n} \alpha_j(t) Q_n \phi_j \right) dx \, dt.$$

Since $Q_n \phi_j \to 0$ in $L^p(\Omega)$ for each j (see (8.11)) we have the required convergence of $P_n f(u_n)$.

It follows that all terms converge in the dual space of $L^2(0, T; V) \cap L^p(\Omega_T)$. This space is $L^2(0, T; V^*) + L^q(\Omega_T)$ (see Exercise 8.1); the equality

$$du/dt + Au = \chi \tag{8.23}$$

holds in this space. To make this more explicit, we have already used above the fact that $L^2(0, T; V^*) + L^q(\Omega_T) \subset L^q(0, T; H^{-s})$, so certainly (8.23) holds as an equality in this space.

It remains to show that $\chi = f(u)$; this is now straightforward using the strong convergence of u_n to u in $L^2(\Omega_T)$ and Lemma 8.3. Since $u_n \to u$ in $L^2(\Omega_T)$, Corollary 1.12 guarantees that there is a subsequence u_{n_j} such that $u_{n_j}(x, t) \to u(x, t)$ for almost every $(x, t) \in \Omega_T$. It follows, using the continuity of f, that $f(u_{n_j}(x, t)) \to f(u(x, t))$ for almost every $(x, t) \in \Omega_T$. Along with the bound on $f(u_{n_j})$ in $L^q(\Omega_T)$ given in (8.20), we can apply Lemma 8.3 to deduce that $f(u_{n_j}) \rightharpoonup f(u)$ in $L^q(\Omega_T)$. By the uniqueness of weak limits, it follows that $\chi = f(u)$.

To prove continuity of the solution $u(t)$ from $[0, T]$ into $L^2(\Omega)$, observe that $u \in L^2(0, T; V) \cap L^p(\Omega_T)$ and that

$$du/dt = -Au + f(u) \in L^2(0, T; V^*) + L^q(\Omega_T),$$

precisely the dual of $L^2(0, T; V) \cap L^p(\Omega_T)$. One can readily adapt the proof of Theorem 7.2 for this case (see Exercise 8.2) and deduce that $u \in C^0([0, T]; L^2(\Omega))$.

To show that $u(0) = u_0$ we use the same technique employed in the previous chapter in (7.25) and (7.26). Choosing some $\phi \in C^1([0, T]; V \cap L^p(\Omega))$ with $\phi(T) = 0$, observe that $\phi \in L^2(0, T; V) \cap L^p(\Omega_T)$, and so in the "limiting equation"

$$\langle du/dt, v \rangle + a(u, v) = \langle f(u), v \rangle,$$

we can integrate by parts in the t variable to give

$$\int_0^T -\langle u, \phi' \rangle + a(u, \phi)\, ds = \int_0^T \langle f(u(s)), \phi \rangle\, ds + (u(0), \phi(0)).$$

Doing the same in the Galerkin approximations yields

$$\int_0^T -\langle u_n, \phi' \rangle + a(u_n, \phi)\, ds = \int_0^T \langle P_n f(u_n(s)), \phi \rangle\, ds + (u_n(0), \phi(0)).$$

$$(8.24)$$

Since our whole argument relies on showing convergence of the terms in (8.24), we already know that we can take limits to conclude that

$$\int_0^T -\langle u, \phi' \rangle + a(u, \phi)\, ds = \int_0^T \langle f(u(s)), \phi \rangle\, ds + (u_0, \phi(0)),$$

since $u_n(0) = P_n u_0 \to u_0$. Thus $u(0) = u_0$.

To prove uniqueness and continuous dependence, let u_0 and v_0 be in H and consider $w(t) = u(t) - v(t)$. Then

$$\frac{\partial w}{\partial t} + Aw = f(u) - f(v), \qquad w(0) = u_0 - v_0,$$

and multiplying by w and integrating over Ω gives

$$\frac{1}{2}\frac{d}{dt}|w|^2 + \|w\|^2 = (f(u) - f(v), u - v).$$

Note that the bound on f' (8.15) shows that

$$
\begin{aligned}
(f(u) - f(v), u - v) &= \int_\Omega [f(u(x)) - f(v(x))](u(x) - v(x)) \, dx \\
&= \int_\Omega \left(\int_{v(x)}^{u(x)} f'(s) \, ds \right) (u(x) - v(x)) \, dx \\
&\leq \int_\Omega l |u(x) - v(x)|^2 \, dx \\
&= l|u - v|^2.
\end{aligned}
$$

We therefore obtain

$$
\tfrac{1}{2} \frac{d}{dt} |w|^2 \leq l|w|^2,
$$

and integrating this gives

$$
|u(t) - v(t)| \leq |u_0 - v_0| e^{lt}. \tag{8.25}
$$

This is uniqueness if $u_0 = v_0$ and is continuous dependence on initial conditions otherwise. ∎

8.3.1 A Semidynamical System on $L^2(\Omega)$

We have obtained existence and uniqueness of solutions and their continuous dependence on initial conditions. We can therefore use these solutions to define a semidynamical system on the phase space $L^2(\Omega)$ by setting

$$
S(t)u_0 = u(t).
$$

The continuity conditions we have obtained, along with the uniqueness, mean that this semigroup of operators $\{S(t)\}_{t \geq 0}$ satisfies

$$
S(0) = I,
$$

$$
S(t)S(s) = S(s)S(t) = S(s + t), \tag{8.26}
$$

$$
S(t)x_0 \text{ is continuous in } x_0 \text{ and } t.
$$

We call this a C^0 semigroup.

The resulting semidynamical system is

$$
(L^2(\Omega), \{S(t)\}_{t \geq 0}).
$$

We have shown that $L^2(\Omega)$ is an appropriate phase space in which to study the dynamics of the reaction–diffusion equation. It is worth emphasising that we can use this phase space without imposing any very strong conditions on the function f (only (8.14) and (8.15)).

In the next section we investigate strong solutions, and we will see there that, under the current weak conditions on f, it is only in the cases $m = 1$ and $m = 2$ that $H_0^1(\Omega)$ is a sensible alternative choice for a phase space.

8.4 Strong Solutions

We now prove a result similar to Theorem 7.7, and we show that if $u_0 \in V \cap L^p$ then $u(t)$ is in this space for all $t \geq 0$. Once again, we do not have to impose any restrictions on p to obtain the existence of this unique strong solution. However, we will take $f(0) = 0$ to simplify the argument, but this is not in fact necessary (see Exercise 8.3).

Theorem 8.5. *Let* $f(0) = 0$. *If* $u_0 \in V \cap L^p(\Omega)$ *then there exists a unique strong solution*

$$u(t) \in C^0([0, T]; V) \cap L^\infty(0, T; L^p(\Omega)) \cap L^2(0, T; D(A)).$$

Without further restrictions on p we cannot prove, for general m, that the map $u_0 \mapsto u(t)$ is continuous, although we can show this for $m \leq 3$ ($m \leq 2$ in Proposition 8.7, and $m = 3$ in Exercise 8.5). Note, however, that uniqueness follows from our previous Theorem 8.4, since a strong solution is automatically a weak solution.

Proof. We make some more estimates on the equation. First, we take the inner product of (8.17) with Au_n, so that (recall (7.28))

$$\frac{1}{2}\frac{d}{dt}\|u_n\|^2 + |Au_n|^2 = -\int_\Omega P_n f(u_n)\Delta u_n \, dx$$

$$= -\int_\Omega f(u_n)\Delta u_n \, dx$$

$$= -\int_\Omega \sum_{j=1}^m f(u_n)\frac{\partial^2 u_n}{\partial x_j^2} \, dx$$

$$= \int_\Omega \sum_{j=1}^m f'(u_n)\left|\frac{\partial u_n}{\partial x_j}\right|^2 dx + \int_{\partial\Omega} f(u_n)\nabla u_n \cdot n \, dS$$

$$= \int_\Omega \sum_{j=1}^m f'(u_n)\left|\frac{\partial u_n}{\partial x_j}\right|^2 dx,$$

using our assumption that $f(0) = 0$ (we know that $u_n = 0$ on $\partial\Omega$).

Therefore we have

$$\frac{1}{2}\frac{d}{dt}\|u_n\|^2 + |Au_n|^2 \le l\|u_n\|^2. \qquad (8.27)$$

Integrating both sides from 0 to T gives

$$\frac{1}{2}\|u_n(T)\|^2 + \int_0^T |Au_n(s)|^2\,ds \le l\int_0^T \|u_n(t)\|^2\,dt + \frac{1}{2}\|u_0\|^2,$$

and so u_n is uniformly bounded in $L^2(0, T; D(A))$ (and in $L^\infty(0, T; V)$), since we already know from (8.19) that $u_n \in L^2(0, T; V)$.

We now make a further estimate on du_n/dt. This time we multiply (8.17) by $\partial u_n/\partial t$ and integrate. Using (7.28) again and noting that

$$\left(P_n f, \frac{du_n}{dt}\right) = \left(f, P_n\frac{du_n}{dt}\right) = \left(f, \frac{du_n}{dt}\right),$$

we have

$$\left|\frac{du_n}{dt}\right|^2 + \frac{d}{dt}\left(\frac{1}{2}\|u_n\|^2 - \int_\Omega \mathcal{F}(u_n)\,dx\right) = 0, \qquad (8.28)$$

where $\mathcal{F}(s) = \int_0^s f(\sigma)\,d\sigma$. Integrating the above equation from 0 to t gives

$$\int_0^t \left|\frac{du_n}{dt}\right|^2 ds + \frac{1}{2}\|u_n(t)\|^2 - \int_\Omega \mathcal{F}(u_n(t))\,dx \le \frac{1}{2}\|u_0\|^2 + \int_\Omega F(u_n(0))\,dx,$$

and using

$$-\kappa - \tilde{\alpha}_1|s|^p \le \mathcal{F}(s) \le \kappa - \tilde{\alpha}_2|s|^p$$

(from (8.14)), we obtain

$$\int_0^t \left|\frac{du_n}{dt}\right|^2 + \frac{1}{2}\|u_n(t)\|^2 + \tilde{\alpha}_2 \int_\Omega |u_n(t)|^p\,dx$$

$$\le 2\kappa|\Omega| + \frac{1}{2}\|u_0\|^2 + \tilde{\alpha}_1 \int_\Omega |u_0|^p\,dx.$$

It follows that du_n/dt is uniformly bounded in $L^2(0, T; H)$ and that u_n is uniformly bounded in $L^\infty(0, T; L^p(\Omega))$.

By extracting the appropriate subsequence we now have

$$u \in L^\infty(0, T; L^p(\Omega)), \quad u \in L^2(0, T; D(A)), \text{ and } du/dt \in L^2(0, T; H),$$

from which Corollary 7.3 gives $u \in C^0([0, T]; V)$. ∎

Note that the result does not pick out a natural phase space for the problem unless $m \le 2$. We need to take u_0 in both V *and* $L^p(\Omega)$ to ensure continuity into V; but this does not give continuity into $L^p(\Omega)$. In general it is only when

$$p \le 2m/(m-2)$$

that we can use the Sobolev embedding theorem (Theorem 5.31) to guarantee that $V \subset L^p(\Omega)$. Thus it is only in this case that $H_0^1(\Omega)$ becomes a good candidate for a phase space.

For this reason we prove continuity with respect to initial conditions in the V norm only in the cases $m = 1$ and $m = 2$. (A similar result for $m = 3$ is given in Exercise 8.5.)

The idea of the proof is that although $f : \mathbb{R} \to \mathbb{R}$, it induces a map F on functions via

$$F[u](x) = f(u(x)).$$

We investigate the smoothness of this map F, and we show that it is Lipschitz continuous from H^1 into L^2.

Proposition 8.6. *Suppose that $f : \mathbb{R} \to \mathbb{R}$ is a C^1 function satisfying*

$$|f'(s)| \le C(1 + |s|^\gamma) \tag{8.29}$$

for some γ, and let $\Omega \subset \mathbb{R}^m$ with $m = 1$ or 2. Then for all $u, v \in H^1(\Omega)$ we have

$$|F(u) - F(v)|_{L^2(\Omega)} \le C[1 + \|u\|_{H^1} + \|u\|_{H^1}]^{1/2} \|u - v\|_{H^1}. \tag{8.30}$$

Although we would expect $\gamma = p - 2$, with p from (8.14), we are in fact introducing another restriction on the function f in (8.29).

Proof. The proof relies on an application of Hölder's inequality and use of the Sobolev embedding Theorem 5.31, which guarantees that

$$H^1(\Omega) \subset L^p(\Omega)$$

for all p when $m = 1$ or 2. We start with the simple equality

$$|F(u) - F(v)|^2 = \int_\Omega \left(f(u(x)) - f(v(x)) \right)^2 dx$$

and then estimate the right-hand side, using C to denote any constant that does not depend on u and v:

$$|F(u) - F(v)|^2 = \int_\Omega \left(\int_{v(x)}^{u(x)} f'(s) \, ds \right)^2 dx$$

$$\leq C \int_\Omega |u(x) - v(x)|^2 (1 + |u(x)|^\gamma + |v(x)|^\gamma)^2 \, dx$$

$$\leq C \int_\Omega |u(x) - v(x)|^2 (1 + |u(x)|^{2\gamma} + |v(x)|^{2\gamma}) \, dx$$

$$\leq C |u - v|^2_{L^{2p}} \left(1 + |u|^2_{L^{2q\gamma}} + |v|^2_{L^{2q\gamma}} \right).$$

Since we can replace the norm in L^p with a constant times the H^1 norm, we obtain (8.30) as required. ∎

We can use this to prove the continuous dependence on initial data.

Proposition 8.7. *If $m = 1$ or $m = 2$ then, for each fixed $t \geq 0$, the map $u_0 \mapsto u(t)$ is continuous from V into V.*

Proof. We know that $u \in L^2(0, T; D(A))$ and that $du/dt \in L^2(0, T; H)$, so we can use Corollary 7.3 to take the inner product of

$$\frac{d}{dt} w + Aw = f(u) - f(v)$$

with Aw to obtain

$$\tfrac{1}{2} \frac{d}{dt} \|w\|^2 + |Aw|^2 = (F(u) - F(v), Aw)$$

$$\leq |F(u) - F(v)||Aw|.$$

We use (8.30) to bound $|F(u) - F(v)|$ in terms of the H^1 norm of u and v,

$$\tfrac{1}{2} \frac{d}{dt} \|w\|^2 + |Aw|^2 \leq \tfrac{1}{2} C(t) \|u - v\|^2 + \tfrac{1}{2} |Aw|^2,$$

where $C(t) = C(1 + \|u\| + \|v\|) \in L^\infty(0, T)$ since u and v are in $L^\infty(0, T; V)$ (Theorem 8.5). Therefore we have

$$\frac{d}{dt} \|w\|^2 \leq C(t)^2 \|w\|^2$$

from which it follows, by using Gronwall's inequality (Lemma 2.8), that

$$\|w(t)\|^2 \leq \exp\left(\int_0^s C(s)^2 \, ds \right) \|w(0)\|^2,$$

which gives continuous dependence on initial conditions as required. ∎

It follows that for $m = 1$ and $m = 2$ we can take $H_0^1(\Omega)$ as a sensible phase space for the reaction–diffusion equation, and along with the semigroup $\tilde{S}(t) : V \to V$ given by $\tilde{S}(t)u_0 = u(t)$,

$$\left(H_0^1(\Omega), \{\tilde{S}(t)\}_{t \geq 0}\right)$$

is a semidynamical system. Note that the uniqueness of weak solutions shows that in fact this is the restriction of the semidynamical system

$$\left(L^2(\Omega), \{S(t)\}_{t \geq 0}\right)$$

to $H_0^1(\Omega)$. We will see in Chapter 11 that even in this case ($m = 1, 2$) it is still more convenient to work with the weak solutions and the semidynamical system they induce on $L^2(\Omega)$ than with these more regular solutions.

Exercises

8.1 Show that the dual space of $X \cap Y$ (with norm $\|u\|_X + \|u\|_Y$) is $X^* + Y^*$. [This shows that the dual space of $L^2(0, T; V) \cap L^p(\Omega_T)$ is $L^2(0, T; V^*) + L^q(\Omega_T)$.]

8.2 Adapt the argument of Theorem 7.2 to show that if $u \in L^2(0, T; H^1) \cap L^p(\Omega_T)$ and $du/dt \in L^2(0, T; H^{-1}) + L^q(\Omega_T)$ then $u \in C^0([0, T]; L^2)$.

8.3 In the case that $f(0) \neq 0$, show that we still have u_n uniformly bounded in $L^2(0, T; D(A))$, as in Theorem 8.5. [Hint: Use the trace theorem (Theorem 5.35) and the result $\|u\|_{H^2(\Omega)} \leq C|Au|$ from Theorem 6.16.]

8.4 Show that if $m = 3$ then the conclusion of Proposition 8.6 holds provided that $\gamma \leq 2$.

8.5 If $m = 3$, a version of the proof of Proposition 8.6 shows that

$$|F[u] - F[v]| \leq C(1 + |Au| + |Av|)^{1/2}\|u - v\|^{1/2}|Au - Av|^{1/2}. \tag{8.31}$$

Use this, along with the fact that $u, v \in L^2(0, T; D(A))$, to show that the map $u_0 \mapsto u(t)$ is continuous from V into V for each fixed $t \geq 0$.

8.6 A different method of proving existence and uniqueness relies on applying a contraction mapping argument to the integral form of

$$du/dt + Au = f(u) \tag{8.32}$$

(cf. proof of ODE existence, Theorem 2.3). By considering the derivative with respect to s of

$$g(s) = e^{-A(t-s)}u(s),$$

where $u(t)$ is a strong solution of (8.32) and e^{-At} is the exponential operator defined in Exercise 3.11, show that $u(t)$ satisfies

$$u(t) = e^{-At}u_0 + \int_0^t e^{-A(t-s)} f(u(s)) \, ds. \qquad (8.33)$$

This is known as the "variation of constants formula." (Hint: Use the result of Exercise 3.11.)

Notes

It is possible to treat more complicated models than the simple scalar equation we have covered here. For multicomponent equations, for example

$$\frac{\partial u}{\partial t} - \Delta u = f(u, v),$$

$$\frac{\partial v}{\partial t} - \nu \Delta v = g(u, v),$$

we could follow a similar analysis with appropriate conditions on f and g. Another common problem is for multicomponent equations with invariant regions. For example, if u_i, $i = 1, \ldots, n$ denote the concentrations of n chemicals then we would certainly expect $u_i(x) \geq 0$. Many "popular" models in fact have bounded invariant regions. Different methods can be used to analyse these equations, and Temam (1988) and Marion (1987) both outline approaches to proving existence and uniqueness in this situation. For a more general treatment see Smoller (1983).

We have chosen in this chapter to concentrate mainly on weak solutions of our simple equation, which generate a semigroup on $L^2(\Omega)$. To obtain these weak solutions we need to use the Galerkin method, the application of which as it appears here is taken essentially from Marion (1987). Proofs of the two results from functional analysis in Section 8.1 can be found in Lions (1969), although the proof given of the compactness theorem (Theorem 8.1) is taken from Temam (1984).

We have not produced a satisfactory theory of strong solutions that is valid in dimensions higher than two. In fact we need to make additional assumptions on f to ensure the existence of a semigroup on V in these higher dimensions; for $m = 3$ we need to impose

$$|f'(s)| \leq c(1 + |s|^\gamma) \qquad \text{with} \qquad 1 \leq \gamma \leq 2,$$

which in fact ensures that Proposition 8.6 still holds, as shown in Exercise 8.4. Properties of the semigroup on $H_0^1(\Omega)$ are discussed in more detail in Hale (1988).

Evans (1998) treats the problem of this chapter by using a contraction mapping argument based on Theorem 7.7 and Proposition 8.6. If $v(t) \in C^0([0, T]; V)$ then

$[h(t)](x) = f(v(x, t)) \in L^2(0, T; H)$, and so Theorem 7.7 shows that the equation

$$du/dt + Au = h(t)$$

has a solution $u \in C^0([0, T]; V)$. By showing that the map $v \mapsto u$ is a contraction on a subset of $C^0([0, T]; V)$ one can deduce the existence of a strong solution.

Henry (1984), Taira (1995), and Cazenave & Haraux (1998) use a different contraction mapping approach, based on the fractional power spaces introduced at the end of Chapter 3, and on the "variation of constants formula" from Exercise 8.6, which gives an implicit solution of the equation

$$du/dt + Au = f(u), \qquad u(0) = u_0$$

as

$$u(t) = e^{-At}u_0 + \int_0^t e^{-A(t-s)} f(u(s)) \, ds$$

(cf. the integral form of the ODE $\dot{x} = f(x)$ in Lemma 2.1). The standard version of this approach does not allow us to construct the semigroup on $L^2(\Omega)$ obtained in this chapter (but see Arrieta & Carvalho (1999) for a more refined theory that does).

9

The Navier–Stokes Equations
Existence and Uniqueness

The Navier–Stokes equations are the fundamental model used for problems involving the flow of fluids. Despite their great physical importance, existence and uniqueness results for the equations in the three-dimensional (3D) case are still not known, and only the two-dimensional (2D) situation is amenable to a complete mathematical treatment. This problem of "regularity" for the 3D equations is one of the outstanding issues in the theory of PDEs.

For an incompressible (constant density) fluid, the Navier–Stokes equations determine the velocity, $u(x, t)$, and the scalar pressure, $p(x, t)$, at each point in the domain Ω. For a fluid with constant density (we take this density as $\rho = 1$, and we can always treat this case by an appropriate rescaling of the variables), the equations are

$$u_t - \nu \Delta u + (u \cdot \nabla)u + \nabla p = f(x, t), \qquad \nabla \cdot u = 0, \qquad (9.1)$$

where the parameter $\nu > 0$ is the kinematic viscosity. The evolution equation is essentially Newton's second law $F = ma$, and $\nabla \cdot u = 0$ is the mass conservation equation $\nabla \cdot (\rho u) = 0$ for an incompressible fluid. For a much more thorough treatment of the physical background leading to these equations, see Doering & Gibbon (1995).

The pressure term in the evolution equation is such as to ensure that $\nabla \cdot u$ remains zero as the flow evolves in time. We will see that by restricting to a phase space in which all elements are divergence-free we can (in some sense) remove this pressure term and so simplify the analysis.

We will consider Equation (9.1) in the most simple case, on a domain $Q = [0, L]^m$ (with $m = 2$ or 3), and impose periodic boundary conditions. Although this corresponds to no physically realistic situation,* the mathematical treatment

---

* cf. Temam (1985): "... it is interesting to consider another boundary condition which has no physical meaning."

234

is much simplified. Nonetheless, the problems that obstruct a 3D existence proof for Dirichlet boundary conditions are also present in this case, and all the analysis we will perform below can be extended to the more physically relevant case of Dirichlet boundary conditions on a smooth domain Ω ($u = 0$ on $\partial\Omega$).

We will also choose our initial conditions and the forcing term so that we can work in a space of periodic functions with zero integral, allowing us to make use of the Poincaré inequality (see Lemma 5.40)

$$|u| \le \lambda_1^{-1/2} |\nabla u|, \quad \text{with} \quad \lambda_1 = \frac{4\pi^2}{L^2}. \tag{9.2}$$

We will assume that

$$\int_Q u_0(x)\,dx = 0 \quad \text{and} \quad \int_Q f(x,t) = 0 \quad \text{for all} \quad t \ge 0.$$

It follows from the evolution equation that

$$\frac{d}{dt}\int_Q u_i\,dx = \int_Q \left[\sum_j \nu D_j(D_j u_i) - u_j D_j u_i - D_i p + f \right] dx.$$

Since $\int_Q D_j g = 0$ when g is periodic, after an integration by parts the right-hand side reduces to

$$\int_Q \sum_j (D_j u_j) u_i \, dx,$$

which is zero since $\nabla \cdot u = 0$. Therefore we have

$$\int_Q u(x,t)\,dx = 0 \tag{9.3}$$

for all $t \ge 0$, and we can work consistently in spaces such as $\dot{L}^2(Q)$.

Even with these simplifying assumptions, the problem of existence and uniqueness of solutions in the 3D case is still unresolved in general (but has been dealt with successfully in the presence of certain symmetries; see Notes at the end of this chapter).

9.1 The Stokes Operator

We first analyse the "Stokes problem," which we obtain by neglecting all the nonlinear and time-dependent terms in (9.1),

$$-\nu\Delta u + \nabla p = f, \qquad \nabla \cdot u = 0. \tag{9.4}$$

Once again, the problem consists of two coupled equations, one involving the velocity u and the pressure p, and the other (the incompressibility condition) involving only the velocity.

Since we are considering the case of periodic boundary conditions, we can use the simple Fourier series approach that we used for the Laplacian in Chapter 6, bearing in mind that in this case the function u is not a scalar, but a vector-valued function.

We first define a space of smooth functions that incorporates the periodicity and the divergence-free condition,

$$\mathbb{V} = \left\{ u \in \left[\dot{C}_{\mathrm{p}}^{\infty}(Q) \right]^{m} : \nabla \cdot u = 0 \right\};$$

that is, $\mathbb{V}$ consists of m-component divergence-free vectors, each component of which is in $\dot{C}_{\mathrm{p}}^{\infty}(Q)$. Since we will be using spaces such as $\mathbb{V}$ consisting of m-vectors throughout this chapter, we will use the notation

$$\mathbb{L}^{2}(Q) = [L^{2}(Q)]^{m} \qquad \text{and} \qquad \mathbb{H}_{\mathrm{p}}^{k}(Q) = \left[H_{\mathrm{p}}^{k}(Q) \right]^{m},$$

where, for example, the norm of $(u_1, \ldots, u_m) \in \mathbb{L}^2(Q)$ is

$$\|u\|_{\mathbb{L}^2(Q)}^2 = \sum_{j=1}^{m} |u_j|_{L^2(Q)}^2.$$

We now take the inner product of the first equation in (9.4) with an element v of $\mathbb{V}$, to obtain

$$v a(u, v) + \int_{Q} \nabla p \cdot v \, dx = (f, v),$$

where $a(u, v)$ is the bilinear form we used before:

$$a(u, v) = \int_{Q} \nabla u \cdot \nabla v \, dx.$$

Now, if we integrate the p term by parts we obtain

$$\int_{Q} \nabla p \cdot v \, dx = \int_{Q} p(\nabla \cdot v) \, dx = 0, \tag{9.5}$$

since $v \in \mathbb{V}$ and so $\nabla \cdot v = 0$. The pressure term has dropped out, and we are left with

$$v a(u, v) = (f, v)$$

for all $v \in \mathbb{V}$. Now, we know that $\mathbb{V}$ is dense in

$$V = \left\{ u \in \mathbb{H}_{\mathrm{p}}^{1}(Q) : \nabla \cdot u = 0 \right\}, \tag{9.6}$$

and so we in fact want (9.7) to hold for all $v \in V$. In this form, observe that the equation makes sense for $f \in \mathbb{H}^{-1}(Q)$,

$$va(u, v) = \langle f, v \rangle. \tag{9.7}$$

Defining the linear "Stokes operator" A from V into V^* by

$$\langle Au, v \rangle = a(u, v) \qquad \text{for all} \qquad v \in V,$$

we obtain

$$vAu = f.$$

Equation (9.7) is in exactly the same form as we derived for Poisson's equation in Chapter 6, and so it follows as there that for every $f \in \mathbb{H}^{-1}(Q)$ there exists a unique solution u that lies in V.

To investigate further regularity for the Stokes problem we want to consider the smoothness of u when $f \in \dot{\mathbb{L}}^2(Q)$. As in Section 6.3, we will use the Fourier expansion. We take $v = 1$ and solve

$$Au = f,$$

expanding f as

$$f = \sum_{k \in \mathbb{Z}^m} e^{2\pi i k \cdot x / L} f_k,$$

where now each $f_k \in \mathbb{R}^m$, and we have $f_0 = 0$ and $\sum_k |f_k|^2 < \infty$ (since $f \in \dot{\mathbb{L}}^2(Q)$). If we similarly expand u as

$$u = \sum_{k \in \mathbb{Z}^m} e^{2\pi i k \cdot x / L} u_k,$$

(with $u_k \in \mathbb{R}^m$) and p as

$$p = \sum_{k \in \mathbb{Z}^m} e^{2\pi i k \cdot x / L} p_k$$

(note that p is a scalar, so $p_k \in \mathbb{R}$), we have

$$-\Delta u = \frac{4\pi^2}{L^2} \sum_{k \in \mathbb{Z}^m} e^{2\pi i k \cdot x / L} |k|^2 u_k$$

and

$$\nabla p = \frac{2\pi}{L} \sum_{k \in \mathbb{Z}^m} i k e^{2\pi i k \cdot x / L} p_k.$$

Equating coefficients in (9.4) now gives

$$\frac{4\pi^2|k|^2}{L^2}u_k - \frac{2\pi ik}{L}p_k = f_k, \tag{9.8}$$

and the divergence-free condition ($\nabla \cdot u = 0$) becomes

$$k \cdot u_k = 0. \tag{9.9}$$

Taking the scalar product of (9.8) with k gives an expression for p_k when $k \neq 0$ (see also Exercise 9.1):

$$p_k = \frac{L}{2\pi i}\frac{(f_k \cdot k)}{|k|^2}, \tag{9.10}$$

and so one obtains the following expression for u_k when $k \neq 0$:

$$u_k = \frac{L^2}{4\pi^2|k|^2}\left(f_k - \frac{k(k \cdot f_k)}{|k|^2}\right).$$

To fix u and p we set $u_0 = 0$ and $p_0 = 0$, which is fine since we required $f_0 = 0$ in the first place (and we want $u_0 \in V$).

It follows (cf. Theorem 6.5) that

$$\|u\|^2_{\dot{\mathbb{H}}_p^{s+2}} \leq C\|f\|^2_{\dot{\mathbb{H}}_p^s}, \tag{9.11}$$

and so in particular if $f \in \dot{\mathbb{L}}^2(Q)$ then $u \in \dot{\mathbb{H}}_p^2(Q)$. Thus the domain of A is given by

$$D(A) = \left\{u \in \dot{\mathbb{H}}_p^2(Q) : \nabla \cdot u = 0\right\}.$$

It is easy to see that we also have

$$D(A^{s/2}) = \left\{\dot{\mathbb{H}}_p^s(Q) : \nabla \cdot u = 0\right\} \tag{9.12}$$

and that

$$|A^{s/2}u| = C\|u\|_{\dot{\mathbb{H}}_p^s(Q)}, \qquad u \in D(A^{s/2}). \tag{9.13}$$

Finally, if we define

$$H = \{u \in \dot{\mathbb{L}}^2(Q) : \nabla \cdot u = 0\}, \tag{9.14}$$

that is, the space of all $u \in \dot{\mathbb{L}}^2(Q)$ whose Fourier coefficients satisfy (9.9), then,

if $f \in H$, the solution of the Stokes problem is given by the function u with
Fourier coefficients

$$u_k = \frac{L^2}{4\pi^2|k|^2} f_k.$$

This expression is exactly the same as (6.12), which gives the solution of
Poisson's equation. So we can deduce that

$$Au = -\Delta u \qquad \text{for all} \qquad u \in D(A). \tag{9.15}$$

Note that this particular result (9.15) is true only in the case of periodic boundary
conditions.

We can use the Hilbert–Schmidt theorem (Corollary 3.26) to deduce that A
has a set of orthonormal eigenfunctions w_j (the sine and cosine Fourier modes)
and corresponding eigenvalues λ_j,

$$Aw_j = \lambda_j w_j,$$

which we can order so that

$$\lambda_{j+1} \geq \lambda_j.$$

It follows as in Corollary 6.7 that $w_j \in \dot{C}_p^\infty(Q)$.

9.2 The Weak Form of the Navier–Stokes Equation

We have already chosen a suitable phase space for the full evolutionary Navier–
Stokes equations in our discussion of the Stokes problem, the space H defined
in (9.14). Again we will find that by incorporating the divergence-free condition
into the definition of H we will eliminate the pressure from our equation. We
will use $|\cdot|$ for the norm on H, which is just the $\mathbb{L}^2$ norm.

If $u(x, t)$ is a smooth solution of (9.1), we can take the inner product of (9.1)
with a function $v \in V$. Then

$$\left(\frac{\partial u}{\partial t}, v\right) - \nu \int_Q \Delta u\, v + \int_Q (u \cdot \nabla u)v + \int_Q (\nabla p)v = \int_Q fv.$$

Now, integrating the second term by parts gives the familiar bilinear form

$$\sum_{j=1}^m \int_Q \frac{\partial u}{\partial x_j} \frac{\partial v}{\partial x_j}\, dx = a(u, v) \tag{9.16}$$

(the boundary contribution is zero because of the periodic boundary conditions).

Integrating the pressure term by parts gives zero, as in (9.5). We have thus "removed" the pressure term and obtained

$$\left(\frac{\partial u}{\partial t}, v\right) + va(u, v) + \int_Q (u \cdot \nabla u)v \, dx = \int_Q fv \, dx \qquad \text{for all} \qquad v \in \mathbb{V}.$$

(9.17)

One can check that if $u(t)$ is in $[C_{\mathrm{p}}^2(Q)]^m$ then (9.17) implies that u satisfies the original Navier–Stokes equation (cf. Exercise 6.1).

Dealing with the nonlinear term in a similar way to (9.16), we define a trilinear form b by

$$b(u, v, w) = \sum_{i,j=1}^n \int_Q u_i \frac{\partial v_j}{\partial x_i} w_j \, dx.$$

We can then write (9.17) as

$$\left(\frac{\partial u}{\partial t}, v\right) + va(u, v) + b(u, u, v) = (f, v) \qquad \text{for all} \qquad v \in \mathbb{V}.$$

Defining V as in (9.6),

$$V = \{u \in \mathbb{\dot{H}}^1(Q) : \nabla \cdot u = 0\},$$

we can now use the density of $\mathbb{V}$ in V to require this equation to hold for every $v \in V$,

$$\left(\frac{\partial u}{\partial t}, v\right) + va(u, v) + b(u, u, v) = \langle f, v \rangle \qquad \text{for all} \qquad v \in V.$$

Since functions in V have zero integral over Q, we have the Poincaré inequality (9.2), and so we can use

$$\|u\|^2 = \sum_{j=1}^m \left|\frac{\partial u}{\partial x_j}\right|^2$$

as a norm on V, with the corresponding inner product

$$((u, v)) = \sum_{j=1}^m \left(\frac{\partial u}{\partial x_j}, \frac{\partial v}{\partial x_j}\right).$$

As in the previous two chapters we have $a(u, v) = ((u, v))$.

Now, the linear operator A from V into V^* associated with the bilinear form $((u, v))$,

$$\langle Au, v \rangle = ((u, v)) \qquad \text{for all} \qquad v \in V,$$

(9.18)

is the Stokes operator discussed in Section 9.1. By analogy with this we can define a bilinear operator $B(u, v)$ from $V \times V$ into V^* that is associated with $b(u, v, w)$, setting

$$\langle B(u, v), w \rangle = b(u, v, w) \qquad \text{for all} \qquad w \in V. \tag{9.19}$$

We can then rewrite the equation

$$\left(\frac{\partial u}{\partial t}, v \right) + v((u, v)) + b(u, u, v) = \langle f, v \rangle \qquad \text{for all} \qquad v \in V$$

as

$$\frac{du}{dt} + vAu + B(u, u) = f, \tag{9.20}$$

using (9.18) and (9.19). If we assume that $f \in V^*$, then we can expect that (9.20) will hold as an equality in V^* for almost every $t \in [0, T]$.

To analyse (9.20) we will need various properties of the nonlinear term $b(u, v, w)$. We derive these in the next section before proceeding to the questions of existence and uniqueness of weak solutions.

9.3 Properties of the Trilinear Form

We start off with some elementary identities.

Proposition 9.1. *If $m = 2$ or $m = 3$, then for $u \in H$, $v, w \in V$,*

$$b(u, v, w) = -b(u, w, v), \tag{9.21}$$

whence the orthogonality relation

$$b(u, v, v) = 0. \tag{9.22}$$

For the case $m = 2$ (and only with periodic boundary conditions)

$$b(u, u, Au) = 0 \qquad \text{for all} \qquad u \in D(A) \tag{9.23}$$

and hence (by differentiation)

$$b(v, u, Au) + b(u, v, Au) + b(u, u, Av) = 0 \qquad \text{for all} \qquad u, v \in D(A). \tag{9.24}$$

Proof. Equation (9.21) follows from an integration by parts. Taking $u, v, w \in \mathbb{V}$, we have

$$b(u, v, w) = \int_Q \sum_{i,j=1}^m u_i (D_i v_j) w_j \, dx$$

$$= -\int_Q \sum_{i,j=1}^m D_i (u_i w_j) v_j \, dx$$

$$= -\int_Q \sum_{i,j=1}^m (D_i u_i) w_j v_j + u_i (D_i w_j) v_j \, dx$$

$$= -b(u, w, v),$$

since $\sum_i D_i u_i = \nabla \cdot u = 0$. Equation (9.21) follows immediately using the density of $\mathbb{V}$ in H and V.

We now restrict to the case $m = 2$. To prove (9.23), note that if $u \in D(A)$ then $Au = -\Delta u$ (9.15), and so

$$b(u, u, Au) = -\sum_{i,j=1}^2 \int_Q u_i (D_i u_j) \Delta u_j \, dx$$

$$= -\sum_{i,j,k=1}^2 \int_Q u_i (D_i u_j) D_k^2 u_j \, dx.$$

An integration by parts yields

$$b(u, u, Au) = -\sum_{i,j,k=1}^2 \int_Q u_i D_{ik} u_j D_k u_j \, dx + \sum_{i,j,k=1}^2 \int_Q D_k u_i D_i u_j D_k u_j \, dx.$$

Both integrals vanish, the first because

$$\sum_{i=1}^2 \int_Q u_i D_{ik} u_j D_k u_j \, dx = \tfrac{1}{2} \sum_{i=1}^2 \int_Q u_i D_i (D_k u_j)^2$$

$$= -\tfrac{1}{2} \int_Q (\nabla \cdot u)(D_k u_j)^2 \, dx,$$

and the second because, writing $u_{i,j} = D_j u_i$, we find that the sum is given by

$$\sum_{i,j,k=1}^2 D_k u_i D_i u_j D_k u_j$$

$$= u_{1,1}^3 + u_{1,1} u_{1,2}^2 + u_{1,1} u_{2,1}^2 + u_{1,2} u_{2,1} u_{2,2}$$

$$\quad + u_{2,1} u_{1,2} u_{1,1} + u_{2,2} u_{1,2}^2 + u_{2,1}^2 u_{2,2} + u_{2,2}^3$$

$$= u_{1,1}^3 + u_{2,2}^3 = (u_{1,1} + u_{2,2})(u_{1,1}^2 - u_{1,1} u_{2,2} + u_{2,2}^2)$$

$$= 0,$$

by using $\nabla \cdot u = u_{1,1} + u_{2,2} = 0$ repeatedly.

To obtain (9.24), we use the identity (9.23) with $w = u + \epsilon v$ to get

$$\epsilon\left[b(v, u, Au) + b(u, v, Au) + b(u, u, Av)\right]$$
$$+ \epsilon^2\left[b(u, v, Av) + b(v, u, Av) + b(v, v, Au)\right] = 0;$$

we can equate terms of $O(\epsilon)$ to obtain (9.24). ∎

Now we turn to inequalities for b. The proof of these is straightforward but unexciting, involving repeated application of Hölder's inequality and results from the Sobolev embedding theorem (Theorem 5.31). We omit some of the details from the proof, leaving them as exercises.

Proposition 9.2. *If $m = 2$ or $m = 3$, then*

$$|b(u, v, w)| \le \|u\|_\infty\|v\|\|w|, \qquad u \in L^\infty, \; v \in V, \; w \in H. \tag{9.25}$$

If $u, v, w \in V$ then

$$|b(u, v, w)| \le k \begin{cases} |u|^{1/2}\|u\|^{1/2}\|v\||w|^{1/2}\|w\|^{1/2}, & m = 2, \\ |u|^{1/4}\|u\|^{3/4}\|v\||w|^{1/4}\|w\|^{3/4}, & m = 3, \end{cases} \tag{9.26}$$

and if $u \in V$, $v \in D(A)$, and $w \in H$,

$$|b(u, v, w)| \le k \begin{cases} |u|^{1/2}\|u\|^{1/2}\|v\|^{1/2}|Av|^{1/2}|w|, & m = 2, \\ \|u\|\|v\|^{1/2}|Av|^{1/2}|w|, & m = 3. \end{cases} \tag{9.27}$$

Proof. All the inequalities follow from an application of the generalised version of Hölder's inequality proved in Exercise 1.8,

$$\left|\int_Q fgh\,dx\right| \le \|f\|_{L^p}\|g\|_{L^q}\|h\|_{L^r},$$

with $p^{-1} + q^{-1} + r^{-1} = 1$. Applied to $b(u, v, w)$ this yields

$$|b(u, v, w)| \le \sum_{i,j=1}^m \int_Q |u_i(D_i v_j)w_j|\,dx$$
$$\le \sum_{i,j=1}^m \|u_i\|_{L^p}\|D_i v_j\|_{L^q}\|w_j\|_{L^r}.$$

Now, choose $(p, q, r) = (\infty, 2, 2)$ to obtain (9.25) after a further application of the Cauchy–Schwarz inequality. If we choose $(p, q, r) = (4, 2, 4)$ and use

"Ladyzhenskaya's inequalities" (see Lemma 5.27 for $m = 2$ and Exercise 9.2 for $m = 3$)

$$\|u_i\|_{L^4} \leq \begin{cases} C|u_i|^{1/2}\|u_i\|^{1/2}, & m = 2, \\ C|u_i|^{1/4}\|u_i\|^{3/4}, & m = 3, \end{cases} \tag{9.28}$$

we obtain (9.26).

Finally, to obtain (9.27) with $m = 2$ take $(p, q, r) = (4, 4, 2)$ and use (9.28) again, along with the Cauchy–Schwarz inequality (Exercise 9.3). When $m = 3$, take $(p, q, r) = (6, 3, 2)$ to obtain

$$|b(u, v, w)| \leq \sum_{i,j=1}^{3} \|u_i\|_{L^6}\|D_i v_j\|_{L^3}|w_j|.$$

Applying the Sobolev embedding $H^1 \subset L^6$ from Theorem 5.31 and using the result

$$\|u\|_{L^3} \leq C|u|^{1/2}\|u\|^{1/2}$$

from Exercise 5.7, we obtain

$$|b(u, v, w)| \leq k \sum_{i,j=1}^{3} \|u_i\| |D_i v_j|^{1/2}\|D_i v_j\|^{1/2}|w_j|.$$

So we have

$$|b(u, v, w)| \leq k\|u\|\|v\|^{1/2}\|v\|_{H^2}^{1/2}|w|.$$

Finally, we use $\|v\|_{H^2} = C|Av|$ (9.13) to obtain (9.27). ∎

9.4 Existence of Weak Solutions

With the above preparations behind us, we now begin to investigate existence and uniqueness for the Navier–Stokes equations.

We will show the following result, valid for both $m = 2$ and $m = 3$. Note that the theorem says nothing about uniqueness, and the solution need not be continuous into H. However, we do obtain *weak continuity* into H, that is, for every $\phi \in H$,

$$\lim_{t \to t_0}(u(t) - u(t_0), \phi) = 0. \tag{9.29}$$

Theorem 9.3 (Weak solutions). *Let $f \in L^2_{\text{loc}}(0, T; V^*)$. Then if $u_0 \in H$, there exists a weak solution $u(t)$ of*

$$du/dt + \nu Au + B(u, u) = f$$

such that, for any $T > 0$,

$$u \in L^\infty(0, T; H) \cap L^2(0, T; V),$$

and the equation holds as an equality in $L^p(0, T; V^*)$, with $p = 2$ if $m = 2$ and $p = 4/3$ if $m = 3$. Furthermore, the solution is weakly continuous into H as in (9.29).

Proof. We look at the finite-dimensional equation obtained by keeping only the first n Fourier modes, the n-dimensional Galerkin approximation,

$$u_n = \sum_{j=1}^{n} u_{nj}(t) w_j.$$

The equation for u_n is

$$\frac{du_n}{dt} + \nu Au_n + P_n B(u_n, u_n) = P_n f, \qquad (9.30)$$

where, as in the previous two chapters, P_n is the projection onto the first n Fourier modes,

$$P_n x = \sum_{j=1}^{n} (x, w_j) w_j.$$

As before, we try to find a bound on $|u_n|$ uniform in n.

To bound $|u_n|$ we take the inner product of (9.30) with u_n, obtaining

$$\frac{1}{2} \frac{d}{dt} |u_n|^2 + \nu(Au_n, u_n) + (P_n B(u_n, u_n), u_n) = \langle P_n f, u_n \rangle.$$

Noting that (since $u_n \in P_n H$)

$$(P_n B(u_n, u_n), u_n) = (B(u_n, u_n), P_n u_n)$$
$$= (B(u_n, u_n), u_n) = b(u_n, u_n, u_n),$$

and using (9.18) and the orthogonality property (9.22), we can write this as

$$\frac{1}{2} \frac{d}{dt} |u_n|^2 + \nu \|u_n\|^2 = \langle f, u_n \rangle \leq \|f\|_* \|u_n\|.$$

Using Young's inequality on the right-hand side, we obtain

$$\frac{1}{2}\frac{d}{dt}|u_n|^2 + v\|u_n\|^2 \le \frac{v}{2}\|u_n\|^2 + \frac{\|f\|_*^2}{2v},$$

so that

$$\frac{d}{dt}|u_n|^2 + v\|u_n\|^2 \le \frac{\|f\|_*^2}{v}.$$

Integrating both sides between 0 and t yields

$$|u_n(t)|^2 + v\int_0^t \|u_n(s)\|^2\,ds \le |u_n(0)|^2 + \frac{\|f\|_{L^2(0,t;V^*)}^2}{v}.$$

Since $|u_n(0)| = |P_n u_0| \le |u_0|$ (Lemma 7.5), we have the bounds

$$\sup_{t\in[0,T]}|u_n(t)|^2 \le K = |u_0|^2 + \frac{\|f\|_{L^2(0,T;V^*)}^2}{v}$$

and

$$\int_0^T \|u_n(s)\|^2\,ds \le K/v$$

uniformly in n. Thus u_n is bounded uniformly (in n) in

$$L^\infty(0,T;H) \qquad \text{and} \qquad L^2(0,T;V). \tag{9.31}$$

These uniform bounds allow us to use the Alaoglu compactness theorem (Theorem 4.18) to find a subsequence (which we shall relabel u_n) such that

$$u_n \overset{*}{\rightharpoonup} u \quad \text{in} \quad L^\infty(0,T;H)$$

and, extracting a further subsequence with Corollary 4.19 (and relabelling again) we have

$$u_n \rightharpoonup u \quad \text{in} \quad L^2(0,T;V),$$

with

$$u \in L^\infty(0,T;H) \cap L^2(0,T;V).$$

Finally, we need to obtain bounds on the derivatives, du_n/dt. Here we find a difference between $m = 2$ and $m = 3$. For $m = 2$ we can show that du_n/dt is uniformly bounded in $L^2(0,T;V^*)$, whereas for $m = 3$ we can obtain a bound

only in $L^{4/3}(0, T; V^*)$. We set $p = 2$ in the case $m = 2$ and $p = 4/3$ if $m = 3$.
Since

$$du_n/dt = -\nu Au_n - P_n B(u_n, u_n) + P_n f,$$

we need to show that each term on the right-hand side is uniformly bounded
in $L^p(0, T; V^*)$. This follows for Au_n since u_n is uniformly bounded in
$L^2(0, T; V)$ and A is a continuous linear operator from V into V^*. Clearly $P_n f$
is also bounded in this sense, since we have assumed that $f \in L^2(0, T; V^*)$.
It remains only to verify the same kind of bound for $P_n B(u_n, u_n)$, and this is
where the difference arises between the cases $m = 2$ and $m = 3$.

The following bounds on $\| B(u, u) \|_*$ are a consequence of (9.26) from Proposition 9.2:

$$\| B(u, u) \|_* \leq \begin{cases} k|u| \|u\|, & m = 2, \\ k|u|^{1/2} \|u\|^{3/2}, & m = 3 \end{cases}$$

(see Exercise 9.4). Since (see Lemma 7.5)

$$\| P_n B(u, v) \|_* \leq \| B(u, v) \|_*$$

we have

$$\| P_n B(u_n, u_n) \|_{L^2(0,T;V^*)}^2 \leq \int_0^T \| B(u_n(s), u_n(s)) \|_*^2 \, ds,$$

so that for $m = 2$

$$\| P_n B(u_n, u_n) \|_{L^2(0,T;V^*)} \leq k \int_0^T |u_n(s)|^2 \|u_n(s)\|^2 \, ds$$

$$\leq k \|u_n\|_{L^\infty(0,T;H)}^2 \|u_n\|_{L^2(0,T;V)}^2,$$

and for $m = 3$

$$\| P_n B(u_n, u_n) \|_{L^{4/3}(0,T;V^*)}^{4/3} \leq k \int_0^T |u_n(s)|^{2/3} \|u_n(s)\|^2 \, ds$$

$$\leq k \|u_n\|_{L^\infty(0,T;H)}^{2/3} \|u_n\|_{L^2(0,T;V)}^2.$$

Since u_n is uniformly bounded in $L^\infty(0, T; H)$ and $L^2(0, T; V)$ [see (9.31)],
$P_n B(u_n, u_n)$ is uniformly bounded (in n) in $L^2(0, T; V^*)$ if $m = 2$ and $L^{4/3}$
$(0, T; V^*)$ if $m = 3$. This gives the same bounds on du_n/dt:

$$du_n/dt \quad \text{is uniformly bounded in} \quad \begin{cases} L^2(0, T; V^*), & m = 2, \\ L^{4/3}(0, T; V^*), & m = 3. \end{cases}$$

We can now use the compactness theorem (Theorem 8.1) to guarantee that there is a subsequence $\{u_n\}$ (after relabelling) that converges to u strongly in $L^2(0, T; H)$, and this will in turn give us weak-* convergence of the nonlinear term in $L^p(0, T; V^*)$,

$$B(u_n, u_n) \overset{*}{\rightharpoonup} B(u, u) \qquad \text{in} \qquad L^p(0, T; V^*). \tag{9.32}$$

Indeed, if $w \in L^q(0, T; V)$ then

$$\int_0^T b(u_n, u_n, w)\,dt = -\int_0^T b(u_n, w, u_n)\,dt$$

$$= -\sum_{i,j=1}^m \int_0^T \int_Q (u_n)_i (D_i w_j)(u_n)_j \, dx \, dt.$$

Now we show (9.32). We have

$$\int_0^T b(u_n, u_n, w) - b(u, u, w)\,dt = \sum_{i,j=1}^m \int_0^T \int_Q [(u_n)_i - u_i](D_i w_j)u_j$$

$$+ (u_n)_i (D_i w_j)[(u_n)_j - u_j]\,dx\,dt.$$

So we need to consider expressions of the form

$$E_n = \int_0^T \int_Q (v_n - v)w v_n \, dx \, dt,$$

where $v_n \to v$ in $L^2(0, T; H)$, $w \in L^q(0, T; H)$, and v_n is uniformly bounded in $L^\infty(0, T; H)$. Since

$$\|wv_n\|_{L^2(0,T;H)} \le \|w\|_{L^2(0,T;H)} \|v_n\|_{L^\infty(0,T;H)}$$

it follows that $E_n \to 0$. Thus $B(u_n, u_n)$ converges weakly-* to $B(u, u)$ in $L^p(0, T; V^*)$ as required.

We can now show that $P_n B(u_n, u_n)$ converges weakly-* to $B(u, u)$ in $L^p(0, T; V^*)$ by using a similar argument to the one we used in the linear parabolic case (see Exercise 9.5). We have therefore shown convergence of all the terms in $L^p(0, T; V^*)$, and we have a solution $u \in L^2(0, T; V) \cap L^\infty(0, T; H)$ that satisfies

$$du/dt + \nu Au + B(u, u) = f \tag{9.33}$$

as an equality in $L^p(0, T; V^*)$. Lemma 7.4 shows that this is the same as V satisfying (9.33) in V^* for almost every $t \in [0, T]$.

To show that the solution has $u(0) = u_0$, as in the previous two chapters, we choose a test function $\phi \in C^1([0, T]; V)$ with $\phi(T) = 0$, and we compare the result of taking the inner product of (9.33) with ϕ and integrating by parts,

$$-\int_0^T (u(t), \phi'(t)) \, dt + v \int_0^T ((u(t), \phi(t))) \, dt + \int_0^T b(u(t), u(t), \phi(t)) \, dt$$

$$= (u(0), \phi(0)) + \int_0^T \langle f(t), \phi(t) \rangle \, dt,$$

with the result of taking the limit of a similar process applied to the Galerkin approximations,

$$-\int_0^T (u_n(t), \phi'(t)) \, dt + v \int_0^T ((u_n(t), \phi(t))) \, dt$$

$$+ \int_0^T b(u_n(t), u_n(t), \phi(t)) \, dt = (u_n(0), \phi(0)) + \int_0^T \langle f(t), \phi(t) \rangle \, dt,$$

which converges to

$$-\int_0^T (u(t), \phi'(t)) \, dt + v \int_0^T ((u(t), \phi(t))) \, dt + \int_0^T b(u(t), u(t), \phi(t)) \, dt$$

$$= (u_0, \phi(0)) + \int_0^T \langle f(t), \phi(t) \rangle \, dt,$$

since $u_n(0) = P_n u_0 \to u(0)$. As before, it follows that $u(0) = u_0$.

Finally, we need to show the weak continuity into H. Since (9.33) holds for almost every $t \in [0, T]$ as an equality in V^*, take the inner product of (9.33) with a fixed $v \in V$ to obtain

$$(du/dt, v) + v((u, v)) + b(u, u, v) = \langle f, v \rangle.$$

Now integrate this between t_0 and t, using Proposition 7.1, to give

$$(u(t) - u(t_0), v) = v \int_{t_0}^t ((u(s), v)) \, ds + \int_{t_0}^t b(u(s), u(s), v) \, ds$$

$$= \int_{t_0}^t \langle f, v \rangle \, ds, \qquad (9.34)$$

valid for any $0 < t_0 < t$. It follows from the inequalities for b (9.26) and the bounds in (9.31) that for a fixed $v \in V$, $t \mapsto b(u(t), u(t), v) \in L^1(0, T; \mathbb{R})$ for each $T > 0$. Similarly, $((u, v))$ and $\langle f, v \rangle$ are both in $L^1(0, T; \mathbb{R})$, and so it follows from (9.34) that

$$\lim_{t \to t_0} (u(t) - u(t_0), \phi) = 0 \qquad \text{for all} \qquad \phi \in V. \qquad (9.35)$$

Since V is dense in H, (9.35) also holds for all $\phi \in H$. ∎

    This theorem is essentially the full extent of the rigorous results available for the equations in three dimensions (but see the Notes at the end of this chapter). To proceed further with no extra assumptions we will have to restrict ourselves to the equations in the 2D case.

9.5 Unique Weak Solutions in Two Dimensions

In the two-dimensional case we can obtain much better results: continuity of the solution into H and uniqueness of the weak solution.

Theorem 9.4 (Unique Solutions in Two Dimensions). *If $m = 2$ then the solution $u(t)$ of the Navier–Stokes equation*

$$du/dt + \nu Au + B(u, u) = f$$

satisfies

$$u \in C^0([0, T]; H)$$

and depends continuously on the initial condition u_0. In particular the solution is unique.

Proof. The continuity into H follows from $u \in L^2(0, T; V)$ and $du/dt \in L^2(0, T; V^*)$ obtained above for $m = 2$, by using Theorem 7.2.

    We have just proved continuity of the solution $u(t; u_0)$ with respect to t. Continuity in u_0 amounts to a uniqueness result. The inequalities for $b(u, v, w)$ play an important rôle in the proof, and this is what distinguishes the analysis possible for the 2D and 3D problems.

    We consider two solutions u and v of (9.20) and write the equation for their difference $w = u - v$. Then w satisfies

$$dw/dt + \nu Aw + B(u, u) - B(v, v) = 0,$$

which we rewrite, using the bilinearity of B, as

$$B(u - v, u) + B(v, u - v) = B(w, u) + B(v, w),$$

as

$$dw/dt + \nu Aw + B(w, u) + B(v, w) = 0.$$

If we take the inner product of this equation with w by using Theorem 7.2 and use the orthogonality property of b (9.22), we obtain

$$\frac{1}{2}\frac{d}{dt}|w|^2 + \nu\|w\|^2 = -b(w, u, w). \tag{9.36}$$

Thus, using (9.26), we have

$$\frac{1}{2}\frac{d}{dt}|w|^2 + \nu\|w\|^2 \leq |b(w, u, w)|$$

$$\leq k|w|\|w\|\|u\|$$

$$\leq \frac{\nu}{2}\|w\|^2 + \frac{k^2}{2\nu}|w|^2\|u\|^2,$$

and so

$$\frac{d}{dt}|w|^2 + \nu\|w\|^2 \leq \frac{k^2}{\nu}\|u\|^2|w|^2. \tag{9.37}$$

Neglecting the term $\nu\|w\|^2$, we see that

$$\frac{d}{dt}\left\{ \exp\left(-\int_0^t \frac{k^2}{\nu}\|u(s)\|^2\, ds\right)|w(t)|^2 \right\} \leq 0.$$

We can rewrite this as

$$|w(t)|^2 \leq \exp\left(\int_0^t \frac{k^2}{\nu}\|u(s)\|^2\, ds\right)|w(0)|^2, \tag{9.38}$$

and since Theorem 9.3 guarantees that $u \in L^2(0, T; V)$, the integral in the exponential is finite. Thus if we have $w(0) = 0$ then $w(t) = 0$ for all $t \geq 0$, which gives uniqueness of the solution. ∎

Notice also that from the uniqueness expression (9.38) we have a Lipschitz separation property of solutions:

$$|u(t) - v(t)| \leq L(T)|u_0 - v_0|, \qquad 0 \leq t \leq T. \tag{9.39}$$

This will be useful later.

The results of this section show that, when f is independent of t, we can define a semidynamical system on H,

$$\left(H, \{S_H(t)\}_{t\geq 0}\right),$$

where $S_H(t)u_0 = u(t)$ and $S_H(t)$ is a C^0 semigroup [that is, $S_H(t)$ satisfies the properties in (8.28)]:

$$S_H(0) = I,$$

$$S_H(t)S_H(s) = S_H(s)S_H(t) = S_H(s+t),$$

$$S_H(t)x_0 \text{ is continuous in } x_0 \text{ and } t.$$

We will see in the next section that V is also a suitable phase space for this problem.

9.6 Existence of Strong Solutions in Two Dimensions

In the 2D case, we can obtain smoother solutions if we take f smoother and require the initial condition to be in V rather than in H. These are strong solutions, of the same type that we obtained for reaction–diffusion equations in Theorem 8.5. Such strong solutions will play a very important rôle in our analysis of the 3D equations later on, and we shall see in Section 9.7 that if a strong solution of the 3D equations exists, then it is unique.

Theorem 9.5 (Strong Solutions). *If $m = 2$, $u_0 \in V$, and $f \in L^2_{\text{loc}}(0, \infty; H)$, then there is a unique solution of*

$$du/dt + \nu Au + B(u, u) = f \qquad \text{(as an equality in } L^2(0, T; H))$$

that satisfies

$$u \in L^\infty(0, T; V) \cap L^2(0, T; D(A)) \qquad (9.40)$$

and in fact $u \in C^0([0, T]; V)$. Furthermore, the solutions depend continuously on the initial condition u_0.

Note that uniqueness of these solutions follows from the uniqueness of weak solutions, since a strong solution is also a weak solution. However, to obtain the continuity into V and with respect to the initial condition we will have to do a bit more work to find some better bounds on u.

Proof. Take the inner product of the finite-dimensional Galerkin approximation (9.30) with Au_n, so that

$$\tfrac{1}{2}\frac{d}{dt}\|u_n\|^2 + \nu|Au_n|^2 + (P_n B(u_n, u_n), Au_n) = (f, Au_n).$$

As before, we note that A commutes with P_n since $Aw_j = \lambda_j w_j$, and so

$$(P_n B(u_n, u_n), Au_n) = (B(u_n, u_n), Au_n) = b(u_n, u_n, Au_n),$$

and by (9.23) this expression is equal to zero in the 2D periodic case, so we have

$$\tfrac{1}{2}\frac{d}{dt}\|u_n\|^2 + \nu|Au_n|^2 \le |f|\,|Au_n|.$$

Applying Young's inequality as before we obtain

$$\frac{d}{dt}\|u_n\|^2 + \nu|Au_n|^2 \le \frac{|f|^2}{\nu}.$$

Now, just as before, we integrate both sides between 0 and t to find

$$\|u_n(t)\|^2 + \nu \int_0^t |Au_n(s)|^2\,ds \le \|u_n(0)\|^2 + \frac{|f|^2_{L^2(0,t;H)}}{\nu},$$

and, since Lemma 7.5 shows that $\|u_n(0)\| \le \|u_0\|$, we have

$$\sup_{t\in[0,T]} \|u_n(t)\|^2 \le K = \|u_0\|^2 + \frac{|f|^2_{L^2(0,T;H)}}{\nu}$$

and

$$\int_0^T |Au_n(s)|^2\,ds \le K/\nu.$$

Thus u_n is uniformly bounded (in n) in $L^\infty(0, T; V)$ and $L^2(0, T; D(A))$. We extract a subsequence by using Corollary 4.19 such that

$$u_n \overset{*}{\rightharpoonup} u \quad \text{in} \quad L^\infty(0, T; V)$$

and

$$u_n \rightharpoonup u \quad \text{in} \quad L^2(0, T; D(A)),$$

for some

$$u \in L^\infty(0, T; V) \cap L^2(0, T; D(A)).$$

Arguments similar to those in Theorem 9.3 show that du_n/dt is uniformly bounded in $L^2(0, T; H)$, and so, with a further subsequence,

$$du_n/dt \rightharpoonup du/dt \quad \text{in} \quad L^2(0, T; H).$$

Corollary 7.3 now shows that $u \in C^0([0, T]; V)$.

The terms du_n/dt, Au_n, and $P_n f$ all converge weakly in $L^2(0, T; H)$. To show that the nonlinear term $P_n B(u_n, u_n)$ converges in the same sense we once more appeal to Theorem 8.1, which this time shows us that $u_n \to u$ strongly in $L^2(0, T; V)$. This is sufficient to show the required convergence for the nonlinear term, and so we can deduce that (9.20) holds as an equality in $L^2(0, T; H)$ (and so an equality in H for almost every $t \in [0, T]$).

For continuity with respect to initial conditions, once again we set $w = u - v$ and consider

$$dw/dt + \nu Aw + B(u, u) - B(v, v) = f.$$

Taking the inner product with Aw (this makes sense because of Corollary 7.3) we obtain (using inequality (9.27))

$$\frac{1}{2} \frac{d}{dt} \|w\|^2 + \nu |Aw|^2 = b(v, v, Aw) - b(u, u, Aw)$$

$$\leq k \left[|w|^{1/2} \|w\|^{1/2} \|u\|^{1/2} |Au|^{1/2} |Aw| + |v|^{1/2} \|v\|^{1/2} \|w\|^{1/2} |Aw|^{3/2} \right]$$

$$\leq C \left[\|w\|^2 \|u\| |Au| + \|v\|^4 \|w\|^2 \right] + \frac{\nu}{2} |Aw|^2,$$

where we have used Young's inequality in the last line. Neglecting the term in $|Aw|^2$ we have

$$\frac{1}{2} \frac{d}{dt} \|w\|^2 \leq C \left[\|u\| |Au| + \|v\|^4 \right] \|w\|^2,$$

and so

$$\|w(t)\|^2 \leq \exp\left(C \int_0^t \|u(s)\| |Au(s)| + \|v(s)\|^4 \right) \|w(0)\|^2.$$

Continuity with respect to initial conditions follows since u and v are both strong solutions and are therefore bounded in both $L^\infty(0, T; V)$ and $L^2(0, T; D(A))$. ∎

When f does not depend on time we can therefore also define a semidynamical system on V,

$$\left(V, \{S_V(t)\}_{t \geq 0} \right).$$

Because solutions are unique, this is the restriction of the semidynamical system $S_H(t)$ to V, and so we denote both simply by $S(t)$. We will make particular use of this dynamical system on V when we discuss the 3D equations, mainly because of the result in the next section.

9.7 Uniqueness of 3D Strong Solutions

To end this chapter, we show that strong solutions of the 3D equations are unique in the class of weak solutions:

Theorem 9.6. *Let u be a strong solution of the 3D Navier–Stokes equations,*

$$u \in L^\infty(0, T; V) \cap L^2(0, T; D(A)),$$

as in (9.40). Then u is unique in the class of all weak solutions.

Exercise 9.6 shows that the assumption $u \in L^4(0, T; V)$ is also sufficient to ensure uniqueness in the class of weak solutions.

Proof. Once again we consider the equation for the difference of two solutions, $w = u - v$, where u is a strong solution and v is a weak solution. We take the inner product with w and obtain as before (9.36)

$$\tfrac{1}{2}\frac{d}{dt}|w|^2 + v\|w\|^2 = -b(w, u, w). \tag{9.41}$$

We now apply the 3D inequality (9.27) for $b(w, u, w)$ to obtain

$$\tfrac{1}{2}\frac{d}{dt}|w|^2 + v\|w\|^2 \le k\|w\|\|u\|^{1/2}|Au|^{1/2}|w|$$

$$\le \frac{v}{2}\|w\|^2 + \frac{k^2}{2v}\|u\||Au||w|^2.$$

Rewriting this as

$$\frac{d}{dt}|w|^2 + v\|w\|^2 \le \frac{k^2}{v}\|u\||Au||w|^2,$$

and ignoring the $v\|w\|^2$ term (as before), gives

$$|w(t)|^2 \le \exp\left(\int_0^t \|u(s)\||Au(s)|\,ds\right)|w(0)|^2. \tag{9.42}$$

Since

$$\int_0^t \|u(s)\||Au(s)|\,ds \le \|u\|_{L^2(0,T;V)}\|u\|_{L^2(0,T;D(A))},$$

and both these quantities are finite, we have uniqueness. ∎

Note that (9.42) once again ensures the Lipschitz property of solutions (9.39).

9.8 Dynamical Systems Generated by the 2D Equations

We end by summarising the results obtained above. When f is independent of time, the 2D Navier–Stokes equations can be used to generate a semidynamical system either on H (if $f \in V^*$) or on V (if $f \in H$). In the 3D case we have no way to define a dynamical system, since we cannot prove uniqueness of weak solutions nor existence of strong solutions.

We will see in the next chapter how to use the concept of the global attractor to help investigate the long-term behaviour of the semidynamical systems arising in the 2D equations.

Exercises

9.1   The techniques in this chapter all involve transforming the equation by projecting it onto a space of divergence-free functions so that p drops out. Show that the pressure can be recovered by solving the equation

$$\Delta p = \nabla \cdot f.$$

In fact we have already obtained the solution of this somewhere in this chapter. Where? What happens if $f \in H$?

9.2   Prove the 3D Ladyzhenskaya inequality for $u \in H_p^1(Q)$,

$$\|u\|_{L^4} \le C|u|^{1/4}\|u\|^{3/4}.$$

(Hint: Use Hölder's inequality and the embedding $H_p^1 \subset L^6$.)

9.3   Use the Cauchy–Schwarz inequality twice to show that

$$\left| \sum_{i,j=1}^{m} a_i b_{i,j} c_j \right| \le \left(\sum_{i=1}^{m} |a_i|^2 \right)^{1/2} \left(\sum_{i,j=1}^{m} |b_{i,j}|^2 \right)^{1/2} \left(\sum_{i=1}^{m} |c_i|^2 \right)^{1/2}.$$

$$(9.43)$$

9.4   Use the bounds in (9.26) to deduce the following bounds on $\|B(u,u)\|_*$:

$$\|B(u,u)\|_* \le \begin{cases} k|u|\|u\|, & m=2, \\ k|u|^{1/2}\|u\|^{3/2}, & m=3. \end{cases}$$

9.5   Fill in the argument of Theorem 9.3 discussed before Equation (9.33): use the argument from the linear parabolic case (see Section 7.4.3) to show that $P_n B(u_n, u_n)$ converges weakly to $B(u,u)$ in $L^p(0, T; V^*)$ ($p = 2$ if $m = 2$; $p = 4/3$ if $m = 3$).

9.6   Show that if $u \in L^4(0, T; V)$ is a solution of the 3D Navier–Stokes equations then it is unique in the class of weak solutions.

Notes

This style of analysis of the Navier–Stokes equations goes back to Leray (1933, 1934a,b) and Hopf (1951), who essentially proved all the results here. The technicalities in the Dirichlet boundary condition case are only minimally more involved, and this case is treated in detail in Ladyzhenskaya (1963), Lions (1969), Temam (1984), and Constantin & Foias (1988). The periodic boundary condition case forms the subject of Temam (1985) and Doering & Gibbon (1995). Recent technical texts are Galdi (1994a,b) and Lions (1994, 1996).

Doering & Gibbon (1995) is a particularly readable introduction to the subject that uses Fourier series to avoid the language of Sobolev spaces, but nonetheless it treats the problems of proving existence and uniqueness in considerable detail.

There are some existence and uniqueness results in three dimensions in the presence of various symmetries; see Ladyzhenskaya (1963), Ukhovskii & Yudovitch (1995), and Mahalov *et al.* (1990).

Part III

Finite-Dimensional Global Attractors

10

The Global Attractor
Existence and General Properties

We have now seen how to use the solutions of some ordinary and partial differential equations (ODEs/PDEs) to define dynamical and semidynamical systems on various phase spaces. For ODEs $\dot{x} = f(x)$ with Lipschitz nonlinearities we can define a dynamical system on $\mathbb{R}^m$; for reaction–diffusion equations we saw how to define a semidynamical system on $L^2(\Omega)$ and, if $m = 1$ or $m = 2$, on $H_0^1(\Omega)$; and for the 2D Navier–Stokes equations we could define a semidynamical system on either H (essentially $\mathbb{L}^2(Q)$) or V (essentially $\mathbb{H}^1(Q)$). We will discuss this in more detail in Section 10.1 below.

One of the main insights from the theory of dynamical systems is that a reduction in the possible complexity of the dynamics is obtained if we are content to study the long-term asymptotic behaviour of solutions. In the context of fluid mechanics, we can think of this as concentrating on the phenomena of "fully developed turbulence," rather than on the transient behaviour of the fluid flow.

In this chapter we introduce the concept of the global attractor for a dynamical system: this is a compact subset of the phase space that attracts all the trajectories. As such, we can expect the set of solutions that lie in the attractor to cover all possible "eventual" dynamical behaviours of the system (we make this precise in Proposition 10.14). We will give a general result that can be used to prove the existence of a global attractor in a variety of systems, and then we discuss some of its properties. Apart from one simple ODE example (the Lorenz equations) we will delay all other, PDE, examples to Chapters 11 and 12.

10.1 Semigroups

For ODEs we defined a solution operator $T(t)$, and we could do this for any $t \in \mathbb{R}$, giving a group of transformations. For many PDEs (including the

261

examples we studied in the last two chapters) it is only sensible to consider solutions for $t \geq 0$ (witness (10) in the Introduction).

Our standing assumption in this chapter is that we can find a phase space H (usually a Hilbert space L^2 or some related space), such that for $u_0 \in H$ the equation has a unique solution $u(t; u_0)$ for all positive times. In this case, we can define a C^0 semigroup of solution operators $S(t) : H \to H$ by

$$S(t)u_0 = u(t; u_0).$$

These have the properties given in (8.26),

$$S(0) = I,$$

$$S(t)S(s) = S(s)S(t) = S(s + t), \qquad (10.1)$$

$$S(t)u_0 \text{ is continuous in } u_0 \text{ and } t,$$

and we consider the semidynamical system

$$\left(H, \{S(t)\}_{t \geq 0}\right).$$

The joint continuity of $S(t)$ in both u_0 and t means that solutions vary continuously in a uniform way with respect to the initial conditions "about a compact set." For any compact set K, we have

$$|S(t)u_0 - S(t)v_0| \leq \delta_K(T, |u_0 - v_0|) \qquad \text{for all} \qquad u_0 \in K, \qquad (10.2)$$

where $\delta_K(t, d)$ has $\delta_K(t, 0) = 0$, $\delta_K(0, d) = d$, and δ_K is nondecreasing in both t and d (for the ODE case (2.24) we had $\delta_K(t, d) = de^{Lt}$). (For a proof of (10.2) see Exercise 10.1.)

10.2 Dissipation

We have already seen how important it is to obtain bounds on the solutions to ensure that they exist for all time (in Chapter 2 for ODEs and in the previous three chapters for various PDEs). We think (imprecisely) of an equation's being "dissipative" if all solutions are eventually bounded, provided that this bound is uniform over all trajectories. If we want to be more precise (and we do) then there are various ways of making a more formal definition. First we will examine various notions of dissipation in the context of ODEs.

We will begin by considering the weakest useful notion of dissipation. We say that $S(t)$ is *point dissipative* if there exists a bounded set $B \subset H$ such that for every x_0 there is a $t_0(x_0)$ such that

$$S(t)x_0 \in B \qquad \text{for all} \qquad t \geq t_0(x_0). \tag{10.3}$$

This definition says that each trajectory is eventually bounded and that this bound is the same for any trajectory.

In $\mathbb{R}^m$, this is equivalent to the stronger property of $S(t)$ being *bounded dissipative*. For this, we require the existence of a bounded set B that absorbs all initial conditions starting within any *bounded set X* in a uniform time: for each bounded set X there is a time $t_1(X)$ such that

$$S(t)X \subset B \qquad \text{for all} \qquad t \geq t_1(X). \tag{10.4}$$

We call B an "absorbing set," and we can think of this definition as giving some tolerance for error in our knowledge of the initial conditions. Even if we know only that u_0 is within ϵ of some u^*, we can still guarantee that $u(t) \in B$ for all $t \geq t_1(B(u^*, \epsilon))$.

We show the equivalence of point and bounded dissipativity in $\mathbb{R}^m$ in the following proposition. Note how the compactness of closed bounded sets in $\mathbb{R}^m$ is central to the argument.

Proposition 10.1. *Suppose that $S(t)$ is a point dissipative semigroup on $\mathbb{R}^m$. Then $S(t)$ is also bounded dissipative. Indeed, for any $\epsilon > 0$ the set*

$$B_\epsilon = \bigcup_{0 \leq t < \infty} S(t)\overline{N}(B, \epsilon)$$

[where B is the set from (10.3)] is a bounded, positively invariant set that absorbs any bounded set X in some time $t_1(X)$.

A set Y is *positively invariant* if $S(t)Y \subseteq Y$ for all $t \geq 0$. Note that an absorbing set is not necessarily positively invariant, although one can always obtain a positively invariant set from an absorbing set (see Exercise 10.2).

Proof. Since $B_\epsilon \supset B$ it must absorb points, and it is clearly positively invariant ($S(t)B_\epsilon \subseteq B_\epsilon$) by definition. Since $S(t)$ is point dissipative, for any $z \in \overline{N}(B, \epsilon)$ there is a $t_0(z)$ such that $S(t)z \in B$ for all $t \geq t_0(z)$ (using (10.3)). Since the solutions depend continuously on initial conditions, there is an open neighbourhood of each point, $N(z)$, such that

$$S(t_0(z))N(z) \subset N(B, \epsilon).$$

Now take a finite cover of $\overline{N}(B, \epsilon)$ by a set of such neighbourhoods $N(z_i)$, and set $t^* = \max_i t_0(z_i)$. Then, since

$$B_\epsilon = \bigcup_{0 \le t \le t^*} S(t)\overline{N}(B, \epsilon),$$

B_ϵ is clearly both closed and bounded.

Now suppose that X is a bounded set. Arguing as above, we find that there is a finite cover of X by open neighbourhoods $N(x_i)$ such that

$$S(t_0(x_i))N(x_i) \subset N(B, \epsilon),$$

and then

$$S(t)N(x_i) \subset B_\epsilon \qquad \text{for all} \qquad t \ge t_0(x_i).$$

Setting $t_1(X) = \max_i t_0(x_i)$, we obtain

$$S(t)X \subset B_\epsilon$$

for all $t \ge t_1(X)$. ∎

This result shows that in $\mathbb{R}^m$ the time $t(x_0)$ in (10.3) can be chosen uniformly for initial conditions in any bounded set X as in (10.4). Since bounded sets in $\mathbb{R}^m$ have compact closure, we could rewrite our definition of a bounded dissipative semigroup as follows. We omit the word "bounded" since this is the notion of dissipation we will use from now on.

Definition 10.2. *A semigroup is* dissipative *if it possesses a compact absorbing set B; that is, for any bounded set X there exists a $t_0(X)$ such that*

$$S(t)X \subset B \qquad \text{for all} \qquad t \ge t_0(X).$$

We will show in the next two chapters that the scalar reaction–diffusion equation considered in Chapter 8 and the 2D Navier–Stokes equation treated in Chapter 9 are both dissipative in this sense.

We have incorporated some compactness into our definition of a dissipative semigroup, and this will allow us to get around the fact that in general bounded sets are not compact in infinite-dimensional phase spaces.

10.3 Limit Sets and Attractors

We now want to find a recipe for constructing an attracting set for our semigroup. We start off by considering general notions of limit sets from the standard theory of dynamical systems. We then show that the ω-limit set of the absorbing set B gives an attracting set with several desirable properties.

10.3.1 Limit Sets

The ω-limit set of a set X consists of all the limit points of the orbit of X,

$$\omega(X) = \{y : \exists\, t_n \to \infty,\ x_n \in X \text{ with } S(t_n)x_n \to y\}. \qquad (10.5)$$

This can also be characterised as

$$\omega(X) = \bigcap_{t \geq 0} \overline{\bigcup_{s \geq t} S(s)X} \qquad (10.6)$$

(see Exercise 10.4). The set $\omega(X)$ in some sense captures all the recurrent dynamics of the orbit through X.

Our construction of attractors will be based on the following fundamental result concerning such limit sets. Note that compactness properties of the semigroup feature in a central way in the conditions. A set X is *invariant* if

$$S(t)X = X \qquad \text{for all} \qquad t \geq 0. \qquad (10.7)$$

In particular if you start in X you stay in X. Equation (10.7) also implies that the whole of X is important in the dynamics, since no part of X "disappears" as we run the dynamics on X forward in time.

Proposition 10.3. *Let $X \subset H$. If, for some $t_0 > 0$, the set*

$$\overline{\bigcup_{t \geq t_0} S(t)X} \qquad (10.8)$$

is compact, then $\omega(X)$ is nonempty, compact, and invariant.

Proof. First we use the characterisation (10.6). Since for $t \geq t_0$ the sets

$$\overline{\bigcup_{s \geq t} S(s)X}$$

are a sequence of nonempty compact sets decreasing as t increases, their intersection $(\omega(X))$ is nonempty and compact.

To show invariance, suppose that $x \in \omega(X)$. Then, using (10.5), we find that
there exist sequences $\{t_n\}$ and $\{x_n\}$ with $t_n \to \infty$ and $x_n \in X$ such that

$$S(t_n)x_n \to x,$$

and so

$$S(t)S(t_n)x_n = S(t + t_n)x_n \to S(t)x,$$

since $S(t)$ is continuous. So $S(t)\omega(X) \subset \omega(X)$. To show equality, for $t_n \geq$
$t + t_0$ [t_0 from (10.8)], the sequence $S(t_n - t)x_n$ is in the set (10.8) and so
possesses a convergent subsequence

$$S(t_{n_j} - t)x_{n_j} \to y,$$

and so $y \in \omega(X)$. But since $S(t)$ is continuous (10.1),

$$x = \lim_{j \to \infty} S(t)S(t_{n_j} - t)x_{n_j} = S(t)y,$$

and so $\omega(X) \subset S(t)\omega(X)$. Thus $S(t)\omega(X) = \omega(X)$ for all $t \geq 0$. ∎

10.3.2 The Global Attractor

We now discuss what manner of limit set we can expect to give us the best
information about the asymptotic dynamics. In the next section we prove an
existence result that shows that the set we choose [$\omega(B)$] does indeed have all
the properties we require.

Since in a dissipative system all trajectories eventually enter and stay in B,
one might expect that

$$\Lambda(B) = \bigcup_{x \in B} \omega(x) \tag{10.9}$$

would capture "all the asymptotic dynamics." However, there are some problems with this set.

Consider the example shown in Figure 10.1(a) (from Hale, 1988), where the
ω-limit set of every point is an equilibrium. In this case,

$$\Lambda_0(B) = \{a, b, c\},$$

a set of three isolated points. However, a small perturbation can produce a
periodic orbit Γ, as in Figure 10.1(b), so that now

$$\Lambda_\epsilon(B) = \{a, b, c, \Gamma\}.$$

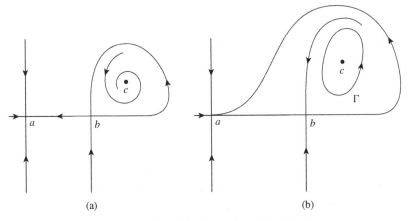

Figure 10.1. Phase portraits of two planar dynamical systems.

The periodic orbit is produced by bifurcation from the homoclinic orbits, a well-known source of global bifurcations (Glendinning, 1994). Since $\Lambda_0(B)$ does not take into account the homoclinic orbit, the appearance of Γ produces a sudden "explosion" in the set $\Lambda_\epsilon(B)$. (See also Exercise 10.3.)

The set $\mathcal{A} = \omega(B)$ turns out to be a more convenient way to study the asymptotic dynamics. These sets are shown in Figure 10.2 for the dynamics in Figure 10.1. Observe that attractor $\mathcal{A}_0$ includes not only the homoclinic orbit but all points in its interior; under perturbation $\mathcal{A}_\epsilon$ enlarges only slightly. By including more information we have obtained a much more stable object, and we will show later in this chapter (Theorem 10.16) that, in general, the global attractor (defined in this way) cannot "explode."

However, it can shrink drastically – if one considers a perturbation as in Figure 10.3 one can see that the attractor $\mathcal{A}_\epsilon$ is now homeomorphic to a line. We could argue that in this case the set $\Lambda(B)$ behaves better than $\omega(B)$; but a reduction in the size of the attractor is clearly less problematic (from an

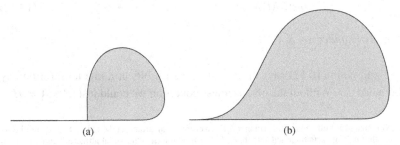

Figure 10.2. The global attractors for the dynamical systems illustrated in Figure 10.1.

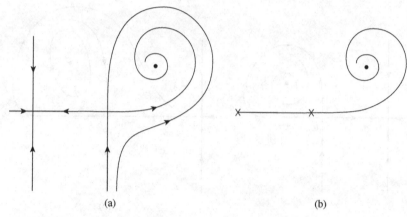

Figure 10.3. (a) Another planar dynamical system and (b) its global attractor.

analytical and dynamical point of view) than a sudden explosion. (See also Exercise 10.3.)

When B is an absorbing set as in Definition 10.2, then

$$\omega(B) = \bigcap_{t \geq 0} S(t)B \tag{10.10}$$

(see Exercise 10.5). Exercise 10.6 discusses some other properties of these sets.

In the next section we will show that $\omega(B)$ gives a set that is compact, connected, invariant, and attracts all bounded sets – the global attractor.*

Definition 10.4. *The global attractor $\mathcal{A}$ is the maximal compact invariant set*

$$S(t)\mathcal{A} = \mathcal{A} \quad \text{for all} \quad t \geq 0 \tag{10.11}$$

and the minimal set that attracts all bounded sets:

$$\text{dist}(S(t)X, \mathcal{A}) \to 0 \quad \text{as} \quad t \to \infty, \tag{10.12}$$

for any bounded set $X \subset H$.

Expression (10.12) says that $\mathcal{A}$ attracts all orbits at a rate uniform on any bounded set. Without the compactness condition we could just take $\mathcal{A} = H$.

---

* Note that $\mathcal{A}$ is both "maximal" (in that if Y is a compact invariant set then $\mathcal{A} \supseteq Y$) and "minimal" (in that if Z attracts bounded sets then $\mathcal{A} \subseteq Z$). The terms "maximal attractor" and "minimal attractor" are both used in the literature; "global attractor" should avoid any possible confusion.

The distance in (10.12) is the semidistance between two sets,

$$\text{dist}(X, Y) = \sup_{x \in X} \inf_{y \in Y} |x - y|.$$

Note that this distance does *not* define a metric – indeed, if $\text{dist}(X, Y) = 0$ then one has only $X \subset Y$. [To obtain a metric on subsets of H, we need to use the symmetric Hausdorff metric,

$$\text{dist}_{\mathcal{H}}(X, Y) = \max(\text{dist}(X, Y), \text{dist}(Y, X)). \tag{10.13}$$

(One can show that the space of all compact subsets of $\mathbb{R}^m$ is a complete space when endowed with this metric; see Exercise 10.7.)]

10.4 A Theorem for the Existence of Global Attractors

We now show that $\omega(B)$ is the global attractor, provided that $S(t)$ is dissipative and B is an absorbing set. We already know that $\omega(B)$ is nonempty, compact, and invariant, so it will remain only to show that it attracts trajectories as in (10.12).

Theorem 10.5. *If $S(t)$ is dissipative and B is a compact absorbing set then there exists a global attractor $\mathcal{A} = \omega(B)$. If H is connected then so is $\mathcal{A}$.*

Proof. That $\omega(B)$ is nonempty, compact, and invariant follows immediately from Proposition 10.3, as the dissipativity assumption ensures that (10.8) holds; indeed,

$$\overline{\bigcup_{t \geq t_0} S(t)B}$$

is a subset of B (and so compact) if $t_0 \geq t_0(B)$, from Definition 10.2. Furthermore, $\mathcal{A}$ is clearly the maximal compact invariant set since if Y is compact and invariant then $S(t)Y = Y$, and, choosing $t \geq t_0(Y)$, we get $Y \subset B$. Thus $\omega(Y) = Y \subset \omega(B) = \mathcal{A}$. (This argument also shows that $\mathcal{A}$ is the maximal *bounded* invariant set.)

To show that $\mathcal{A}$ attracts bounded sets we argue by contradiction. Indeed, if not, then there is a bounded set X, a $\delta > 0$, and a sequence $\{t_n\}$ with $t_n \to \infty$ such that

$$\text{dist}(S(t_n)X, \mathcal{A}) \geq \delta.$$

It follows that there are $x_n \in X$ with

$$\text{dist}(S(t_n)x_n, \mathcal{A}) \geq \delta/2.$$

As X is bounded, $S(t_n)x_n \in B$ for n large enough. As B is compact, there is a subsequence with

$$S(t_{n_j})x_{n_j} \to \beta \in B,$$

and

$$\text{dist}(\beta, \mathcal{A}) \geq \delta/2. \qquad (10.14)$$

But

$$\beta = \lim_{j \to \infty} S(t_{n_j})x_{n_j}$$

$$= \lim_{j \to \infty} S(t_{n_j} - t_0(X))S(t_0(X))x_{n_j},$$

and, setting $\beta_j = S(t_0(X))x_{n_j}$, notice that $\beta_j \in B$ and thus $\beta \in \mathcal{A}$, contradicting (10.14).

$\mathcal{A}$ is clearly the minimal set with this property, since it must attract itself, and $S(t)\mathcal{A} = \mathcal{A}$ for all $t \geq 0$.

Now suppose that $\mathcal{A}$ is not connected. Then there exist two open sets O_1 and O_2 such that $\omega(B) \subset O_1 \cup O_2$, $\omega(B) \cap O_i \neq \emptyset$, and $O_1 \cap O_2 = \emptyset$. Since B is bounded, B is contained in a ball U that is a connected set and $\omega(U)$ is clearly equal to $\omega(B)$. Since $S(t)$ is continuous $S(t)U$ is connected.

Thus $O_i \cap S(t)U \neq \emptyset$ and $O_1 \cup O_2$ does not cover $S(t)U$. So for each $t > 0$ there is an $x_t \in S(t)U$ with $x_t \notin O_1 \cup O_2$. Consider $\{x_n\}$, $n \in \mathbb{Z}^+$. Then for $n > t_0(U)$, $x_n \in B$ so $x_{n_j} \to x$ with $x \notin O_1 \cup O_2$. But we know that $\mathcal{A}$ attracts bounded sets, so $\text{dist}(x_n, \mathcal{A}) \to 0$ as $n \to \infty$, which is a contradiction. ∎

Importantly, we can show that if the semigroup $S(t)$ is injective on $\mathcal{A}$ (we make this precise in the statement of the next theorem) then the dynamics, restricted to $\mathcal{A}$, actually define a dynamical system; that is, $S(t)|_{\mathcal{A}}$ makes sense for all $t \in \mathbb{R}$. This is one good reason for investigating the dynamics on the attractor.

Theorem 10.6. *If the semigroup is injective on $\mathcal{A}$, so that*

$$S(t)u_0 = S(t)v_0 \in \mathcal{A} \qquad \text{for some} \qquad t > 0 \qquad \Rightarrow \qquad u_0 = v_0,$$

$$(10.15)$$

then every trajectory on $\mathcal{A}$ is defined for all $t \in \mathbb{R}$, and (10.11) holds for all $t \in \mathbb{R}$. In particular,

$$\left(\mathcal{A}, \{S(t)\}_{t \in \mathbb{R}}\right)$$

is a dynamical system.

Proof. For each $u \in \omega(B)$ we know that $u \in S(t)\omega(B)$, and so there exists a unique $v \in \omega(B)$ with $S(t)v = u$. We define $S(-t)u = v$ to give $S(t)$ for all $t \in \mathbb{R}$, and hence we obtain (10.11) for $t < 0$ also. Since $\mathcal{A} = \omega(B)$ is compact, it follows from Exercise 10.8 that $S(-t)$ as defined here is continuous on $\mathcal{A}$. Thus $S(t)$ is a continuous map from $\mathcal{A}$ into itself for all $t \in \mathbb{R}$. Finally, it follows that $S(t)|_{\mathcal{A}}$ satisfies properties in (10.1) for all $t, s \in \mathbb{R}$. ∎

We will prove the injectivity property (10.15) for the reaction–diffusion and Navier–Stokes equations in the next two chapters.

10.5 An Example – The Lorenz Equations

Before treating various properties of the global attractor, we will consider a simple example for which it is straightforward to prove the existence of such a set $\mathcal{A}$. The Lorenz equations are a 3D system of ODEs, introduced by Lorenz in 1963, which have become a standard example in the theory of finite-dimensional dynamical systems (see Sparrow, 1982, for more details). The equations are

$$\frac{dx}{dt} = -\sigma x + \sigma y,$$

$$\frac{dy}{dt} = rx - y - xz,$$

$$\frac{dz}{dt} = xy - bz,$$

with σ, r, and b all positive. Since bounded sets are compact in $\mathbb{R}^3$, all we need to show is the existence of a bounded absorbing set. In fact, we can use Proposition 10.1, and all we need is to show a uniform bound on each individual solution. We will show that a large enough sphere centred on $(0, 0, r + \sigma)$ is absorbing.

Consider

$$V(x, y, z) = x^2 + y^2 + (z - r - \sigma)^2,$$

which satisfies

$$\frac{dV}{dt} = -2\sigma x^2 - 2y^2 - 2bz^2 + 2b(r + \sigma)z$$
$$= -2\sigma x^2 - 2y^2 - b(z - r - \sigma)^2 - bz^2 + b(r + \sigma)^2$$
$$\leq -\alpha V + b(r + \sigma)^2,$$

where $\alpha = \min(2\sigma, 2, b)$. Then by (2.21) from the Gronwall inequality (Lemma 2.8), we have

$$V(t) \leq \frac{2b(r + \sigma)^2}{\alpha}$$

when t is large enough.

It follows by Theorem 10.5 that the Lorenz equations have a global attractor. However, as we will see below (Theorem 10.10), this attractor contains more than the "butterfly" so familiar from popular chaos writings.

10.6 Structure of the Attractor

We now want to examine the attractor itself in more detail, first investigating its structure. We show that it consists of all complete bounded orbits and contains the unstable manifolds of all fixed points and periodic orbits. This gives us a better idea of the kind of dynamics we can expect to understand if we restrict our attention to the attractor.

Theorem 10.7. *All complete bounded orbits lie in $\mathcal{A}$. If $S(t)$ is injective on $\mathcal{A}$ [as in (10.15)] then $\mathcal{A}$ is the union of all the complete bounded orbits.*

A "complete" orbit $u(t)$ is a solution of the PDE (or ODE) that is defined for all $t \in \mathbb{R}$. In general we do not expect the solutions of a PDE to lie on a complete orbit, since we cannot define $S(t)$ for $t < 0$.

Proof. Assume that $\mathcal{O}$ is not contained in $\mathcal{A}$; then for some $\epsilon > 0$ there is a point $x \in \mathcal{O}$ with $x \notin N(\mathcal{A}, \epsilon)$. However, since $\mathcal{A}$ attracts bounded sets, for t large enough

$$\text{dist}(S(t)z, \mathcal{A}) < \epsilon \qquad \text{for all} \qquad z \in \mathcal{O}. \qquad (10.16)$$

Since $\mathcal{O}$ is a complete orbit, $x = S(t)\tilde{x}$ for some $\tilde{x} \in \mathcal{O}$; (10.16) now gives a contradiction.

If $x \in \mathcal{A}$ and $S(t)$ is injective then the orbit through x is defined for all $t \in \mathbb{R}$ (Theorem 10.6) and is contained in $\mathcal{A}$ since $\mathcal{A}$ is invariant. Thus $\mathcal{A}$ consists of only complete bounded orbits. ∎

Note from the proof that we use only the injectivity to show that every point in $\mathcal{A}$ lies on a complete bounded orbit. Even without the injectivity we know that every complete bounded orbit lies in $\mathcal{A}$.

To investigate the structure of the attractor further, we need to recall the definition of stable and unstable manifolds. If z is a fixed point, then

Definition 10.8. *The* unstable manifold of z *is the set*

$$W^u(z) = \{u_0 \in H : S(t)u_0 \text{ defined for all } t, S(-t)u_0 \to z \text{ as } t \to \infty.\}$$

Similarly,

Definition 10.9. *The* stable manifold of z *is the set*

$$W^s(z) = \{u_0 \in H : S(t)u_0 \to z \text{ as } t \to \infty.\}$$

Note that if the semigroup is injective, the first condition in Definition 10.8 [$S(t)u_0$ is defined for all t] is satisfied automatically since $z \in \mathcal{A}$ (using Theorem 10.7).

One can define the unstable manifold of a general invariant set X just as in Definition 10.8, replacing "$S(-t)u_0 \to z$" with

$$\text{dist}(S(-t)u_0, X) \to 0.$$

Now, the unstable manifolds of all invariant sets (in particular of all the fixed points and periodic orbits) are contained in the attractor:

Theorem 10.10. *If X is a compact invariant set, then*

$$W^u(X) \subset \mathcal{A}$$

Proof. Let $u \in W^u(X)$. Then by definition (Definition 10.8) u lies on the complete orbit $Y = \cup_{t \in \mathbb{R}} u(t)$. As $t \to -\infty$ we know that $\text{dist}(u(t), X) \to 0$, and as $t \to \infty$ we know that $\text{dist}(u(t), \mathcal{A}) \to 0$, so the orbit $u(t)$ is bounded. Thus u lies on a complete bounded orbit, and by Theorem 10.7, $u \in \mathcal{A}$. ∎

Thus the commonly drawn picture of the attractor for the Lorenz equations is not in fact the whole of the global attractor found in Section 10.5. There are unstable fixed points at the centre of the "eyes," and their unstable manifolds will fill out these eyes.

10.6.1 Gradient Systems and Lyapunov Functions

Some semigroups have particularly simple asymptotic dynamics. These are systems that possess a Lyapunov functional, termed gradient (or strictly "gradient-like") systems. This Lyapunov functional gives a powerful way of "managing" the dynamics and proving much stronger results than those of the previous section. For example, we will show here that the attractors for such systems consist entirely of the unstable manifolds of the fixed points. We will see in the next chapter that scalar reaction–diffusion equations are one example of a gradient system.

First, we give the definition.

Definition 10.11. *A Lyapunov function for $S(t)$ on a positively invariant set $X \subset H$ is a continuous function $\Phi : X \to \mathbb{R}$ such that*

(i) for each $u_0 \in X$ the function $t \mapsto \Phi(S(t)u_0)$ is nonincreasing and
(ii) if $\Phi(S(\tau)u) = \Phi(u)$ for some $\tau > 0$, then u is a fixed point of $S(t)$.

This is a strong definition of a Lyapunov function. We could consider weaker Lyapunov functions that do not satisfy (ii), but for simplicity we consider only the above case. We will be able to construct such a strong Lyapunov function for the reaction–diffusion example in the next chapter.

Gradient systems are a good example of a situation in which the attractor contains much more information than the set $\Lambda(B)$ defined in (10.9). Note that (ii) in the definition implies that the system can have no periodic orbits. It is then perhaps unsurprising that the only limit sets of individual trajectories that are possible are fixed points (or unions of fixed points). We denote the set of all fixed points by $\mathcal{E}$.

Proposition 10.12. *Suppose that $S(t)$ has a Lyapunov function on $\tilde{B}$, where $\tilde{B}$ is a positively invariant absorbing set. Then $\omega(u_0) \subset \mathcal{E}$ for every $u_0 \in H$. In particular if H is connected and $\mathcal{E}$ is discrete then $\omega(u_0) \in \mathcal{E}$.*

Proof. For each $u_0 \in H$, there exists a t_0 such that $u_1 = S(t_0)u_0 \in \tilde{B}$. Since $\omega(u_0) = \omega(u_1)$, we consider the trajectory starting at u_1, $u(t) = S(t)u_1$. Then

$u(t) \in \tilde{B}$ for all $t \geq 0$, and

$$\omega(u_0) = \bigcap_{s>0} \overline{\{u(t) : t \geq s\}}$$

is nonempty, compact, connected, and invariant (Proposition 10.3). Now, Φ is constant on $\omega(u_0)$, since

$$\Phi|_{\omega(u_0)} = \lim_{t \to +\infty} \Phi(u(t)) = \inf_{t \in \mathbb{R}} \Phi(u(t)). \qquad (10.17)$$

Φ is bounded below since $u(t)$ is a subset of the compact set $\mathcal{A}$, and as $\Phi(u(t))$ is a nonincreasing function as $t \to +\infty$ the limit in (10.17) exists. Thus $\omega(u_0)$ consists only of stationary points [Definition 10.11, part (ii)].

Since $\omega(u_0)$ is connected if H is (by the same argument that gives connectedness of $\mathcal{A}$ in Theorem 10.5) it follows that if the fixed points of $S(t)$ are discrete then $\omega(u_0)$ consists of precisely one element of $\mathcal{E}$. ∎

The structure of the attractor follows from a very similar argument.

Theorem 10.13. *Suppose that $S(t)$ has a Lyapunov function on $\mathcal{A}$. Then*

$$\mathcal{A} = W^u(\mathcal{E}). \qquad (10.18)$$

Furthermore, if H is connected and $\mathcal{E}$ is discrete then

$$\mathcal{A} = W^u(\mathcal{E}) = \bigcup_{z \in \mathcal{E}} W^u(z) \qquad (10.19)$$

and also

$$\mathcal{A} = W^s(\mathcal{E}) = \bigcup_{z \in \mathcal{E}} W^s(z). \qquad (10.20)$$

Proof. Theorem 10.10 shows that $W^u(\mathcal{E}) \subset \mathcal{A}$ [and $W^u(z) \subset \mathcal{A}$ for all $z \in \mathcal{E}$]. To show that $\mathcal{A} \subset W^u(\mathcal{E})$, apply the argument of the previous proposition, replacing $\omega(u_0)$ with

$$\gamma = \bigcap_{s<0} \overline{\{u(t) : t < s\}},$$

where $u(t) = S(t)u_0$, $u_0 \in \mathcal{A}$. This shows that $\gamma \subset \mathcal{E}$, and (10.18) follows.

If the stationary points are discrete and γ is connected, then $\gamma = z$ for some $z \in \mathcal{E}$. This gives (10.19). A very similar argument gives (10.20). ∎

When $\mathcal{E}$ is discrete, the double equality given in the theorem means that in this case the attractor is made up of points u_0 with

$$\lim_{t \to -\infty} S(t)u_0 = z_1, \qquad \lim_{t \to +\infty} S(t)u_0 = z_2, \qquad z_1, z_2 \in \mathcal{E},$$

that is, heteroclinic orbits joining two stationary points. Thus the structure of such attractors can be fully specified by a list of which stationary points are joined to each other. This *connection problem* has received much attention, and techniques to establish which pairs of stationary points are in fact joined to each other have been developed for certain systems (see for example Fusco & Rocha, 1991, and Fiedler & Rocha, 1996).

10.7 How the Attractor Determines the Asymptotic Dynamics

We now make precise the idea that the dynamics "on the attractor" serve to determine all possible long-time dynamics of individual trajectories. The proposition says that after a sufficiently long time (τ) any trajectory $u(t)$ of the original equation will look like some trajectory on the attractor for a long time.

Proposition 10.14. *Given a trajectory $u(t) = S(t)u_0$, $\epsilon > 0$ and $T > 0$, there exists a time $\tau = \tau(\epsilon, T) > 0$ and a point $v_0 \in \mathcal{A}$ such that*

$$|u(\tau + t) - S(t)v_0| \le \epsilon \qquad \text{for all} \qquad 0 \le t \le T.$$

Proof. Since trajectories depend continuously on the initial conditions as in (10.2), given $\epsilon, T > 0$ there exists a $\delta(\epsilon, T)$ such that

$$u_0 \in \mathcal{A} \quad \text{and} \quad |u_0 - v_0| \le \delta(\epsilon, T) \qquad \text{imply that}$$

$$|u(t) - v(t)| \le \epsilon, \qquad t \in [0, T]. \tag{10.21}$$

Now, since $\mathcal{A}$ is the global attractor, the trajectory $u(t)$ tends towards $\mathcal{A}$; thus there exists a time τ and a point $v_0 \in \mathcal{A}$ such that

$$\text{dist}(u(\tau), \mathcal{A}) = |u(\tau) - v_0| \le \delta.$$

We now consider the trajectory $v(t)$ on $\mathcal{A}$ with $v(0) = v_0$. Then, the two trajectories $u(t)$ [seen as a trajectory starting at the point $u(\tau)$] and $v(t) = S(t)v_0$ satisfy, by (10.21),

$$|u(\tau + t) - S(t)v_0| \le \epsilon. \qquad \blacksquare$$

To follow the chosen trajectory $u(t)$ for a longer time, one has to skip from one trajectory on $\mathcal{A}$ to another. The following result is a straightforward corollary of Proposition 10.14 (for a proof see Exercise 10.9).

Corollary 10.15. *Given a solution $u(t)$, there exists a sequence of errors $\{\epsilon_n\}_{n=1}^{\infty}$, with*

$$\epsilon_n \to 0,$$

an increasing sequence of times $\{t_n\}_{n=1}^{\infty}$, with

$$t_{n+1} - t_n \to \infty \qquad as \qquad n \to \infty,$$

and a sequence of points $\{v_n\}_{n=1}^{\infty}$, with $v_n \in \mathcal{A}$, such that

$$|u(t) - S(t - t_n)v_n| \leq \epsilon_n \qquad for\ all \qquad t_n \leq t \leq t_{n+1}.$$

Furthermore, the jumps $|v_{n+1} - S(t_{n+1} - t_n)v_n|$ decrease to zero.

This result shows that the chosen trajectory is shadowed closer and closer and for longer and longer times by portions of trajectories that lie within $\mathcal{A}$. Although elementary, this result is instructive. Indeed, it shows exactly how the dynamics on $\mathcal{A}$ can be said to determine the asymptotic behaviour of trajectories on H. A trivial example is the 3D system

$$dx/dt = z(x - y),$$
$$dy/dt = z(y - x),$$
$$dz/dt = -\lambda z|z|,$$

which has $z \equiv 0$ as an attractor (not "the global attractor," however, since it is not compact), on which the dynamics are trivial ($\dot{x} = \dot{y} = 0$). However, trajectories that start with $z \neq 0$ approach algebraically slowly according to

$$z(t) = \frac{z_0}{1 + \mu|z_0|t}.$$

Thus for an initial condition (r, θ_0, z_0) (where the x and y have been turned into polar coordinates), $\theta(t)$ is given by

$$\theta(t) = \theta_0 + \frac{1}{\mu}\ln(1 + \mu|z_0|t).$$

The trajectory is "determined" by that on $z \equiv 0$ inasmuch as it remains nearly constant (within an ϵ error) for longer and longer time intervals. Without a result such as Corollary 10.15, this interpretation is not necessarily an obvious one.

10.8 Continuity Properties of the Attractor

As the final topic, we show as promised in Section 10.3.2 that the attractor cannot "explode," a property known as upper semicontinuity. Although this is relatively easy result to prove, it is essential in guaranteeing some stability for our notion of an attractor. We then discuss when it is possible to show that the attractor does not implode.

10.8.1 Upper Semicontinuity

We treat a semigroup $S_0(t)$ with a global attractor $\mathcal{A}_0$ and a family of perturbed semigroups $S_\eta(t)$ with attractors $\mathcal{A}_\eta$. Such a situation could arise (for example) by making small changes to the right-hand side of an ODE or changing the forcing term in the Navier–Stokes equations. We first prove upper semicontinuity. Since $\mathrm{dist}(\mathcal{A}_\eta, \mathcal{A}_0) < \epsilon$ means that $\mathcal{A}_\eta$ lies within a small ϵ-neighbourhood of $\mathcal{A}_0$, it is clear that the result of the following theorem means that $\mathcal{A}_0$ cannot explode.

Theorem 10.16. *Assume that for $\eta \in [0, \eta_0)$ the semigroups S_η each have a global attractor A_η and that there exists a bounded set X such that*

$$\bigcup_{0 \leq \eta < \eta_0} A_\eta \subset X.$$

If in addition the semigroups S_η converge to S_0 in that, for each $t > 0$, $S_\eta(t)x \to S_0(t)x$ uniformly on bounded subsets Y of H,

$$\sup_{u_0 \in Y} |S_\eta(t)u_0 - S_0(t)u_0| \to 0 \qquad as \qquad \eta \to 0,$$

then

$$\mathrm{dist}(\mathcal{A}_\eta, \mathcal{A}_0) \to 0 \qquad as \qquad \eta \to 0.$$

Proof. Given $\epsilon > 0$, we first show that $S_\eta(t)X \subset N(\mathcal{A}_0, \epsilon)$ for some $t > 0$ and all $\eta \leq \eta(\epsilon)$. Now, since $\mathcal{A}_0$ attracts X there exists a time t such that

$$S(t)X \subset N(\mathcal{A}_0, \epsilon/2).$$

Then, for η sufficiently small ($\leq \eta(\epsilon)$, say), we can ensure that

$$\sup_{x \in X} |S_\eta(t)x - S_0(t)x| < \epsilon/2.$$

Thus in fact, for $\eta \leq \eta(\epsilon)$, since $A_\eta \subset X$,

$$A_\eta = S_\eta A_\eta \subset S_\eta(t)X \subset N(\mathcal{A}_0, \epsilon),$$

and it follows that dist$(\mathcal{A}_\eta, \mathcal{A}_0) \leq \epsilon$. ∎

This is the most that we can expect to be true in general, as the example discussed in Section 10.3.2 demonstrates.

10.8.2 Lower Semicontinuity

The result that would give full continuity would be that

$$\text{dist}(\mathcal{A}_0, \mathcal{A}_\eta) \to 0 \qquad \text{as} \qquad \eta \to 0.$$

This is not known except in a few special cases. The most general result holds for a generalisation of gradient (Lyapunov) systems, inspired by the result of Theorem 10.13 on the structure of the attractor.

Theorem 10.17 (Stuart & Humphries, 1996). *Let the assumptions of Theorem 10.16 hold and in addition let $\mathcal{A}_0$ be given by the closure of the unstable manifolds of a finite number of fixed points, so that*

$$\mathcal{A}_0 = \bigcup_{z \in \mathcal{E}} \overline{W^u(z)}.$$

Provided that the unstable manifolds vary continuously with η near $\eta = 0$ in some neighbourhood of each fixed point, then the attractor is lower semicontinuous:

$$\text{dist}(\mathcal{A}_0, \mathcal{A}_\eta) \to 0 \qquad as \qquad \eta \to 0. \tag{10.22}$$

It follows that the attractor is continuous in the Hausdorff metric:

$$\text{dist}_{\mathcal{H}}(\mathcal{A}_\eta, \mathcal{A}_0) \to 0 \qquad as \qquad \eta \to 0.$$

Proof. We need to prove (10.22), i.e. that within ϵ of any point in $\mathcal{A}_0$ there is a point in $\mathcal{A}_\eta$ for all $\eta \leq \eta^*$, with $\eta^* > 0$. Since $\mathcal{A}_0$ is compact this reduces to showing that there are points in $\mathcal{A}_\eta$ within ϵ of some finite set of points

$\{x_k\}_{k=1}^N$ in $\mathcal{A}_0$. Since $\mathcal{A}_0$ is the closure of the unstable manifolds, there are points $\{y_k\}_{k=1}^N$, lying in the unstable manifolds in $\mathcal{A}_0$, with

$$|x_k - y_k| \le \epsilon/2.$$

We write $y_k = S_0(t_k)z_k$, with each z_k within a small neighbourhood of the fixed points.

Now choose $\delta > 0$ such that

$$|z_k - u| \le \delta \quad \Rightarrow \quad |S_0(t_k)z_k - S_0(t_k)u| \le \epsilon/4$$
$$\text{for all} \quad k = 1, \dots, N,$$

using the continuity of S_0, and then pick η small enough ($\eta \le \eta_1$) such that

$$|S_0(t_k)u - S_\eta(t_k)u| \le \epsilon/4 \quad \text{for all} \quad u \in N(\mathcal{A}_0, \delta), \ k = 1, \dots, N.$$

Since the local unstable manifolds perturb smoothly, provided that η is small enough ($\eta \le \eta^* \le \eta_1$) there are points z_k^η in the unstable manifolds within $\mathcal{A}_\eta$ that satisfy

$$\left|z_k^\eta - z_k\right| < \delta.$$

It follows that

$$
\begin{aligned}
\left|S_\eta(t_k)z_k^\eta - y_k\right| &= \left|S_\eta(t_k)z_k^\eta - S_0(t_k)z_k\right| \\
&\le \left|S_\eta(t_k)z_k^\eta - S_0(t_k)z_k^\eta\right| + \left|S_0(t_k)z_k^\eta - S_0(t_k)z_k\right| \\
&\le \epsilon/4 + \epsilon/4 = \epsilon/2.
\end{aligned}
$$

Therefore

$$\left|S_\eta(t_k)z_k^\eta - x_k\right| \le \epsilon,$$

and since $S_\eta(t_k)z_k^\eta \in \mathcal{A}_\eta$, we have obtained lower semicontinuity. Coupled with the result of Theorem 10.16, this proves the continuity of $\mathcal{A}_\eta$ at $\eta = 0$ in the Hausdorff metric. ∎

10.9 Conclusion

In this chapter we have introduced the global attractor and shown that it exists for dissipative systems (those that have a compact absorbing set). We have also seen some of its basic structural properties and investigated briefly how the attractor serves to determine the asymptotic dynamics. In the next two chapters we will verify the conditions necessary to prove the existence of global attractors for reaction–diffusion equations (Chapter 11) and the Navier–Stokes equations (Chapter 12). We then go on to address various questions about the dynamics on these sets.

Exercises

10.1 Show that if $f : H \rightarrow H$ is continuous and K is a compact set, then f is "uniformly continuous near K," in that there is a $\delta_K(\epsilon)$ such that $u \in K$, $v \in H$, and $|u - v| \leq \delta_K(\epsilon)$ implies that $|f(u) - f(v)| \leq \epsilon$.

10.2 Suppose that $S(t)$ is a dissipative semigroup with a bounded absorbing set B. Show that

$$\overline{\bigcup_{t \geq 0} S(t)B} \tag{10.23}$$

is a bounded, positively invariant set that is also absorbing (cf. Proposition 10.1). If B is compact, show that (10.23) is also compact.

10.3 Another problem with the set $\Lambda(B)$ defined in (10.9) is that

$$\Lambda(B) \neq \Lambda[\Lambda(B)]$$

in general. Verify this by finding $\Lambda(B)$ and $\Lambda^2(B)$ for the example

$$\dot{r} = r(1 - r),$$
$$\dot{\theta} = (1 - r)^2 + \theta^2.$$

10.4 Prove that the characterisations of $\omega(X)$ given in (10.5) and (10.6) are equivalent.

10.5 Show that (10.10) holds, i.e. that

$$\omega(B) = \bigcap_{t \geq 0} \overline{S(t)B}.$$

Show also that, if $T > 0$,

$$\omega(B) = \bigcap_{n=1}^{\infty} \overline{S(nT)B}.$$

10.6 Show that if X is bounded and $X \supset Y$ then $\omega(X) \supset \omega(Y)$. Deduce that if Y is an absorbing set then $\omega(X) = \omega(Y)$.

10.7   Show that the space $\mathcal{K}(\mathbb{R}^m)$ consisting of all compact subsets of $\mathbb{R}^m$ is complete when equipped with the Hausdorff metric (10.13). [Hint: From a Cauchy sequence $\{K_j\}$ in $\mathcal{K}(\mathbb{R}^m)$ first take a subsequence such that

$$\mathrm{dist}_{\mathcal{H}}(K_j, K_i) \le j^{-1} \qquad \text{for all} \qquad i \ge j, \tag{10.24}$$

and then consider the candidate limit set

$$K_\infty = \bigcap_{j=1}^{\infty} \overline{\bigcup_{i=j}^{\infty} K_i}.$$

Show that K_∞ is compact, and then using (10.24) deduce that

$$\mathrm{dist}_{\mathcal{H}}(K_j, K_\infty) \to 0.]$$

10.8   Suppose that X is compact and that $f : X \to Y$ is injective. Show that $f^{-1} : f(X) \to X$ is continuous.

10.9   Prove Corollary 10.15.

10.10   The following property is proved in Babin & Vishik (1992, Chapter 2, Theorem 1.1); they call it "uniform continuity for all $t \ge 0$ modulo the attractor." We will say that the semigroup $S(t)$ is C^0 *uniformly continuous* if (10.2) holds for u_0 in any *bounded* (rather than just compact) set X. Prove that if $S(t)$ is a uniformly C^0 semigroup then

$$\sup_{t \ge 0} \, \mathrm{dist}_{\mathcal{H}}(S(t)B_1 \cup \mathcal{A}, S(t)B_2 \cup \mathcal{A}) \to 0$$

as $\mathrm{dist}_{\mathcal{H}}(B_1, B_2) \to 0$, where B_1 and B_2 are bounded subsets of H. [Recall the definition of the Hausdorff metric $\mathrm{dist}_{\mathcal{H}}$ given in (10.13).]

Notes

There are many versions of Theorem 10.5 in the literature, the first probably that of Billotti & LaSalle (1971). The statement given here, along with its proof, can be found in Temam (1988). See also Theorem 2.1 and Lemma 2.1 in Chapter 2 of Babin & Vishik (1992) and Ladyzhenskaya's "semigroups of class $\mathcal{K}$" (1991).

In many applications (although not those considered in this book) the result in this form is too restrictive, and Temam also gives another result under weaker conditions. Note first that the use of the dissipation in the proof of the theorem is to ensure that the condition of Proposition 10.3 applies, namely that

$$\bigcup_{t \ge t_0} S(t)B \tag{10.25}$$

is compact for some t_0. To generalise the conditions of the theorem, we assume that the semigroup $S(t)$ can be decomposed into one part that preserves the property in (10.25) and one part that decays to zero.

In this case, one need only assume that the absorbing set is *bounded*, rather than compact. More precisely,

Theorem 10.18. *If the semigroup $S(t)$ can be decomposed as*

$$S(t) = S_1(t) + S_2(t),$$

such that for any bounded set X there exists a $t_0(X)$ giving

$$\overline{\bigcup_{t \geq t_0} S_1(t)X} \qquad \text{compact in } H,$$

and

$$r_X(t) = \sup_{\varphi \in X} |S_2(t)\varphi| \to 0 \qquad \text{as} \qquad t \to \infty,$$

and if there exists a bounded *absorbing set B, then $\omega(B)$ is the global attractor as in Theorem 10.5.*

There are other equivalent conditions in the literature. For example, Ladyzhenskaya (1991) calls a semigroup "asymptotically compact" if for any bounded sequence $\{x_k\}$ and $t_k \to \infty$

$$\{S(t_k)x_k\}_k \qquad \text{has a convergent subsequence in } H.$$

She shows that an asymptotically compact semigroup with a bounded absorbing set has a global attractor. Babin & Vishik (1992, Chapter 2, Theorem 2.2) and Hale (1988) prove the existence of a global attractor under the condition that there is a compact set K such that for any bounded set B

$$\text{dist}(S(t)B, K) \to 0 \qquad \text{as} \qquad t \to \infty.$$

Goubet & Moise have shown that these three conditions are equivalent (the proof is given in Temam, 1997).

Theorem 10.7 is taken from Stuart & Humphries (1996), as is the lower semicontinuity theorem. Although their book is primarily concerned with analysis of numerical methods in $\mathbb{R}^m$ using a dynamical systems approach, it contains many results that are also interesting in the PDE case. The other results of Section 10.6 are taken from Temam (1988).

The tracking property is proved in Langa & Robinson (1999), where they also give conditions that ensure that the attractor is simply connected (this is not true in general).

Continuity properties are investigated in various texts, including those by Hale (1988), Temam (1988), and Stuart & Humphries (1996). The original treatment of the problem in this setting was in the paper of Hale *et al.* (1988), with a lower semicontinuity result in Hale & Raugel (1989). Related theories have been developed for Morse–Smale systems (Smale, 1967; Hale *et al.*, 1984) and using Conway's attractor–repellor pairs (e.g., Hurley, 1983; Smoller, 1983).

Finally, one can generalise the theory in various ways to cover nonautonomous semi-dynamical systems (Haraux, 1988; Sell, 1967; Vishik, 1992) and the "cocyles" generated by stochastic PDEs (Crauel & Flandoli, 1994; Crauel *et al.*, 1995; Schmalfuß, 1992).

11

The Global Attractor
for Reaction–Diffusion Equations

In this chapter we will apply Theorem 10.5 to reaction–diffusion equations. We prove the existence of an attractor in L^2 and show that the equation gives rise to a gradient system (Definition 10.11). It follows from Theorem 10.13 that the structure of this attractor can be well understood.

In order to apply Theorem 10.5 we have to prove the existence of a compact absorbing set. The usual method is to start by finding an absorbing set in L^2, and then to use this result to help find an absorbing set in H^1. Since $H^1 \subset\subset L^2$, this gives a compact absorbing set in L^2. As in Chapters 7–9, we will denote the norm in L^2 by $|\cdot|$, the norm in H_0^1 by $\|\cdot\|$, and the norm in H^{-1} by $\|\cdot\|_*$.

We will treat the equation

$$\frac{\partial u}{\partial t} - \Delta u = f(u), \qquad u|_{\partial\Omega} = 0 \tag{11.1}$$

under the same conditions that we imposed in Chapter 8 to ensure that solutions generate a semidynamical system on $L^2(\Omega)$. Namely, Ω must be "sufficiently smooth" (see Section 7.1), and f must satisfy the conditions

$$-k - \alpha_1|s|^p \le f(s)s \le k - \alpha_2|s|^p, \qquad p > 2, \tag{11.2}$$

and

$$f'(s) \le l,$$

for all $s \in \mathbb{R}$. As in Theorem 8.5 we also take $f(0) = 0$ to simplify the algebra.

11.1 Absorbing Sets and the Attractor

To prove the existence of a global attractor we need to show that there are absorbing sets in L^2 and in H_0^1.

285

11.1.1 An Absorbing Set in L^2

It is straightforward to find an absorbing set in L^2.

Proposition 11.1. *The reaction–diffusion equation has an absorbing set in L^2; there is a constant ρ_H and a time $t_0(|u_0|)$ such that, for the solution $u(t) = S(t)u_0$,*

$$|u(t)| \leq \rho_H \qquad \text{for all} \qquad t \geq t_0(|u_0|). \qquad (11.3)$$

In addition there is a constant I_V such that

$$\int_t^{t+1} \|u(s)\|^2 \, ds \leq I_V \qquad \text{for all} \qquad t \geq t_0(|u_0|). \qquad (11.4)$$

The integral bound in (11.4) will help us to find the absorbing set in V in the next section. For an analysis valid in the case of periodic boundary conditions, see Exercise 11.1.

Proof. We write the equation as

$$\frac{du}{dt} + Au = f(u), \qquad (11.5)$$

and we take the inner product with u (using Theorem 7.2) to obtain, using (11.2),

$$\frac{1}{2}\frac{d}{dt}|u|^2 + \|u\|^2 = \int_\Omega f(u(x))u(x) \, dx$$

$$\leq \int_\Omega k - \alpha_2 |u|^p \, dx \qquad (11.6)$$

[cf. (8.18)]. Now we can drop the term in $|u|^p$ and use the Poincaré inequality

$$|u| \leq \lambda_1^{-1/2} \|u\|$$

(where λ_1 is the first eigenvalue of the Laplacian on Ω) to obtain

$$\frac{d}{dt}|u|^2 + 2\lambda_1 |u|^2 \leq 2k|\Omega|.$$

The Gronwall inequality (Lemma 2.8) gives

$$|u(t)|^2 \leq |u_0|^2 e^{-2\lambda_1 t} + \frac{k|\Omega|}{\lambda_1}(1 - e^{-2\lambda_1 t}).$$

It follows that if

$$t \geq t_0(|u_0|) \equiv \frac{1}{2\lambda_1} \ln \frac{\lambda_1 |u_0|^2}{k|\Omega|}$$

then

$$|u(t)|^2 \leq \rho_H^2 = \frac{2k}{\lambda_1} |\Omega|.$$

To deduce the integral bound on $\|u\|^2$, we return to (11.6). Dropping the term in $|u|^p$ we are left with

$$\tfrac{1}{2} \frac{d}{dt} |u|^2 + \|u\|^2 \leq k|\Omega|,$$

which we can integrate between t and $t+1$ to obtain

$$\int_t^{t+1} \|u(s)\|^2 \, ds \leq k|\Omega| + \tfrac{1}{2}|u(t)|^2.$$

This shows that

$$\int_t^{t+1} \|u(s)\|^2 \, ds \leq k|\Omega| + \tfrac{1}{2}\rho_H^2 \qquad \text{for all} \qquad t \geq t_0(|u_0|),$$

which is (11.4), with $I_V = k|\Omega| + \rho_H^2$. ∎

11.1.2 An Absorbing Set in H_0^1

We first give a heuristic derivation of the existence of an absorbing set in H_0^1. Ideally we would take the inner product of (11.5) with Au to obtain [cf. (8.26)]

$$\tfrac{1}{2} \frac{d}{dt} \|u\|^2 + |Au|^2 \leq l\|u\|^2 \tag{11.7}$$

on $(0, T)$. Integrating (11.7) between s and t $(t - 1 \leq s < t)$ gives

$$\|u(t)\|^2 \leq 2l \int_s^t \|u(\xi)\|^2 \, d\xi + \|u(s)\|^2.$$

Now, if we integrate this last equation with respect to s between $t - 1$ and t we obtain

$$\|u(t)\|^2 \leq (l + 1) \int_{t-1}^t \|u(s)\|^2 \, ds,$$

and we can now use the bound from (11.4) to deduce that

$$\|u(t)\|^2 \le (l+1)\big[k|\Omega| + \rho_H^2\big] + C \tag{11.8}$$

provided that $t \ge t_0(|u_0|) + 1$. (Although this may seem like a trick, the method can be generalised as the "uniform Gronwall lemma" and applied in many different examples; see Exercise 11.2.)

However, we do not know that u is smooth enough to justify the computations we have just performed. Corollary 7.3, which would allow us to take the inner product with Au to write (11.7), requires $u \in L^2(0, T; D(A))$ and $du/dt \in L^2(0, T; L^2)$, which we do not have for our weak solutions. A calculation like that above, which assumes that we have sufficient regularity, is referred to as a "formal" calculation. Although it usually gives the right answer, it really needs to be justified rigorously before we can believe its conclusions. To do this we will use the Galerkin expansion again and the following lemma.

Lemma 11.2. *Let $V \subset\subset H$, with dual V^*. Suppose that $\{u_n\}$ is uniformly bounded in $L^\infty(0, T; V)$,*

$$\operatorname{ess\,sup}_{t \in [0,T]} \|u_n(t)\| \le C, \tag{11.9}$$

and that $u_n \rightharpoonup u$ in $L^2(0, T; V)$; then

$$\operatorname{ess\,sup}_{t \in [0,T]} \|u(t)\| \le C. \tag{11.10}$$

Furthermore, if $u \in C^0([0, T]; H)$ then in fact

$$\sup_{t \in [0,T]} \|u(t)\| \le C. \tag{11.11}$$

We will apply the theorem below with $V = H_0^1(\Omega)$ and $H = L^2(\Omega)$.

Proof. First we prove (11.10). Since $\{u_n\}$ is uniformly bounded in $L^\infty(0, T; V)$ it has a subsequence $\{u_{n_j}\}$ that converges weakly-* to some $v \in L^\infty(0, T; V)$. We have, for all $\phi \in L^1(0, T; V^*)$,

$$\left| \int_0^T \langle u_{n_j}(t), \phi(t) \rangle \, dt \right| \le C \|\phi\|_{L^1(0,T;V^*)},$$

by using (11.9). Since $u_{n_j} \rightharpoonup v$ in $L^\infty(0, T; V)$ means that

$$\int_0^T \langle u_{n_j}(t), \phi(t) \rangle \, dt \to \int_0^T \langle v(t), \phi(t) \rangle \, dt$$

for all such ϕ, it follows that

$$\|v\|_{L^\infty(0,T;V)} \le C,$$

[i.e. the limit v satisfies (11.10)]. Since $L^2(0, T; V^*) \subset L^1(0, T; V^*)$, it follows that $u_{n_j} \rightharpoonup v$ in $L^2(0, T; V)$. By the uniqueness of weak limits, $u = v$ and so u satisfies (11.10).

To deduce (11.11), suppose that there exists a $t_0 \in [0, T]$ such that $\|u(t_0)\| > C$. Then, since $\|u(t)\| \le C$ almost everywhere, we can find a sequence $t_j \to t_0$ such that $\|u(t_j)\| \le C$ for every $j \ne 0$. Since $u(t_j)$ is bounded in V, it follows that some subsequence converges weakly to v with $\|v\| \le C$. Since $u(t_j) \to u(t_0)$ strongly in H, it follows that $v = u(t_0)$, and so in fact we do have $\|u(t_0)\| \le C$, and (11.11) follows. ∎

We now use the lemma to prove, rigorously, the existence of an absorbing set in $H_0^1(\Omega)$.

Proposition 11.3. *The reaction–diffusion equation has an absorbing set in H_0^1; there is a constant ρ_V and a time $t_1(|u_0|)$ such that*

$$\|u(t)\| \le \rho_V \quad \text{for all} \quad t \ge t_1(|u_0|).$$

Proof. We perform the same estimates as above, but work with the truncated Galerkin equations

$$du_n/dt + Au_n = P_n f(u_n), \qquad u_n(0) = P_n u_0$$

instead of the full PDE. We note that the calculations of Proposition 11.1 can be followed identically to show that

$$|u_n(t)| \le \rho_H \quad \text{for all} \quad t \ge t_0(|u_0|),$$

since $|u_n(0)| \le |u_0|$. Also, we can obtain (11.7) with u replaced by u_n, and then the equivalent of (11.8),

$$\|u_n(t)\| \le (1+l)\left[k|\Omega| + \rho_H^2\right] \quad \text{for all} \quad t \ge t_0(|u_0|) + 1,$$

a bound uniform in n. If we take

$$\rho_V = (1+l)\left[k|\Omega| + \rho_H^2\right] \quad \text{and} \quad t_1(|u_0|) = t_0(|u_0|) + 1,$$

then we can write, for any $T > t_1$,

$$\|u_n\|_{L^\infty(t_1,T;V)} \le \rho_V.$$

Since $u_n \rightharpoonup u$ in $L^2(0, T; V)$ and $u \in C^0([0, T]; H)$, it follows using Lemma 11.2 that

$$\|u(t)\| \leq \rho_V \qquad \text{for all} \qquad t \geq t_1(|u_0|). \qquad \blacksquare$$

It is easy to see that what we have done in the proof of this proposition is to perform exactly the same calculations as in the formal case and then to take limits. Because of this, most research papers and monographs tend to skip over this detail with a remark such as "this is a formal calculation, which can be made rigorous using a Galerkin truncation . . . ." Of course, this makes life easier for researchers, but we will tread carefully in this chapter. When we move on to the Navier–Stokes equations, however, we will adopt the simpler approach, relying on formal estimates and leaving the fully rigorous version for Galerkin enthusiasts.

11.1.3 The Global Attractor

Using the absorbing set in H_0^1, we can deduce the existence of a global attractor.

Theorem 11.4. *The reaction–diffusion equation has a connected global attractor $\mathcal{A}$.*

Proof. A bounded set in H_0^1 is compact in L^2. The result follows from Theorem 10.5, since L^2 is connected. $\blacksquare$

11.2 Regularity Results

We have shown that the global attractor is a bounded subset of both L^2 and H_0^1. However, in many situations we can do even better and show higher regularity of solutions in the attractor. In this section we will show that the attractor is bounded in L^∞, and also in H^2. This will enable us to prove (in Section 11.4) the existence of a Lyapunov functional for the equation, indicating that we have a gradient system. The results in the previous chapter then apply to limit its structure.

11.2.1 A Bound in L^∞

We start with the L^∞ bound. To prove the theorem we will use a truncated version of the function $u(x) - M$ for some appropriate constant M. For $u(x) \in L^2$

we define

$$u_+(x) = \begin{cases} u(x), & u(x) > 0, \\ 0, & \text{otherwise,} \end{cases}$$

and similarly

$$u_-(x) = \begin{cases} u(x), & u(x) < 0, \\ 0, & \text{otherwise.} \end{cases}$$

Clearly, if $u \in L^2(\Omega)$ then u_+ and u_- are in $L^2(\Omega)$, with

$$|u_+| \le |u| \qquad \text{and} \qquad |u_-| \le |u|.$$

Furthermore, if $u \in H^1(\Omega)$ then so are u_+ and u_-, as shown in the following lemma.

Lemma 11.5. *If $u \in H^1(\Omega)$ then so are u_+ and u_-, with*

$$\|u_+\|_{H^1} \le \|u\|_{H^1} \qquad \text{and} \qquad \|u_-\|_{H^1} \le \|u\|_{H^1}. \tag{11.12}$$

In fact,

$$Du_+(x) = \begin{cases} Du(x), & u(x) > 0, \\ 0, & \text{otherwise} \end{cases} \tag{11.13}$$

and

$$Du_-(x) = \begin{cases} Du(x), & u(x) < 0, \\ 0, & \text{otherwise.} \end{cases}$$

Proof. We prove the lemma for u_+; the result for u_- follows immediately since $u_- = -(-u)_+$. Consider the function u_ϵ defined by

$$u_\epsilon(x) = \begin{cases} (u(x)^2 + \epsilon^2)^{1/2} - \epsilon, & u(x) \ge 0, \\ 0, & \text{otherwise.} \end{cases}$$

Now, $u_\epsilon \in H^1(\Omega)$, with

$$u_\epsilon'(x) = \begin{cases} u'(x)\dfrac{u(x)}{(u(x)^2+\epsilon^2)^{1/2}}, & u(x) > 0, \\ 0, & \text{otherwise.} \end{cases}$$

Clearly $u_\epsilon \to u_+$ in L^2 and is uniformly bounded in H^1 by $\|u\|_{H^1}$. Since u_ϵ is uniformly bounded in the Hilbert space H^1 it has a weakly convergent subsequence, converging to some v. Since $u_\epsilon \to u_+$ in L^2, it follows that $v = u_+$, and so $u_+ \in H^1$, bounded as in (11.12). Noting that

$$u'_\epsilon(x) \to \begin{cases} Du(x), & u(x) > 0, \\ 0, & \text{otherwise}, \end{cases} \qquad \text{pointwise}$$

the equality (11.13) follows using Lebesgue's dominated convergence theorem.

∎

It follows immediately that

$$(Au, u_+) = a(u, u_+) = (Du, Du_+) = |Du_+|^2,$$

and we use this below.

Theorem 11.6. *The global attractor of the reaction–diffusion equation (11.1) is uniformly bounded in $L^\infty(\Omega)$, with*

$$\|u\|_\infty \le \left(\frac{k}{\alpha_2}\right)^{1/p} \qquad \text{for all} \qquad u \in \mathcal{A}.$$

Proof. Recall that f is bounded [from (11.2)] according to

$$f(s)s \le k - \alpha_2|s|^p.$$

It follows that

$$f(s) \le 0 \qquad \text{when} \qquad s \ge M = (k/\alpha_2)^{1/p}. \qquad (11.14)$$

If we multiply (11.1) by $(u(x) - M)_+$ and integrate over Ω we obtain

$$\frac{1}{2}\frac{d}{dt}\int_\Omega (u(x) - M)_+^2\, dx + \int_\Omega |\nabla(u - M)_+|^2\, dx = \int_\Omega f(u)(u - M)_+\, dx$$
$$\le 0,$$

using (11.14). Using Poincaré's inequality it follows that

$$\frac{d}{dt}\int_\Omega (u(x) - M)_+^2\, dx \le -C\int_\Omega (u(x) - M)_+^2\, dx,$$

and so

$$\int_\Omega (u(x) - M)_+^2\, dx \le e^{-Ct}\int_\Omega (u_0(x) - M)_+^2\, dx.$$

Since the attractor is bounded in L^2 and for any $v \in \mathcal{A}$ there exists a u_0 such that $v = S(t)u_0$, we have

$$\int_\Omega (u - M)^2_+ \, dx = 0$$

for all $u \in \mathcal{A}$. A similar argument can be applied to $(u + M)_-$ (see Exercise 11.4), which shows that

$$\|u\|_{L^\infty} \leq M \qquad \text{for all} \qquad u \in \mathcal{A}. \qquad \blacksquare$$

Note that this result does not tell us that there is an absorbing set in L^∞, but only that each solution on $\mathcal{A}$ is uniformly bounded. It is in fact possible to show that there is an L^∞ absorbing set, but the calculations are significantly more involved than those here (see Notes).

11.2.2 A Bound in $H^2(\Omega)$

We now use the L^∞ bound to deduce that the attractor is bounded in $H^2(\Omega)$. Again, note that we do not obtain an $H^2(\Omega)$ absorbing set (although this would follow from the existence of an absorbing set in L^∞, we avoided the lengthy proof of this above). Because of this, we will not prove the existence of a global attractor for the dynamical system

$$\left(H_0^1(\Omega), \{S(t)\}_{t \geq 0} \right),$$

which we could define for $m = 1$ and $m = 2$, as we saw in Chapter 8. This will not trouble us, since by treating weak solutions we include all the strong solutions too.

The rigorous argument for the H^2 bound is based on the following formal calculation. We use the equation

$$Au = -du/dt + f(u). \qquad (11.15)$$

The idea is to show that the right-hand side is bounded in L^2 and then use the theory of elliptic regularity to show that u must be bounded in H^2. Since $u \in L^\infty$, certainly $f(u) \in L^2$. We need to show that du/dt is also bounded in L^2 on $\mathcal{A}$. We will write $u_t = du/dt$ to simplify the presentation.

The formal calculations to obtain the bound on u_t are as follows. First, we estimate $\int_0^t |u_t|^2$. Multiply

$$u_t + Au = f(u) \qquad (11.16)$$

by u_t to obtain

$$|u_t|^2 + \tfrac{1}{2}\frac{d}{dt}\|u\|^2 = f(u)u_t = \frac{d}{dt}\int_\Omega F(u(x,t))\,dx, \qquad (11.17)$$

where $\mathcal{F}(s)$ is defined as before [near (8.28)], $\mathcal{F}(s) = \int_0^s f(\sigma)\,d\sigma$. Now integrate from 0 to t to give

$$\int_0^t |u_t|^2\,dt + \tfrac{1}{2}\|u(t)\|^2 = \tfrac{1}{2}\|u_0\|^2 + \int_\Omega F(u(x,t))\,dx - \int_\Omega F(u(x,0))\,dx.$$

Since $\mathcal{A}$ is bounded in H_0^1 and in L^∞, this gives

$$\int_0^t |u_t|^2\,dt + \tfrac{1}{2}\|u(t)\|^2 \le K, \qquad (11.18)$$

for some K.

Now we obtain the required bound on u_t in L^2. Differentiate (11.16) to give

$$\frac{d}{dt}u_t + Au_t = f'(u)u_t$$

and take the inner product with $t^2 u_t$ (this deals with the lack of smoothness in u_0 at $t=0$) to give

$$(t^2 u_t, \partial_t u_t) = t^2(u_t, \Delta u_t) + t^2(f'(u)u_t, u_t),$$

and so

$$\tfrac{1}{2}\frac{d}{dt}|tu_t|^2 - t|u_t|^2 + t^2\|u_t\|^2 \le t^2 l|u_t|^2.$$

Integrating between 0 and t again gives

$$|tu_t|^2 + \int_0^t t^2\|u_t\|^2\,ds \le \int_0^t (s + ls^2)|u_t|^2\,ds,$$

and since $(s + ls^2)$ is bounded on $[0, 1]$, we obtain, setting $t = 1$,

$$|u_t(1)|^2 \le (1+l)\int_0^1 |u_t|^2\,ds \le (1+l)K. \qquad (11.19)$$

Using (11.18) this gives an L^2 bound on $u_t(1)$, uniform over all the attractor.

Since any $u \in \mathcal{A}$ is given as $S(1)v$ for some $v \in \mathcal{A}$, it follows that $f(u) - du/dt$ is uniformly bounded in L^2 over all of $\mathcal{A}$. Since Au is uniformly bounded in L^2 it follows that u is uniformly bounded in H^2.

To make this rigorous we have to be careful in two ways. Just as before, the calculations we have made assume a smoothness of u and u_t that is not guaranteed by our existence theorem, so we should really calculate with the Galerkin approximations and take a careful limit. Second, Equation (11.15) only has to hold for almost every t, so we will have to exercise some care there too.

Theorem 11.7. *The global attractor of the reaction–diffusion Equation (11.1) is bounded in $H^2(\Omega)$.*

Proof. We know that the equality

$$Au = -du/dt + f(u) \qquad (11.20)$$

holds for a.e. t along a trajectory. Since $u \in L^\infty$, so is $f(u)$, and so certainly $f(u) \in L^\infty(0, T; L^2)$.

The bounds we obtain for u_t above can be obtained by using exactly the same manipulations on the Galerkin truncations, although we write the final conclusion (11.19) in a form to which we can apply Lemma 11.2:

$$\|du_n/dt\|_{L^\infty(1-\epsilon, 1+\epsilon; L^2)} \le (2 + l)K.$$

Arguing as before, we take the limit to find that

$$|du/dt(1)| \le (2 + l)K.$$

It follows, using (11.20), that $u(t) \in L^\infty(0, T; D(A))$.

Since the trajectory is continuous into L^2, we can use the argument of Lemma 11.2 to show that in fact we have a bound on $|Au(t)|$ that is uniform for all $0 < t < T$, and so the attractor is uniformly bounded in H^2. ∎

11.2.3 Further Regularity

With more sophisticated arguments, similar to those needed to prove the existence of an absorbing set in L^∞, it is possible to show that the regularity of the attractor increases as f becomes more regular. We will give only a statement of the result. A proof is outlined in Marion (1989).

Theorem 11.8. *If Ω is a bounded C^∞ domain and f is a C^∞ function, then the global attractor is a bounded subset of $H^k(\Omega)$ for every $k \ge 0$. In particular, if $u \in \mathcal{A}$ then $u \in C^\infty(\overline{\Omega})$.*

11.3 Injectivity on $\mathcal{A}$

We have seen that in general it is not possible to use the solutions of a PDE to define $S(t)$ for all $t \in \mathbb{R}$. However, for some examples it is possible to prove an injectivity, or "backwards uniqueness," property: if $S(t)u_0$ and $S(t)v_0$ are two trajectories starting from u_0 and v_0 at $t = 0$,

$$S(T)u_0 = S(T)v_0, \qquad \text{for some } T > 0, \qquad \Rightarrow \qquad u_0 = v_0.$$

One can view this as saying that if there is a solution that goes backwards in time, from u_0 to $u(-T)$, then there can only be one such solution.

We will prove this property for solutions lying on the attractor. We can then use Theorem 10.6 to show that

$$\left(\mathcal{A}, \{S(t)\}_{t \in \mathbb{R}} \right)$$

is a dynamical system.

The proofs in this section are essentially technical and can easily be skipped on first reading. We first prove a result under some abstract assumptions, which we will then check for the reaction–diffusion equation, and later for the 2D Navier–Stokes equations. The first step is a lemma that treats the quotient of the norms in V and H. Note that we can identify H with H^* by using the Riesz representation theorem (Theorem 4.9).

Lemma 11.9. *Let H and V be Hilbert spaces, where V^* is the dual of V, with $V \subset\subset H \simeq H^* \subset V^*$. Suppose that*

$$w \in L^\infty(0, T; V) \cap L^2(0, T; D(A))$$

satisfies

$$\frac{dw}{dt} + Aw = h(t, w(t))$$

as an equality in $L^2(0, T; H)$, where A is a bounded linear operator from V into V^*, and

$$|h(t, w(t))| \le k(t)\|w(t)\| \tag{11.21}$$

with $k(t) \in L^2(0, T)$. If we write

$$\Lambda(t) = \frac{\|w(t)\|^2}{|w(t)|^2}$$

then we have

$$\Lambda(t) \le \Lambda(0) \exp\left(2 \int_0^t k^2(s)\,ds\right).$$

We will take w to be the difference of two solutions when we apply the lemma.

Proof. Differentiating $\Lambda(t)$ gives

$$\frac{1}{2}\frac{d\Lambda}{dt} = \frac{((w',w))}{|w|^2} - \frac{\|w\|^2}{|w|^4}(w',w)$$

$$= \frac{(w',Aw)}{|w|^2} - \frac{\Lambda}{|w|^2}(w',w)$$

$$= \frac{1}{|w|^2}(w',Aw-\Lambda w)$$

$$= \frac{1}{|w|^2}(h-Aw,Aw-\Lambda w).$$

Now, $(Aw,w) = \Lambda(w,w)$, so

$$\frac{1}{2}\frac{d\Lambda}{dt} = -\frac{|Aw-\Lambda w|^2}{|w|^2} + \frac{1}{|w|^2}(Aw-\Lambda w,h).$$

Using the Cauchy–Schwarz and Young inequalities on the last term gives

$$\frac{1}{2}\frac{d\Lambda}{dt} \le -\frac{1}{2}\frac{|Aw-\Lambda w|^2}{|w|^2} + \frac{1}{2}\frac{|h|^2}{|w|^2},$$

so that, using the assumption on h (11.21), we get

$$\Lambda' + \frac{|Aw-\Lambda w|^2}{|w|^2} \le 2k^2\Lambda.$$

Dropping the second term gives

$$\frac{d\Lambda}{dt} \le 2k^2\Lambda,$$

and the result follows using Gronwall's inequality (Lemma 2.8). ∎

We use this to prove the following "backwards uniqueness" result.

Theorem 11.10. *Let* $w(t)$ *satisfy the assumptions from the previous lemma. If* $w(T) = 0$ *for some* $T > 0$, *then* $w(t) = 0$ *for all* $0 \le t \le T$.

Proof. Suppose not; then $w(t_0) \neq 0$ for some $t_0 \in [0, T)$. It follows from the assumptions on w that $dw/dt \in L^2(0, T; L^2)$, and so (using Corollary 7.3) we know that $w \in C^0([0, T]; H^1)$. Therefore, by continuity, we must have $w(t) \neq 0$ for some interval $(t_0, t_0 + \epsilon)$. We can denote by t_1 the largest time for which

$$|w(t)| \neq 0 \quad \text{on} \quad [t_0, t_1).$$

Clearly, $w(t_1) = 0$.

On the interval $[t_0, t_1)$ we can consider the function $t \mapsto \log |w(t)|$, and, differentiating, we have

$$
\begin{aligned}
\frac{d}{dt} \log \frac{1}{|w|} &= -\frac{1}{2} \frac{d}{dt} \log |w|^2 \\
&= -\frac{(w', w)}{|w|^2} \\
&= -\frac{(h - Aw, w)}{|w|^2} \\
&= \Lambda - \frac{(h, w)}{|w|^2} \\
&\leq \Lambda + k\Lambda^{1/2},
\end{aligned}
$$

by using the bound on h in (11.21). An application of Young's inequality produces

$$\frac{d}{dt} \log \frac{1}{|w|} \leq 2\Lambda + k^2.$$

Integrating this between t_0 and $t \in [t_0, t_1)$ yields

$$
\begin{aligned}
\log \frac{1}{|w(t)|} &\leq \log \frac{1}{|w(t_0)|} + \int_{t_0}^{t} (2\Lambda(s) + k(s)^2) \, ds \\
&\leq \log \frac{1}{|w(t_0)|} + \int_{t_0}^{T} (2\Lambda(s) + k(s)^2) \, ds.
\end{aligned}
$$

Since $k \in L^2(0, T)$ and so is Λ (by the previous lemma), this gives a uniform bound on $1/|w(t)|$ over $[t_0, T)$, a contradiction. ∎

To apply this abstract theorem to the reaction–diffusion equation is relatively straightforward.

Theorem 11.11. *The reaction–diffusion equation (11.1) has the injectivity property on the attractor: if $u(t)$ and $v(t)$ are two trajectories on $\mathcal{A}$ with $u(T) = v(T)$ for some $T > 0$ then $u(t) = v(t)$ for all $0 \leq t \leq T$.*

Proof. The equation for $w = u - v$ is

$$\frac{dw}{dt} + Aw = f(u) - f(v).$$

That $w \in L^{\infty}(0, T; V)$ is clear since the attractor is bounded in V; similarly, that $w \in L^2(0, T; D(A))$ follows since the attractor is also bounded in H^2 (Theorem 11.7). Since $h(t, w(t)) = f(u(t)) - f(v(t))$, we can use the bound on f in L^{∞} to write

$$
\begin{aligned}
|f(u) - f(v)|^2 &= \int_{\Omega} |f(u) - f(v)|^2 \, dx \\
&= \int_{\Omega} \left| \int_{v(x)}^{u(x)} f'(s) \, ds \right|^2 dx \\
&\leq \int_{\Omega} \|f'\|_{\infty}^2 |u(x) - v(x)|^2 \, dx \\
&\leq \|f'\|_{\infty}^2 |u - v|_{L^2}^2,
\end{aligned}
\tag{11.22}
$$

where $\|f'\|_{\infty}$ is the bound on $f'(u)$ that is uniform over all $u \in \mathcal{A}$. This inequality shows that

$$|h(t, w(t))| \leq \|f'\|_{\infty} |w(t)|,$$

which gives (11.21). ∎

Corollary 11.12. *The restriction of the semigroup $\{S(t)\}_{t \geq 0}$ to $\mathcal{A}$ gives rise to a dynamical system*

$$\left(\mathcal{A}, \{S(t)\}_{t \in \mathbb{R}} \right),$$

where the norm on $\mathcal{A}$ inherited from $L^2(\Omega)$ is used.

11.4 A Lyapunov Functional

Recall that a Lyapunov functional for a set $\mathcal{A}$ is a functional $\Phi : \mathcal{A} \mapsto \mathbb{R}$ that is

(i) continuous on $\mathcal{A}$,

(ii) nonincreasing along trajectories [$\Phi(u(t))$ is nonincreasing as a function of t].

Furthermore,

(iii) if $\Phi(u(t)) = \Phi(u_0)$ for some $t > 0$ then u_0 is a fixed point.

Proposition 11.13. *The functional*

$$\Phi(u) = \int_\Omega \tfrac{1}{2}|\nabla u|^2 - \mathcal{F}(u)\,dx,$$

where $\mathcal{F}(s) = \int_0^s f(\sigma)\,d\sigma$, is a Lyapunov functional on $\mathcal{A}$ for the reaction–diffusion equation (11.1).

Proof. We saw above (11.17) that

$$\frac{d}{dt}\Phi(u) = -|u_t|^2,$$

so that Φ is clearly nonincreasing along trajectories, and if $\Phi(u(T)) = \Phi(u_0)$ we must have $u(t) \equiv u_0$ for all $0 \le t \le T$, and so u_0 must be a fixed point. It remains only to check the continuity of $\Phi(u)$ on $\mathcal{A}$. Now, the first part of the integral is just $\|u\|^2$, and from the definition of A we have

$$\|u\|^2 = ((u, u)) = (Au, u) \le |Au||u| \le |u|\|u\|_{H^2}.$$

It follows that $\|u\|^2$ is Lipschitz from $L^2(\Omega)$ into $\mathbb{R}$. For the second term, we can write [cf. (11.22)]

$$\left|\int_\Omega \mathcal{F}(u) - \mathcal{F}(v)\right|dx = \int_\Omega \left|\int_{v(x)}^{u(x)} f(s)\,ds\,dx\right|$$

$$\le \int_\Omega \|f\|_\infty |u(x) - v(x)|\,dx$$

$$\le \|f\|_\infty |u - v|_{L^2},$$

so that the second term is also Lipschitz into $\mathbb{R}$. ∎

We now know from Theorem 10.13 that the attractor is given by

$$\mathcal{A} = W^u(\mathcal{E}), \tag{11.23}$$

where $\mathcal{E}$ is the set of fixed points, which are the solutions of the equation

$$-\Delta u = f(u(x)).$$

If $\mathcal{E}$ is discrete then we can further simply (11.23) and deduce that the attractor in this case consists of the unstable manifolds of the fixed points,

$$\mathcal{A} = \bigcup_{z \in \mathcal{E}} W^u(z).$$

We will now investigate a particular example in which we can show that the equilibrium points are indeed discrete and say something more about the structure of the attractor.

11.5 The Chaffee–Infante Equation

This is a much-studied example on a one-dimensional domain $[0, \pi]$, with the simple nonlinear term $u - u^3$:

$$u_t - u_{xx} = \lambda(u - u^3), \qquad u(0) = u(\pi) = 0. \tag{11.24}$$

We have also included a parameter λ that enables us to adjust the relative balance of the diffusion term and the nonlinear term. For small λ we would expect the diffusion to dominate and the behaviour to be very simple. Indeed, we can write

$$\frac{1}{2}\frac{d}{dt}|u|^2 + \|u\|^2 = \lambda \int_\Omega |u|^2 - |u|^4 \, dx,$$

and so

$$\frac{1}{2}\frac{d}{dt}|u|^2 + \|u\|^2 \le \lambda|u|^2.$$

It follows from the Fourier sine expansion of u that $\|u\| \ge |u|$ (the constant in Poincaré's inequality is 1), and so this becomes

$$\frac{1}{2}\frac{d}{dt}|u|^2 \le (1 - \lambda)|u|^2. \tag{11.25}$$

Clearly, if $\lambda < 1$ then $u(t) \to 0$ as $t \to \infty$ and there is just one stable fixed point, $u \equiv 0$. For λ larger we would expect the behaviour, and hence the attractor, to become more complicated.

11.5.1 Stationary Points

We want to investigate the stationary points of the equation, solutions of

$$-u_{xx} = \lambda(u - u^3), \qquad u(0) = u(\pi) = 0. \tag{11.26}$$

Taking $v = u_x$ we can write a coupled pair of differential equations in place of (11.26):

$$u_x = v,$$
$$v_x = -\lambda(u - u^3). \tag{11.27}$$

We can now apply phase plane ideas, treating x as the time variable. It follows from (11.27) that, on any trajectory,

$$\tfrac{1}{2}|u_x|^2 + \lambda\left(\tfrac{1}{2}u^2 - \tfrac{1}{4}u^4\right) = \text{constant}.$$

We write this as

$$|u_x|^2 + \lambda\left(u^2 - \tfrac{1}{2}u^4\right) = \lambda E, \tag{11.28}$$

where E is a constant.

The phase portrait for (11.27) is given in Figure 11.1. To satisfy the boundary conditions in (11.26) we need a trajectory that starts on the v axis at $x = 0$ and moves back onto the v axis when $x = \pi$. Clearly, the only possible trajectories in the phase diagram are those that are on closed orbits about the origin. These correspond to values of $0 \le E < \tfrac{1}{2}$; we will investigate therefore the time it takes to traverse half of such an orbit, as indicated by the heavy black portion in the figure.

For a given value of E, the "velocity" in the u coordinate is u_x, and this is given using (11.28) as

$$u_x = \sqrt{\lambda}\left(E - u^2 + \tfrac{1}{2}u^4\right)^{1/2}.$$

Now, a trajectory with this fixed value of E starts at $u = 0$, $v = \sqrt{\lambda E}$ and moves around clockwise until it strikes the u axis at a value $u = u_0$, where

$$u_0^2 - \tfrac{1}{2}u_0^4 = E.$$

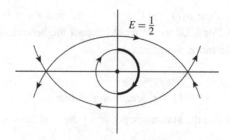

Figure 11.1. The phase portrait for Equation (11.27).

The "time" $t(E)$ it has taken to reach this point is then given by

$$t(E) = \frac{1}{\sqrt{\lambda}} \int_0^{u_0} \left(E - u^2 + \tfrac{1}{2}u^4\right)^{-1/2} du.$$

The properties of the fixed points of Equation (11.24) follow from the following properties of the integral $t(E)$:

(i) As $E \to \frac{1}{2}, t(E) \to \infty$.
(ii) As $E \downarrow 0^+, t(E) \to \pi/\sqrt{\lambda}$.
(iii) $t(E)$ is strictly increasing as a function of E.

In particular, it follows that

$$\frac{\pi}{\sqrt{\lambda}} < t(E) < \infty.$$

To obtain a solution of (11.26) from a trajectory of (11.27), we need to be on a trajectory where $nt(E) = \pi$ for some integer n; we circle around $n/2$ times, ending up back on the v axis. To find the number of fixed points we therefore have to find the number of values of E for which $nt(E) = \pi$.

First, if $\lambda < 1$ then $\frac{\pi}{\sqrt{\lambda}} > \pi$, and so the only trajectory that fulfils our criteria is the origin. This corresponds to the steady state $u \equiv 0$, and we label this state ϕ_0.

Now, if $1 < \lambda < 2^2$, the values of E are bounded below by $\pi/2$ but do include the value π. Therefore there are two new fixed points, corresponding to orbits that perform a half loop, and we call these $\phi_1^{\pm}$.

For $2^2 < \lambda < 3^2$ we now have orbits that loop around $3/2$ times, since one of the orbits has $t(E) = \pi/3$; these are $\phi_2^{\pm}$. We now have five fixed points.

This argument now provides us with a full set of fixed points.

Proposition 11.14. *If $n^2 < \lambda \le (n+1)^2$ then there are $2n+1$ fixed points, ϕ_0, and n pairs, $\phi_1^{\pm}, \ldots, \phi_n^{\pm}$. The function $\phi_j^{\pm}$ has j zeros in $(0, \pi)$.*

Theorem 10.13 therefore applies, and we know that the attractor is given as the union of the unstable manifolds of these fixed points. To obtain some more information, we would have to investigate the stability of these fixed points. Rather than do this in detail, we will investigate only the linearisation near the zero state. This at least gives some indication of the stability of the fixed points that bifurcate from $u \equiv 0$ as λ crosses through square integer values.

11.5.2 Bifurcations around the Zero State

We end with a very heuristic investigation of the stability of the fixed point at zero, assuming that approaches valid in the finite-dimensional setting remain so here. We give references in the notes for rigorous treatments.

To study the stability of a stationary point of the equation

$$du/dt = G(u),$$

where $G(u)$ is in general some unbounded nonlinear operator, we consider the problem

$$\frac{d}{dt}(u^* + \epsilon v) = G(u^* + \epsilon v),$$

where ϵv is a small perturbation away from the fixed point u^*, and so we study the linearised equation

$$\frac{d}{dt}v = DG(u^*)v.$$

We investigate this by looking for a solution $v(t) = we^{\sigma t}$, which gives an eigenvalue problem for w,

$$DG(u^*)w = \sigma w.$$

If all eigenvalues have $\mathrm{Re}(\sigma) > 0$ then the fixed point is stable; otherwise it is unstable.

For our example, we are interested in the solutions of

$$w_{xx} - \lambda(1 - 3u^2)w = \sigma w.$$

(The validity of this linearisation will be considered in Chapter 13.) If we choose the fixed point $u \equiv 0$ this becomes the simple equation

$$w_{xx} - \lambda w = \sigma w.$$

Since we already know that the eigenfunctions of the problem

$$\phi_{xx} = \mu\phi, \qquad \phi(0) = \phi(\pi) = 0$$

are simply $\phi_n = \sin nx$ with $\mu_n = n^2$, the eigenvalues about the origin are

$$\sigma_n = \lambda - n^2.$$

When $\lambda < 1$ the origin is stable [as we saw in (11.25)]. Each time λ^2 passes through an integer value, another unstable direction appears. We therefore have successive saddle node bifurcations from the origin that produce all our pairs of fixed points.

In fact, one can show that, once they have appeared, the first bifurcating pair $\phi_1^{\pm}$ is stable, but that all the other fixed points are unstable. This enables us to give an indication of the structure of the attractor.

For $\lambda < 1$ we have a single fixed point. If $1 < \lambda < 4$ we have three fixed points, configured along a line as in Figure 11.2(a). If $4 < \lambda < 9$, two more fixed points have appeared, and the attractor is like the deformed disc in Figure 11.2(b). For $9 < \lambda < 16$ we have yet another two points, and the attractor is now a 3D object, something like Figure 11.2(c). This whole

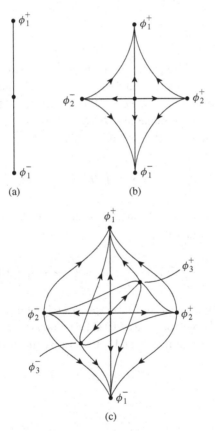

Figure 11.2. Sketches of the attractor for the Chaffee–Infante equation when (a) $1 < \lambda < 4$, (b) $4 < \lambda < 9$, and (c) $9 < \lambda < 16$.

picture can be made rigorous with the appropriate arguments; see Henry (1984).

Note that this discussion indicates that the attractor consists of unions of n-dimensional manifolds when $n^2 < \lambda < (n+1)^2$, and so we would expect

$$d(\mathcal{A}) \sim \lambda^{1/2}, \tag{11.29}$$

where d is some measure of the dimension of $\mathcal{A}$.

In this example an unusually detailed knowledge of the structure of the attractor is possible. Usually, one is not able to obtain such information, and we have to rely on other methods to investigate the asymptotic dynamics. In the next chapter we treat the Navier–Stokes equations and then go on to discuss various methods that are applicable to general attractors, whether or not they possess a Lyapunov functional.

Exercises

11.1   If we had chosen to perform the analysis of Chapter 8 using periodic boundary conditions, we could have obtained the same results. However, we would not have had the Poincaré inequality, since the equation forces us to work in $L^2(Q)$ rather than $\dot{L}^2(Q)$. So we need a different approach to enable us to prove Proposition 11.1. By using

$$\int_\Omega |u|^2 \, dx \le \left(\int_\Omega |u|^p \, dx \right)^{2/p} |\Omega|^{1-(2/p)} \tag{11.30}$$

[which follows from Hölder's inequality with exponents $p/2$ and $p/(p-2)$] deduce that

$$\frac{1}{2} \frac{d}{dt} |u|^2 + \frac{2\alpha_2}{p} |u|^2 \le \left(\frac{2}{p-2} + k \right) |\Omega|,$$

and so obtain the results of Proposition 11.1 in this case.

11.2   Generalise the method used to get (11.8) to obtain the "uniform Gronwall lemma." Prove that if x, a, and b are positive functions such that

$$dx/dt \le ax + b$$

with

$$\int_t^{t+r} x(s) \, ds \le X, \qquad \int_t^{t+r} a(s) \, ds \le A, \qquad \text{and} \qquad \int_t^{t+r} b(s) \, ds \le B$$

for some $r > 0$ and all $t \geq t_0$, then

$$x(t) \leq \left(\frac{X}{r} + B\right)e^A$$

for all $t \geq t_0 + r$. [Hint: Use the integrating factor

$$\exp\left(-\int_s^t a(\tau)\,d\tau\right)$$

with $t_0 \leq t \leq s \leq t + r$.]

11.3 An alternative approach to finding an absorbing set in $H_0^1(\Omega)$ would be to obtain estimates on u to show that Corollary 7.3 can in fact be applied. Assume that $m \leq 3$, and remember that $f(0) = 0$. Then take the inner product of

$$\frac{du_n}{dt} + Au_n = P_n f(u_n)$$

with $t^2 A u_n$, and show that [cf. analysis leading to (8.26)]

$$\frac{1}{2}\frac{d}{dt}\|tu_n\|^2 - 2t\|u_n\|^2 + t^2|Au_n|^2 \leq lt^2\|u_n\|^2.$$

Deduce that u_n is uniformly bounded in $L^2(t, T; D(A))$ for all $0 < t < T$. Now follow the analysis of Section 11.1.2, multiplying by $t^2(u_n)_t$ rather than $(u_n)_t$, to show that $(u_n)_t$ is uniformly bounded in $L^2(t, T; L^2)$ for all $0 < t < T$. It then follows that $u(t)$ is in $L^2(t, T; D(A))$ and that, for $m \leq 3$, $u_t(t)$ is in $L^2(t, T; L^2)$. Corollary 7.3 can be applied to deduce (11.7) on $(0, T)$ and render the calculations there rigorous.

11.4 Repeat the argument of Theorem 11.6, instead multiplying through by $(u(x) + M)_-$, to deduce that

$$\int_\Omega (u + M)_-^2\,dx = 0 \qquad \text{for all} \qquad u \in \mathcal{A}.$$

Notes

The analysis leading to the existence of a global attractor given here is due essentially to Marion (1987); it also appears in Temam (1988), although the rigorous presentation in Section 11.1.2 is inspired by the existence proofs in Evans (1998). See also Babin & Vishik (1992).

The L^∞ bound in the form here can be found in Temam (1988), and the argument for the properties of u_+ and u_- is similar to that in Gilbarg & Trudinger (1983). Marion

(1987) contains the more involved arguments that yield an L^∞ absorbing set, and hence an absorbing set in H^2. The arguments that give higher regularity (as in Theorem 11.8) are outlined in Marion (1989).

The backwards uniqueness result is taken from Temam (1988); his proof in the reaction–diffusion case is much simplified here since we restricted ourselves to solutions on the attractor rather than considering all solutions. Indeed, without further assumptions on the nonlinear term $f(s)$ a general result on backwards uniqueness is unknown.

The analysis of the Lyapunov functional given here is based on Babin & Vishik (1983; see also 1992). The investigation of the structure of the attractor is based largely on the presentation in Hale (1988) and Henry (1984); the results are originally due to Chaffee & Infante (1974).

Hale (1988) treats the semigroup on $H_0^1(\Omega)$ using other methods, based largely on the Lyapunov functional and his more powerful theorems for the existence of attractors discussed briefly in the Notes at the end of Chapter 10. However, this treatment depends on restricting the function f more than we need to here.

12

The Global Attractor for the Navier–Stokes Equations

In this chapter we will investigate the existence of global attractors for the Navier–Stokes equations. We will first consider the 2D equations, for which standard formal estimates (which can be made rigorous using a Galerkin method and taking limits) prove the existence of absorbing sets in $\mathbb{L}^2$, $\mathbb{H}^1$, and $\mathbb{H}^2$. We therefore can show the existence of an attractor for both the dynamical system on $H \subset \mathbb{L}^2$,

$$H = \left\{ u \in \dot{\mathbb{L}}_p^2(Q) : \nabla \cdot u = 0 \right\},$$

and that on $V \subset \mathbb{H}^1$,

$$V = \left\{ u \in \dot{H}_p^1(Q) : \nabla \cdot u = 0 \right\}.$$

Recall that we use $|u|$, $\|u\|$, and $\|u\|_*$ to denote the norm of u in H, V, and V^* (the dual of V) respectively.

We then go on to consider the 3D equations. Although we could not even show that they generated a semidynamical system in Chapter 9, here we take this as an assumption and prove that this implies the existence of a global attractor.

The global attractors we obtain in this chapter do not possess a Lyapunov functional, so we will be unable to determine their structure in the detailed way we did for reaction–diffusion equations in the previous chapter.

12.1 2D Navier–Stokes Equations

We first show the existence of absorbing sets in increasingly regular spaces. Only the first calculation is not "formal." The others need justifying via the Galerkin method, which we leave to the (over)enthusiastic reader.

We consider throughout this section the dynamical system on H defined by virtue of Theorems 9.3 and 9.4 (see end of Section 9.5). We will see that even if $u_0 \in H$, eventually $u(t)$ is smoother (we will show that $u(t) \in V$ for $t \geq t_0$, and $u(t) \in D(A)$ for $t \geq t_1$).

Proposition 12.1. *If $f \in V^*$ then there is an absorbing set in H: there exists a time $t_0(|u_0|)$, a ρ_H, and an I_V such that*

$$|u(t)| \leq \rho_H \quad and \quad \int_t^{t+1} \|u(s)\|^2 \, ds \leq I_V \qquad (12.1)$$

for all $t \geq t_0(|u_0|)$.

The integral bound on $\|u(t)\|^2$ will be useful in obtaining an asymptotic bound on $\|u(t)\|$ in the next section.

Proof. As in our analysis of the Galerkin truncations of the equation in Chapter 9, we take the inner product of the equation

$$du/dt + \nu Au + B(u, u) = f$$

with u and obtain

$$\tfrac{1}{2} \frac{d}{dt} |u|^2 + \nu \|u\|^2 + b(u, u, u) = \langle f, u \rangle.$$

Since $b(u, u, u) = 0$ (9.22), we have

$$\tfrac{1}{2} \frac{d}{dt} |u|^2 + \nu \|u\|^2 \leq \|f\|_* \|u\|,$$

which after application of Young's inequality yields

$$\frac{d}{dt} |u|^2 + \nu \|u\|^2 \leq \frac{\|f\|_*^2}{\nu}. \qquad (12.2)$$

(Recall that we use $\|f\|_*$ for the norm of f in V^*.) Using the Poincaré inequality (9.2) we have $\|u\|^2 \geq \lambda_1 |u|^2$, and so

$$\frac{d}{dt} |u|^2 + \nu \lambda_1 |u|^2 \leq \frac{\|f\|_*^2}{\nu}.$$

By Gronwall's lemma (Lemma 2.8) we immediately deduce that

$$|u(t)|^2 \leq |u_0|^2 \exp(-\nu \lambda_1 t) + \frac{\|f\|_*^2}{\nu^2 \lambda_1} (1 - e^{-\nu \lambda_1 t}), \qquad (12.3)$$

and so there is a time $t_0(|u_0|)$, which we can take as

$$t_0(|u_0|) = \max\left(-\frac{1}{\nu\lambda_1}\ln\left(\frac{\|f\|_*^2}{\nu^2\lambda_1|u_0|}\right), 0\right)$$

such that for all $t \geq t_0$

$$|u(t)|^2 \leq 2\frac{\|f\|_*^2}{\nu^2\lambda_1} \equiv \rho_H^2. \tag{12.4}$$

The bound on the integral of $\|u(s)\|^2$ follows by integrating (12.2) between t and $t + 1$ to obtain

$$\nu\int_t^{t+1}\|u(s)\|^2\,ds \leq \frac{\|f\|_*^2}{\nu} + |u(t)|^2$$

and then using (12.4). ∎

Note that the results of this proposition also hold when $m = 3$, since we have only used the orthogonality property of b (9.22), rather than any of the bounds (9.25–9.27). However, for $m = 3$ we have to obtain these estimates via the Galerkin process: we cannot apply Theorem 7.2, since we only have $du/dt \in L^{4/3}(0, T; V^*)$.

12.1.2 An Absorbing Set in $\mathbb{H}^1$

We now improve on Proposition 12.1, making use of the integral bound on $\|u(s)\|^2$ to obtain an absorbing set in V. We require more regularity of the forcing term, but no more regularity of u_0. The estimates in the rest of this chapter are formal and strictly require justification via the Galerkin method.

Proposition 12.2. *If $f \in H$ then there is an absorbing set in V: there exists a time $t_1(|u_0|)$, a ρ_V, and an I_A such that*

$$\|u(t)\| \leq \rho_V \quad \text{and} \quad \int_t^{t+1}|Au(s)|^2\,ds \leq I_A \tag{12.5}$$

for all $t \geq t_1(|u_0|)$.

The integral bound on $|Au(s)|$ will help us obtain an absorbing set in $D(A)$ in the next section.

Proof. We take the inner product of the equation with Au and obtain, since $b(u, u, Au) = 0$ (9.23),

$$\frac{d}{dt}\|u\|^2 + \nu|Au|^2 \leq \frac{|f|^2}{\nu}. \tag{12.6}$$

Neglecting the second term* gives

$$\frac{d}{dt}\|u\|^2 \leq \frac{|f|^2}{\nu}.$$

Integrating both sides between s and t, with $t - 1 \leq s < t$, gives

$$\|u(t)\|^2 \leq \|u(s)\|^2 + \frac{|f|^2}{\nu}.$$

Now integrate again, between $s = t - 1$ and $s = t$,

$$\|u(t)\|^2 \leq \int_{t-1}^{t} \|u(s)\|^2 \, ds + \frac{|f|^2}{\nu}.$$

Provided that $t \geq t_1(|u_0|) = t_0(|u_0|) + 1$ we can use the integral bound in (12.1) to deduce that

$$\|u(t)\|^2 \leq \rho_V^2 \equiv I_V + \frac{|f|^2}{\nu},$$

an absorbing set in V.

Now if we integrate (12.6) between t and $t + 1$ we obtain

$$\nu \int_t^{t+1} |Au(s)|^2 \leq \frac{|f|^2}{\nu} + \|u(t)\|^2,$$

which gives (12.5) with $\nu I_A = |f|^2/\nu + \rho_V^2$. ∎

This result shows that there is an attractor for the dynamical system on $H = \mathbb{L}^2$ provided that $f \in H$. Also, we have shown that any solution on the

---

* Since we have $|Au|^2 \geq \lambda_1\|u\|^2$, we could write

$$\frac{d}{dt}\|u\|^2 + \nu\lambda_1\|u\|^2 \leq \frac{|f|^2}{\nu}.$$

It is tempting to do this and use Gronwall's inequality (Lemma 2.8) to derive the inequality

$$\|u(t)\|^2 \leq \|u_0\| \exp(-\nu\lambda_1 t) + \frac{|f|^2}{\nu^2\lambda_1}(1 - e^{-\nu\lambda_1 t})$$

[cf. (12.3)]. However, this gives an estimate of $\|u(t)\|$ that is based on the norm in V of u_0. Since we are only assuming that $u_0 \in H$, this is no help, and we are forced to adopt the uniform Gronwall "trick" we used in the previous chapter.

attractor is in $L^\infty(0, T; V) \cap L^2(0, T; D(A))$ and so must be a strong solution (cf. Theorem 9.5).

Theorem 12.3. *If $f \in H$ then the dynamical system on H generated by the 2D Navier–Stokes equations has a global attractor $\mathcal{A}_H$, and the solutions on $\mathcal{A}_H$ are strong solutions of the original equation.*

Note that in Chapter 9 we showed that the 2D equations generated a semigroup not only on H but also on V. The question whether there is a global attractor in V arises naturally. We have just shown that there is a bounded (but not compact) absorbing set in V, and so we *could* define a set by (cf. (10.6))

$$\mathcal{A}_W = \bigcap_{T \geq 0} \overline{\bigcup_{t \geq T} S(t)B}, \tag{12.7}$$

where B is the absorbing ball in V from Proposition 12.2, but the closure is taken in H. This yields a set $\mathcal{A}_W$ that is bounded (but not compact) in V and attracts all orbits in the norm of H. (That this set is bounded in V follows from the result of Exercise 12.3.) However, a global attractor "in V" has to attract in the norm of V. To obtain a proper global attractor for the semigroup on V we have to find a compact absorbing set in V. We do this by showing that there is an absorbing set in $\mathbb{H}^2$.

12.1.3 An Absorbing Set in $\mathbb{H}^2$

With a little more work we can show that there is also an absorbing set in $D(A)$ by finding an asymptotic bound on $|Au|$ and hence, using the regularity result (9.11) for the Stokes operator, an asymptotic bound on $\|u\|_{\mathbb{H}^2}$. For the dynamical system on H this amounts to a regularity result for the attractor. For the dynamical system on V it shows the existence of a compact absorbing set and therefore an attractor in V.

Proposition 12.4. *If $f \in H$ is independent of t then there is an absorbing set in $D(A)$: there exists a time $t_2(|u_0|)$ and a ρ_A such that*

$$|Au(t)| \leq \rho_A \qquad \text{for all} \qquad t \geq t_2(|u_0|). \tag{12.8}$$

In particular, $\mathcal{A}_H$ is bounded in $\dot{\mathbb{H}}_p^2(Q)$.

Note that if f is independent of time (which we need to prove the existence of a global attractor) then $f \in H$ suffices to obtain an absorbing set in $\mathbb{H}^2$

as in this proposition. If f is allowed to depend on t then we need f to be more regular to obtain asymptotic bounds on $|Au(t)|$. This case is treated in Exercises 12.1 and 12.2.

Proof. The idea is to use estimates on the size of the time derivative u_t to help with estimates on the derivatives of u. Indeed, the first step of the proof is to show that the norm of u_t is related to the norm of Au.

Observe first that the bound (9.27) implies that if $u \in D(A)$ then $B(u, u) \in H$, with

$$|B(u, u)| \le k|u|^{1/2}\|u\||Au|^{1/2}. \tag{12.9}$$

It then follows from the governing equation

$$du/dt + \nu Au + B(u, u) = f$$

that

$$|u_t| \le |Au| + k|u|^{1/2}\|u\||Au|^{1/2} + |f|.$$

An application of Young's inequality gives

$$|u_t| \le \tfrac{3}{2}|Au| + \tfrac{1}{2}k^2|u|\|u\|^2 + |f|,$$

and so, using (12.1) and (12.5), we obtain

$$|u_t| \le c|Au| + c\rho_H\rho_V^2 + |f|$$

once t is large enough. It follows that for such large t we can translate the bound from Proposition 12.2 for

$$\int_t^{t+1} |Au(s)|^2 \, ds$$

into a bound on

$$\int_t^{t+1} |u_t|^2 \, ds \le C_t \equiv C\big[I_A + \rho_H^2\rho_V^4 + |f|^2\big]. \tag{12.10}$$

We now differentiate the equation with respect to t and take the inner product with u_t to obtain

$$\tfrac{1}{2}\frac{d}{dt}|u_t|^2 + \nu\|u_t\|^2 \le |b(u_t, u, u_t)|$$

$$\le k\|u\|\|u_t\|\|u_t\|$$

$$\le \frac{\nu}{2}\|u_t\|^2 + \frac{k^2}{2\nu}\|u\|^2|u_t|^2.$$

It follows that for t large enough

$$\frac{d}{dt}|u_t|^2 \leq \frac{k^2 \rho_V^2}{\nu}|u_t|^2.$$

Using the "uniform Gronwall" trick, we integrate the equation between s and $t + 1$ with $t < s < t + 1$ to get

$$|u_t(t+1)|^2 \leq |u_t(s)|^2 + \frac{k^2 \rho_V^2}{\nu}\int_s^{t+1}|u_t(s)|^2\,ds,$$

and then again between t and $t + 1$ so that

$$|u_t(t+1)|^2 \leq \left(1 + \frac{k^2 \rho_V^2}{\nu}\right)\int_t^{t+1}|u_t(s)|^2\,ds \leq C_t\left(1 + k^2 \rho_V^2/\nu\right), \quad (12.11)$$

using the bound from (12.10).

It now follows that

$$|Au| \leq |u_t| + |B(u,u)| + |f|,$$

which we can rearrange using (12.9) to give

$$|Au| \leq 2|u_t| + k^2 \rho_H \rho_V^2 + 2|f|,$$

and so we have that $|Au(t+1)|$ is bounded, using (12.11). ∎

Since $|u_0| \leq \lambda_1^{-1/2}\|u_0\|$, it follows that if $u_0 \in V$ then there exists a time $\tilde{t}_0(\|u_0\|) = t_1(\lambda_1^{-1/2}\|u_0\|)$ such that

$$\|u(t)\| \leq \rho_V \quad \text{and} \quad |Au(t)| \leq \rho_A \qquad \text{for all} \qquad t \geq \tilde{t}_0(\|u_0\|),$$

so there is a compact absorbing set in V. Therefore the dynamical system on V has a global attractor.

Theorem 12.5. *If $f \in H$ then the dynamical system on V generated by the 2D Navier–Stokes equations has a global attractor $\mathcal{A}_V$.*

12.1.4 Comparison of the Attractors in H and V and Further Regularity Results

If $f \in H$ then Theorem 12.3 and Theorem 12.5 both ensure the existence of a global attractor. It is a natural question whether these attractors coincide. They do, and this is ensured by the regularity in Proposition 12.4.

Lemma 12.6. *If $f \in H$ then $\mathcal{A}_H = \mathcal{A}_V$.*

Proof. The asymptotic bound on $|Au|$ shows that $\mathcal{A}_H$ is bounded in $D(A)$ and therefore compact in V. It is thus a compact invariant set in V. Since the global attractor is the maximal compact invariant set, we must have $\mathcal{A}_H \subset \mathcal{A}_V$.

Since the compact absorbing set in H contains the compact absorbing set in V and the norm in H is weaker than the norm in V, expression (10.10) for $\omega(B)$ shows that $\mathcal{A}_H \supset \mathcal{A}_V$. Thus the two attractors are equal. ∎

It is worth emphasising that it is the regularity result that $\mathcal{A}_H$ is bounded in $D(A)$ that is important in the first part of the proof. Further regularity properties of the attractor can be proved. In fact, the attractor consists of smooth functions on $Q = [0, L]^2$. The argument follows the pattern of Proposition 12.4, but the calculations are more involved, and we will not give the proof here. It can be found in Guillopé (1982) or Temam (1988).

Theorem 12.7. *If $f \in \dot{C}_p^\infty(Q)$ then the global attractor is bounded in $\dot{\mathbb{H}}_p^m(Q)$ for every $m \geq 0$. In particular, if $u \in \mathcal{A}$ then $u \in \dot{C}_p^\infty(Q)$.*

12.1.5 Injectivity on the Attractor

It is now simple to apply Theorem 11.10 to prove the injectivity property on the attractor.

Theorem 12.8. *If $f \in H$ then the 2D Navier–Stokes equations have the injectivity property on the attractor $\mathcal{A}_H$.*

Proof. We have to check the assumptions of Lemma 11.9; if $w = u - v$, with u and v two solutions on $\mathcal{A}$, then we need $w \in L^\infty(0, T; V) \cap L^2(0, T; D(A))$, and if the equation for dw/dt is

$$dw/dt + Aw = h(t, w(t))$$

we need

$$|h(t, w(t))| \leq k(t)\|w(t)\|, \qquad (12.12)$$

with $k(t) \in L^2(0, T)$.

If $f \in H$ then Proposition 12.2 shows that the attractor is bounded in V and that solutions lie in $L^2(0, T; D(A))$. We can then use the inequalities for

$b(u, v, w)$ in (9.25) and (9.27) to show that

$$|h(t, w(t))| = |B(u, u) - B(v, v)|$$
$$\leq |B(u, w)| + |B(w, v)|$$
$$\leq k\left[\|u\|_\infty\|w\| + |w|^{1/2}\|w\|^{1/2}\|v\|^{1/2}|Av|^{1/2}\right].$$

Using the result of Exercise 12.4,

$$\|u\|_\infty \leq C|Au|^{1/2}|u|^{1/2} \qquad \text{for all} \qquad u \in D(A),$$

we find that this becomes

$$|h(t, w(t))| \leq C\left[|u|^{1/2}|Au|^{1/2}\|w\| + |w|^{1/2}\|w\|^{1/2}\|v\|^{1/2}|Av|^{1/2}\right]$$
$$\leq C(|Au| + |Av|)\|w\|.$$

Since $u, v \in L^2(0, T; D(A))$, $h(t, w(t))$ satisfies (12.12), and Theorem 11.10 then implies injectivity. ∎

Corollary 12.9. *The solutions of the 2D Navier–Stokes equations generate a dynamical system*

$$\left(\mathcal{A}, \{S(t)\}_{t\in\mathbb{R}}\right)$$

when restricted to the global attractor.

Note since $\mathcal{A}_H = \mathcal{A}_V$, $S_V(t) = S_H(t)$ on the attractor, and we do not need to distinguish between these two cases in the above corollary.

12.2 The 3D Navier–Stokes Equations

In Chapter 9 we could not show that the 3D Navier–Stokes equations generated unique weak solutions, nor could we show that the strong solutions, which are unique, exist for all time. Trying to investigate the existence of attractors without the guarantee of a sensible semigroup seems futile. However, in this section we will show that if we are prepared to assume that the equations generate a semigroup on V (i.e. if we assume the existence of strong solutions), then the equations must have a global attractor.

The main result proves the existence of an absorbing set in V. A much more straightforward argument (whose proof, along the lines of Proposition 12.4, we relegate to an exercise) can then be used to show that there is also an absorbing set in $D(A)$ and hence a global attractor.

What we are doing here is making a physically reasonable assumption in a mathematically precise way and then deducing an entirely mathematical consequence. It allows us to consider the asymptotic regimes of the "true" Navier–Stokes equations, and thus fully developed turbulence, within a mathematical framework.

The analysis is very much geared to the particular form of the equation, and (as a reading of the proof will make clear) it does not fall into the "standard" framework in which we bound u in H and then bound u in V by what are (essentially) straightforward estimates on the terms in the equation.

Another way to view this theorem, which does not require us to make any "unjustified" assumptions, is as a description of the way in which the 3D Navier–Stokes equations must break down if they are not well-posed. The theorem shows that existence and uniqueness fail only if there is some solution $u(t)$ such that $\|u(t)\|$ becomes infinite in some finite time.

12.2.1 An Absorbing Set in V

Theorem 12.10. *Suppose that the 3D Navier–Stokes equations are well-posed on V, so that for any $f \in H$ and $u_0 \in V$,*

$$du/dt + Au + B(u, u) = f$$

has a strong solution $u(t)$, i.e. a solution u with

$$u \in L^\infty(0, T; V) \cap L^2(0, T; D(A))$$

for all $T > 0$. Then there exists an absorbing set in V.

Note that the strong solution is unique by Theorem 9.6.

Proof. The theorem comes in two parts. First we show that there is a uniform bound on $\|u(T)\|$ for all $\|u_0\| \le M$, for each $T > 0$,

$$\sup_{t \in [0,T]} \sup_{\|u_0\| \le M} \|u(T)\| \le K_T < \infty \qquad \text{for all} \qquad T > 0. \qquad (12.13)$$

This does not preclude the possibility of blowup in infinite time, which would be $K_T \to \infty$ as $T \to \infty$. In the second part of the theorem we show that this possibility can be excluded by using almost the same argument that guarantees (12.13).

So, suppose that (12.13) does not hold. Then there must be sequences $\{u_{0n}\}$ and $\{t_n\}$ with $u_{0n} \in V$, $\|u_{0n}\| \le M$, and $t_n \in [0, T]$, such that

$$\|S(t_n)u_{0n}\| \to \infty \qquad (12.14)$$

as $n \to \infty$. We take a subsequence such that $t_n \to t^* \in [0, T]$, and, using the Alaoglu compactness theorem (Corollary 4.19), we take another such that $u_{0n} \rightharpoonup v_0$ in V. Note that we therefore have $\|v_0\| \leq M$, and since V is compactly embedded in H, we can use Theorem 4.15 to take another subsequence and relabel such that

$$u_n \to v_0 \quad \text{in} \quad H. \tag{12.15}$$

As in our proof of the existence of weak solutions, we can show that the solutions $u_n(t)$ of

$$du_n/dt + \nu A u_n + B(u_n, u_n) = f$$

with $u(0) = u_{0n}$ are uniformly bounded in $L^\infty(0, T; H) \cap L^2(0, T; V)$; taking the limit as $n \to 0$, we obtain a weak solution v of

$$dv/dt + \nu A v + B(v, v) = f$$

with $v(0) = v_0$. However, by assumption, this equation possesses a strong solution $y(t)$. Since this solution is unique in the class of weak solutions, we must have $v(t) = y(t)$, and hence

$$v \in L^\infty(0, T; V) \cap L^2(0, T; D(A)).$$

We now use these regularity properties of v to obtain improved convergence of u_n to v; we will show that $u_n \to v$ strongly in $L^2(0, T; V)$.

Consider the equation for the evolution of the difference $w_n = v - u_n$. Then w_n satisfies

$$\frac{d}{dt} w_n + \nu A w_n + B(w_n, w_n) + B(u, w_n) + B(w_n, v) = 0, \quad w_n(0) = u_{0n} - v_0.$$

Taking the inner product with w_n yields [via (9.22) and (9.26)]

$$\tfrac{1}{2} \frac{d}{dt} |w_n|^2 + \nu \|w_n\|^2 \leq |b(w_n, v, w_n)|$$

$$\leq k |w_n|^{1/2} \|w_n\|^{3/2} \|v\|$$

$$\leq \frac{3\nu}{4} \|w_n\|^2 + \frac{k^4}{4\nu^3} |w_n|^2 \|v\|^4,$$

where we have used Young's inequality with $(p, q) = (4, 4/3)$ in the last line. Therefore

$$\frac{d}{dt} |w_n|^2 + \frac{\nu}{2} \|w_n\|^2 \leq \frac{k^4}{2\nu^3} \|v\|^4 |w_n|^2. \tag{12.16}$$

Forgetting the second term and integrating we obtain

$$|w_n(t)|^2 \leq \exp\left\{ \frac{k^4}{2\nu^3} \int_0^t \|v(s)\|^4 \, ds \right\} |w_n(0)|^2.$$

Since $v \in L^\infty(0, T; V)$ and $w_n(0) \to 0$ in H (by (12.15)), this gives

$$w_n \to 0 \quad \text{in} \quad L^\infty(0, T; H). \tag{12.17}$$

Returning to (12.16) and integrating between 0 and T gives

$$\frac{\nu}{2} \int_0^T \|w_n(s)\|^2 \, ds \leq \frac{k^4}{2\nu^3} \int_0^T \|v(s)\|^4 |w_n(s)|^2 \, ds + |w_n(0)|^2.$$

Since $v \in L^\infty(0, T; V)$, and we have just shown (12.17), using (12.15) we now have that $w_n \to 0$ in $L^2(0, T; V)$ as promised. So

$$u_n \to v \quad \text{in} \quad L^2(0, T; V).$$

Since L^2 convergence on an interval implies that there exists a subsequence converging almost everywhere (Corollary 1.12), we have

$$u_n(s) \to v(s) \quad \text{in } V \quad \text{a.e. } s \in [0, T]. \tag{12.18}$$

Take one of these times s_1 such that the convergence in (12.18) holds; then for n large enough we certainly have

$$\|u_n(s_1)\| \leq M_0 \equiv 1 + \|v\|_{L^\infty(0,T;V)}.$$

We now show that there is some small time τ such that

$$\|u_n(s_1 + t)\| \leq 2(1 + M_0) \quad \text{for all} \quad 0 \leq t \leq \tau. \tag{12.19}$$

This will give us the result.

Consider the equation for the evolution of $\|u(t)\|$,

$$\frac{d}{dt}\|u\|^2 + \nu|Au|^2 = -b(u, u, Au) + (f, Au)$$

$$\leq k\|u\|^{3/2}|Au|^{3/2} + |f||Au|,$$

which after an application of Young's inequality on both terms yields

$$\frac{d}{dt}\|u\|^2 + \nu|Au|^2 \leq \frac{2|f|^2}{\nu} + c\|u\|^6 \leq a + b\|u\|^6, \tag{12.20}$$

for some constants $a, b, c > 0$. One can readily deduce from this equation that there exists a time $\tau(\|u_0\|)$ such that

$$\|u(t)\| \leq 2(1 + \|u_0\|) \qquad \text{for all} \qquad 0 \leq t \leq \tau.$$

Choosing $\tau = \tau(M_0)$ implies (12.19).

Now cover the interval $[0, T]$ with subintervals of length τ, $[s_j, s_j + \tau]$, where each s_j is one of the points where the convergence in (12.18) holds. Then

$$\|u_n(t)\| \leq 2(1 + M_0) \qquad \text{for all} \qquad t \in [0, T],$$

which contradicts (12.14). So we have shown (12.13), that K_T is finite for each $T > 0$.

We now have to exclude the possibility that $K_T \to \infty$ as $T \to \infty$. To do this, we will recast the statement that $K_T \to \infty$ in a form identical to (12.14). Recall first that, using Proposition 12.1, we have the integral bound

$$\int_t^{t+1} \|u(s)\|^2 \leq I_V \qquad \text{for all} \qquad t \geq t_0(|u_0|).$$

Now consider the set of all those s in $[t, t+1]$ for which $\|u(s)\|^2 > 2I_V$, and let σ be the measure of this set. Then

$$2I_V \sigma \leq \int_t^{t+1} \|u(s)\|^2 \, ds \leq I_V,$$

so that $\sigma \leq \frac{1}{2}$. Thus in any interval $[t, t+1]$, the measure of points such that

$$\|u(s)\|^2 \leq 2I_V \tag{12.21}$$

is greater than or equal to $\frac{1}{2}$. In particular, in every interval $[t, t+1]$ there exists at least one point s such that (12.21) holds.

Set $\varrho = \sqrt{2I_V}$. We will show that

$$\sup_{t \geq 0} \sup_{\|u_0\| \leq \varrho} \|u(t)\| < \infty.$$

Indeed, if not, there is a sequence $t_n \to \infty$ and points u_{0n} with $\|u_{0n}\| \leq \varrho$, such that

$$\|S(t_n)u_{0n}\| \to \infty. \tag{12.22}$$

Now consider the interval $[t_n - 1, t_n]$. We know that within this interval there must exist an s_n such that

$$\|u_n(s_n)\| \leq \varrho,$$

by (12.21). So consider the time-shifted solution $v_n(t) = u_n(t - s_n)$. Here v_n is a solution of the 3D equation, with $v_n(0) = v_{0n}$, where $\|v_{0n}\| \leq \varrho$, and (12.22) says that there exist $a_n = t_n - s_n \leq 1$ such that

$$\|v_n(a_n)\| \to \infty.$$

But this is exactly (12.14), which we have already shown cannot occur.

In summary, we know that there is a time s, with $t_0(\|u_0\|) \leq s \leq t_0(\|u_0\|)+1$, such that

$$\|u(s)\| \leq \varrho.$$

We have shown that from this point on there must be some R_V such that

$$\|u(t)\| \leq R_V.$$

So certainly, if we set $t_1(\|u_0\|) = t_0(\|u_0\|) + 1$, then we have

$$\|u(t)\| \leq R_V \qquad \text{for all} \qquad t \geq t_1(\|u_0\|),$$

that is, an absorbing set in V. ■

12.2.2 An Absorbing Set in D(A) and a Global Attractor

As discussed above, just after Theorem 12.3, an absorbing set in V allows one to prove the existence of an attractor that attracts the solutions in V "weakly" (i.e. in the norm of H). To show the existence of a global attractor that attracts in the norm of V, we need to show the existence of an absorbing set in $D(A)$. This can be done with an analysis that closely parallels that of Proposition 12.4, and we leave this as an exercise (Exercise 12.5).

Proposition 12.11. *If the assumptions of Theorem 12.10 hold then there is an absorbing set in $D(A)$.*

We have therefore shown the existence of a global attractor for the 3D equations.

Theorem 12.12. *If the 3D Navier–Stokes equations generate unique strong solutions (as in the statement of Theorem 12.10) then there exists a global attractor in V.*

12.3 Conclusion

We have shown the existence of a global attractor in both H and V for the 2D equations and, under the assumption of regularity, for the 3D equations in V. In the next chapter we will show that in fact these attractors are finite-dimensional subsets of the underlying infinite-dimensional phase space.

Exercises

12.1 Show that in two dimensions

$$b(u, u, A^2u) = b(Au, u, Au) + 2\sum_{j=1}^{2} b(D_ju, D_ju, Au),$$

integrating by parts and using the fact that $\nabla \cdot u = 0$. Deduce that

$$|b(u, u, A^2u)| \le \frac{\nu}{4}|A^{3/2}u|^2 + \frac{9k^2}{\nu}\|u\|^2|Au|^2. \tag{12.23}$$

12.2 For the time-dependent 2D equations, suppose that $f(t) \in L^2_{\text{loc}}(0, \infty; V)$, such that

$$\int_t^{t+1} \|f(s)\|^2\, ds \le M \tag{12.24}$$

for all $t \ge 0$. By taking the inner product of

$$du/dt + \nu Au + B(u, u) = f(t)$$

with A^2u, show that

$$\frac{d}{dt}|Au|^2 + \nu|A^{3/2}u|^2 \le \frac{2\|f\|^2}{\nu} + \frac{C}{\nu}\|u\|^2|Au|^2.$$

Now use the "uniform Gronwall lemma" trick, as in the proof of Proposition 12.2, to show that there is an absorbing set in $D(A)$.

12.3 Use the argument of Lemma 11.2 to show that the set defined in (12.7) is bounded in V.

12.4 Use the Fourier expansion of $u \in D(A)$ to prove the 2D version of Agmon's inequality,

$$\|u\|_\infty \le C|u|^{1/2}|Au|^{1/2}.$$

(Hint: $\|u\|_\infty \le \sum_{k\in\mathbb{Z}^2}|u_k|$; split the sum into two parts, one over $|k| \le \kappa$ and one over $|k| > \kappa$, and then choose κ appropriately.)

12.5   Show the existence of an absorbing set in $\mathbb{H}^2$ for the 3D equations under the assumptions of Theorem 12.10. The first step is to obtain an asymptotic bound on

$$\int_{t_0}^{t_0+1} |Au(s)|^2 \, ds,$$

which you should do using Equation (12.20). Then follow the analysis of Proposition 12.4, using the inequality

$$|B(u, u)| \le k\|u\|^{3/2}|Au|^{1/2}, \tag{12.25}$$

which follows from (9.27) when $m = 3$.

Notes

Although the existence of an absorbing set in V for the 2D equations was first shown (in different terminology) by Foias & Prodi (1967), the proof of the existence of a global attractor for the 2D Navier–Stokes equations was first published by Ladyzhenskaya in 1972 (this paper is in Russian – for an English version see Ladyzhenskaya, 1975), and later, along with many other important results, by Foias & Temam (1979).

The proofs given here of absorbing sets in H and V are from Temam (1985); that of the existence of an absorbing set in $D(A)$ for f independent of time is adapted from Heywood & Rannacher (1982). The case of time-dependent f, treated in Exercises 12.1 and 12.2, is taken from Foias et al. (1983) [the result also follows from the analysis in Guillopé (1982)].

The regularity result of Theorem 12.7 is due to Guillopé (1982) and is reproduced in Temam (1988).

Theorem 12.10 was proved first by Constantin et al.(1985); the proof also appears in Constantin & Foias (1988), and the result as presented here is taken from Temam (1988).

Alternative approaches that attempt to apply the theory of global attractors to the solutions of the 3D equations can be found in Foias & Temam (1987) ("weak attractors," which attract in the weak topology on H) and in Sell (1996) ("trajectory attractors," where each point in the attractor represents a whole trajectory of the original equation).

13

Finite-Dimensional Attractors
Theory and Examples

In the previous two chapters we proved the existence of global attractors for scalar reaction–diffusion equations and for the Navier–Stokes equations. Since these attractors are compact, this guarantees that they are (in some sense) "small" subsets of the original phase space. Indeed, the noncompactness of the unit ball in an infinite-dimensional space implies that these attractors have no interior.

In this chapter we will see that one can show that the dimension of these global attractors is finite, even though they are subsets of infinite-dimensional phase spaces. In turn, results in the three next chapters try to make precise the notion that the asymptotic dynamics of these systems is determined by only a finite number of degrees of freedom.

If the attractor were (a subset of) a smooth manifold then we could use the conventional, topological, dimension of the manifold to bound the dimension of the attractor. However, except in some special cases (which we will cover in Chapter 15), we cannot show that the attractor is of this form, and we need to introduce more widely applicable measures of dimension.

We will give two definitions of dimension, both of which will be useful in what follows. We then discuss how to obtain bounds on the dimension of global attractors. The method relies on finding conditions that guarantee that any n-dimensional volume in the phase space is contracted by the evolution of the flow. If n-dimensional volumes are contracted, then it is reasonable to expect that the attractor can have no n-dimensional subsets and so must have dimension $\leq n$.

We then apply this method to find bounds on the dimensions of the attractors for the scalar reaction–diffusion and 2D Navier–Stokes equations.

13.1 Measures of Dimension

There are in fact many ways of generalising our intuitive notion of "dimension" to treat irregular sets and to allow fractional measures of dimension. With any sensible generalisation, we would still expect that

(i) $d(X) \leq d(Y)$ if $X \subset Y$;
(ii) $d(X) = m$ if X is a nonempty open subset of $\mathbb{R}^m$;
(iii) $d(X \cup Y) = \max(d(X), d(Y))$.

We will consider two generalisations that have these properties: the "fractal" dimension $d_f(X)$ and the "Hausdorff" dimension $d_H(X)$. In what follows we consider subsets of an underlying Banach space H (in most applications H will be a Hilbert space).

13.1.1 The "Fractal" Dimension

Our first measure of dimension is the "fractal" dimension, which we will write as $d_f(X)$. It is based on counting the number of closed balls of a fixed radius ϵ needed to cover X.

We denote the minimum number of balls in such a cover by $N(X, \epsilon)$. If X were a line, we would expect $N(X, \epsilon) \sim \epsilon^{-1}$, if X a surface we would have $N(X, \epsilon) \sim \epsilon^{-2}$, and for a (3D) volume we would have $N(X, \epsilon) \sim \epsilon^{-3}$. So one possible method for obtaining a general measure of dimensions would be to extract the exponent from $N(X, \epsilon)$. Accordingly, we make the following definition. [The fractal dimension is also known as the (upper) box-counting dimension,* the entropy dimension, and the limit capacity.]

Definition 13.1. *If $\overline{X}$ is compact, the* fractal dimension *of X, $d_f(X)$, is given by*

$$d_f(X) = \limsup_{\epsilon \to 0} \frac{\log N(X, \epsilon)}{\log(1/\epsilon)}, \tag{13.1}$$

where we allow the limit in (13.1) to take the value $+\infty$.

---

* At least in the finite-dimensional spaces $\mathbb{R}^m$ Definition 13.1 is equivalent to the "upper box-counting dimension." The idea here is take a grid of cubes of side ϵ that cover $\mathbb{R}^m$, and let $N(X, \epsilon)$ be the number of cubes that intersect X. The dimension is then defined exactly as in (13.1). Although these definitions coincide in $\mathbb{R}^m$ (see Exercise 13.1), the "unit cube" $[0, 1]^\infty$ in an infinite-dimensional Hilbert space has elements with arbitrarily large norm, so one cannot "count boxes" in this context.

Note that it follows from the definition that if $d > d_f(X)$, then, for sufficiently small ϵ,

$$N(X, \epsilon) \leq \epsilon^{-d}. \tag{13.2}$$

Along with (i) and (ii) from Section 13.1, the fractal dimension has the following properties.

Proposition 13.2 (Properties of Fractal Dimension)
  (i) *The fractal dimension is stable under finite unions:*

$$d_f\left(\bigcup_{k=1}^{N} X_k\right) \leq \max_k \, d_f(X_k). \tag{13.3}$$

(ii) *If $f : H \to H$ is Hölder continuous with exponent θ,*

$$|f(x) - f(y)| \leq L|x - y|^{\theta}, \tag{13.4}$$

  then $d_f\big(f(X)\big) \leq d_f(X)/\theta$,
(iii) $d_f(X \times Y) \leq d_f(X) + d_f(Y)$.
(iv) *If $\overline{X}$ is the closure of X in H, then $d_f(\overline{X}) = d_f(X)$.*

Proof
  (i) Since

$$N\left(\bigcup_{k=1}^{N} X_k, \epsilon\right) \leq \sum_{k=1}^{N} N(X_k, \epsilon),$$

it follows from (13.2) that, for any $d > \max_k d_f(X_k)$, for all ϵ small enough we have

$$N(\cup X_k, \epsilon) \leq N\epsilon^{-d},$$

and so $d_f(\cup X_k) \leq d$. Since this holds for all $d > \max_k d_f(X_k)$, (13.3) follows.
  (ii) Take a covering of X by $N(X, \epsilon)$ balls of radius ϵ. The images of these balls under f have diameter at most $L\epsilon^{\theta}$, and they provide a cover of $f(X)$. So

$$N(f(X), L\epsilon^{\theta}) \leq N(X, \epsilon),$$

which yields

$$d_f(f(X)) \le \limsup_{\epsilon \to 0} \frac{\log N(f(X), L\epsilon^\theta)}{-\log L\epsilon^\theta}$$

$$\le \limsup_{\epsilon \to 0} \frac{\log N(X, \epsilon)}{-\log L - \theta \log \epsilon} = \frac{d_f(X)}{\theta}.$$

(iii) Finally, if X is covered by $N(X, \epsilon)$ balls of radius ϵ and Y by $N(Y, \epsilon)$ balls of radius ϵ, then $X \times Y$ is covered by the product of all possible pairs of balls: there are at most $N(X, \epsilon)N(Y, \epsilon)$ such pairs, and the diameter of their product is no more than 2ϵ. Therefore

$$N(X \times Y, 2\epsilon) \le N(X, \epsilon)N(Y, \epsilon),$$

and (iii) follows immediately from the definition.
(iv) If B_k is a covering of X by closed balls of radius ϵ, then B_k must also cover $\overline{X}$, and (iv) follows immediately. ∎

We now calculate the fractal dimension of various simple sets. For some examples the result of Exercise 13.2 simplifies the argument.

Example 13.3. *Let $I_Q = [0, 1] \cap \mathbb{Q}$; then $d_f(I_Q) = 1$.*

Proof. Since $\overline{I_Q} = [0, 1]$, this follows from property (iv) above. ∎

Example 13.4. *Let $G_k = \{n^{-k}\}_{n=1}^\infty$. Then $d_f(X) = 1/(k + 1)$.*

Note that no matter what value of k we choose, we always have $d_f(G_k) < 1$, which is good news since $G_k \subset [0, 1]$.

Proof. Let N be the first value of n for which

$$n^{-k} - (n + 1)^{-k} < \epsilon.$$

It follows that the interval $[0, N^{-k}]$ will have to be entirely covered by balls of radius ϵ. If we include the other $N - 1$ points in the interval $(N^{-k}, 1]$, each of which requires its own ball, we have

$$N(X, \epsilon) = \frac{N^{-k}}{\epsilon} + (N - 1). \tag{13.5}$$

Now, we can estimate this critical value of N using the mean-value theorem, since

$$k(n+1)^{-(k+1)} \leq n^{-k} - (n+1)^{-k} \leq kn^{-(k+1)},$$

and so

$$\left(\frac{k}{\epsilon}\right)^{1/(k+1)} - 1 \leq N \leq \left(\frac{k}{\epsilon}\right)^{1/(k+1)}.$$

The estimate in (13.5) now becomes

$$N(X, \epsilon) = O\left(\epsilon^{-1/(k+1)}\right),$$

and so the result follows. ■

We now consider a similar example in an infinite-dimensional space; the fractal dimension can be made as large as we wish by the choice of k.

Example 13.5. *Let $\{e_n\}_{n=1}^{\infty}$ be an orthonormal basis in a Hilbert space, and let $H_k = \{n^{-k}e_n\}_{n=1}^{\infty}$. Then $d_f(H_k) = 1/k$.*

Proof. For a given $\epsilon > 0$, let N be the first n such that

$$n^{-k} < \epsilon,$$

so that $N \sim \epsilon^{-1/k}$. Now, it is clear that a single ball of radius ϵ and centred at the origin is sufficient to cover all the points $j^{-k}e_j$ with $j \geq N$. However, since

$$|n^{-k}e_n - m^{-k}e_m| = n^{-k} + m^{-k},$$

which is greater than 2ϵ if $n, m < N$, we must have

$$N(X, \epsilon) \sim \epsilon^{-1/k},$$

and so we have $d_f(X) = 1/k$. ■

The example $H_{\log} = \{e_n/\log n\}_{n=1}^{\infty}$ can be treated in a similar way (see Exercise 13.3). For this example, $d_f(H_{\log}) = \infty$, which shows that countable sets can have infinite fractal dimension.

Finally, we treat the well-known example of the Cantor middle third set. This is obtained from the interval [0, 1] by removing the middle third to obtain two

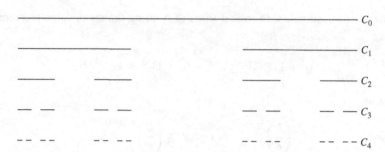

Figure 13.1. The first few stages in the construction of the Cantor set $C = \cap_{n=0}^{\infty} C_n$.

intervals $[0, \frac{1}{3}]$ and $[\frac{2}{3}, 1]$, then removing the middle third of each of these two intervals, and continuing this construction ad infinitum, as in Figure 13.1. More concretely, the set C is the infinite intersection of the sets C_j, where C_j is the set from the jth step in this construction, consisting of 2^j intervals of length 3^{-j}.

The fractal dimension of this set is particularly easy to calculate.

Example 13.6. *The Cantor middle third set C has*

$$d_f(C) = \log 2 / \log 3.$$

Proof. From the construction, C can be covered by 2^j intervals of length 3^{-j}, and this is the minimum such covering. Thus, using the alternative method for calculating fractal dimension provided by Exercise 13.2, we have

$$d_f(C) = \limsup_{j \to \infty} \frac{\log 2^j}{\log 3^j} = \frac{\log 2}{\log 3}. \qquad \blacksquare$$

See also Exercise 13.4.

13.1.2 The Hausdorff Dimension

The fractal dimension is always greater than the second measure of dimension we now introduce (we prove this below in Section 13.1.3). However, the Hausdorff dimension has the advantage that property (i) of Proposition 13.2 also holds for countable unions.

We base this construction on an approximation of the d-dimensional volume of X by a covering of a finite number of balls $B(x_i, r_i)$ with radii $r_i \leq \epsilon$. Note that here, unlike with the fractal dimension, we can take balls with arbitrarily small radii less than ϵ. Suppose that we try to cover a d-dimensional set of

"d-dimensional volume" V with such balls. Then as the diameter of the balls decreases, we would expect

$$\sum_i r_i^d \to \tilde{V} \propto V.$$

More rigorously, we take the best approximation of the volume using such a covering of balls with radii $\leq \epsilon$ and set

$$\mu(X, d, \epsilon) = \inf\left\{ \sum_i r_i^d : r_i \leq \epsilon \text{ and } X \subseteq \cup_i B(x_i, r_i) \right\},$$

where the $B(x_i, r_i)$ are balls with radius r_i. We now define the d-dimensional Hausdorff measure.

Definition 13.7. *The d-dimensional Hausdorff measure of X, $\mathcal{H}^d(X)$, is given by*

$$\mathcal{H}^d(X) = \lim_{\epsilon \to 0} \mu(X, d, \epsilon). \tag{13.6}$$

Since we will allow $\mathcal{H}^d$ to take the value $+\infty$, the limit in (13.6) exists because $\mu(X, d, \epsilon)$ is nonincreasing in ϵ.

The Hausdorff measure is of use in its own right and is a natural generalisation of Lebesgue measure, applicable on manifolds and for noninteger values of d (in fact $\mathcal{H}^m$ is proportional to Lebesgue measure on $\mathbb{R}^m$).

Suppose that we have a smooth d-dimensional manifold. If we take $q > d$ then

$$\sum_i r_i^q \leq \epsilon^{q-d} \sum_i r_i^d \to 0 \qquad \text{as} \qquad \epsilon \to 0,$$

so that $\mathcal{H}^q(X) = 0$ (the conventional volume of a surface is zero), and if we take $p < d$ then

$$\sum_i r_i^p \geq \epsilon^{-(d-p)} \sum_i r_i^d \to \infty \qquad \text{as} \qquad \epsilon \to 0,$$

so that $\mathcal{H}^q(X) = \infty$ (the "length" of a surface is infinite).

We therefore define the Hausdorff dimension of a set X as the smallest value of d for which $\mathcal{H}^d(X)$ is finite.

Definition 13.8. *The* Hausdorff dimension *of a compact set* X, $d_H(X)$, *is defined by*

$$d_H(X) = \inf_{d>0}\{d : \mathcal{H}^d(X) = 0\}.$$

Note that if $d_H(X) = d$ we could have $\mathcal{H}^d(X) = 0$ or $\mathcal{H}^d(X)$ finite but nonzero; either are possible.

The following are useful properties of the Hausdorff dimension, along the lines of Proposition 13.2. Again, (i) and (ii) from Section 13.1 also hold.

Proposition 13.9 (Properties of Hausdorff Dimension)

(i) *The Hausdorff dimension is stable under countable unions:*

$$d_H\left(\bigcup_{k=1}^{\infty} X_k\right) \le \sup_k\ d_H(X_k). \tag{13.7}$$

(ii) *If* $f : H \to H$ *is Hölder continuous with exponent* θ *as in (13.4) then* $d_H\big(f(X)\big) \le d_H(X)/\theta$.
(iii) $d_H(X \times Y) \le d_f(X) + d_H(Y)$.

We have gained significantly over the fractal dimension in property (i), since we now have stability under countable unions, rather than just finite unions. Property (ii) is identical, but note that property (iii) is weaker, since we do not have $d_H(X \times Y) \le d_H(X) + d_H(Y)$ but have to involve the fractal dimension. We have also lost property (iv) entirely.

Proof

(i) Suppose that $d > \sup_k d_H(X_k)$; then

$$\mathcal{H}^d(X_k) = 0$$

for all k. Since (see Exercise 13.5)

$$\mathcal{H}^d\left(\bigcup_{k=1}^{\infty} X_k\right) \le \sum_{k=1}^{\infty} \mathcal{H}^d(X_k), \tag{13.8}$$

it follows that $d_H(\bigcup X_k) \le d$, and we obtain (13.7).
(ii) We consider how $\mathcal{H}^s(X)$ behaves under the image of such maps. If $B(x_i, r_i)$ is a cover of X with $r_i \le \epsilon$, then $B(f(x_i), Lr_i^{\theta})$ is a cover of $f(X)$, with

each $\tilde{r} = Lr^{\theta}$ less than $L\epsilon^{\theta}$. Therefore

$$\sum_i \left[Lr_i^{\theta} \right]^{s/\theta} = L^{s/\theta} \sum_i r_i^s,$$

whence it follows that

$$\mu(f(X), s/\theta, L\epsilon^{\theta}) \leq L^{s/\theta} \mu(X, s, \epsilon).$$

If we let $\epsilon \to 0$ then we get

$$\mathcal{H}^{s/\theta}[f(X)] \leq L^{s/\theta} \mathcal{H}^s(X).$$

The inequality (ii) now follows from the definition of $d_H[f(X)]$.

(iii) Choose $s > d_H(X)$ and $t > d_f(Y)$. Then there exists a $\delta_0 > 0$ such that $N(Y, \delta) \leq \delta^{-t}$ for all $\delta \leq \delta_0$. Now let $\{B(x_i, r_i)\}$ be a cover of X with

$$\sum_i r_i^s < 1 \tag{13.9}$$

[which is possible since $\mathcal{H}^s(X) = 0$ as $s > d_H(X)$]. Now, for each i cover Y with $N(Y, r_i)$ balls of radius r_i, $B(y_{ij}, r_i)$. It follows that $B(x_i, r_i) \times Y$ is covered by $N(Y, r_i)$ products of balls $B(x_i, r_i) \times B(y_{ij}, r_i)$, and so

$$X \times Y \subset \bigcup_i \bigcup_j B(x_i, r_i) \times B(y_{ij}, r_i).$$

Therefore we get

$$\mu(X \times Y, s + t, 2\delta) \leq \sum_i \sum_j (2r_i)^{s+t}$$

$$\leq \sum_i N(Y, r_i) 2^{s+t} r_i^{s+t}$$

$$\leq 2^{s+t} \sum_i r_i^{-t} r_i^{s+t} < 2^{s+t},$$

using (13.9). It follows that $\mathcal{H}^{s+t}(X \times Y) < \infty$ whenever $s > d_H(X)$ and $t > d_f(Y)$, so that $d_H(X \times Y) \leq s + t$. ∎

We now repeat the examples from above. Note that property (i) shows that any countable set X must have $d_H(X) = 0$.

Example 13.10. *Let I_Q, G_k, and H_k be as in Examples 13.3–13.5. Then $d_H(I_Q) = d_H(G_k) = d_H(H_k) = 0$.*

Proof. I_Q, G_k, and H_k are countable. ∎

Example 13.11. $d_H(C) \geq \log 2 / \log 3$.

Proof. We show that $\mathcal{H}^s(C)$, where $s = \log 2/\log 3$, is bounded below by $\frac{1}{2}$. Suppose that we have a cover $(x_i - r_i, x_i + r_i)$ of C. For each i let k be the integer such that

$$3^{-(k+1)} < r_i < 3^{-k}.$$

It follows that $(x_i - r_i, x_i + r_i)$ can intersect at most one of the intervals in C_k (the kth level of the construction of C), since these are at least 3^{-k} apart.

For every $j \geq k$, each $(x_i - r_i, x_i + r_i)$ intersects at most $2^{j-k} = 2^j 3^{-sk} \leq 2^j 3^s r_i^s$ intervals in C_j. Choosing j large enough that $3^{-(j+1)} \leq r_i$ for every i, since the $(x_i - r_i, x_i + r_i)$ intersect all of the 2^j intervals in C_j, we can count intervals to obtain

$$2^j \leq \sum_i 2^j 3^s e_i^s,$$

which gives

$$\sum_i r_i^s \geq 3^{-s} = \tfrac{1}{2}.$$

Since this holds for every cover, we have $\mathcal{H}^s(C) \geq \frac{1}{2}$. It follows that $d_H(C) \geq s$, and so we have the equality given in the statement. ∎

In general it is much harder to obtain lower bounds for the Hausdorff dimension than upper bounds.

13.1.3 Hausdorff versus Fractal Dimension

The definitions of Hausdorff and fractal dimension are not equivalent, as the above examples show. To investigate this further, in this section we write Definition 13.1 in a way that makes comparison with the Hausdorff dimension more straightforward.

Lemma 13.12. *If we define an "approximate fractal measure" as*

$$\mu_f(X, d, \epsilon) = N(X, \epsilon)\epsilon^d,$$

and the "d-dimensional fractal measure"* by

$$m_f(X, d) = \limsup_{\epsilon \to 0} \mu_f(X, d, \epsilon),$$

then

$$d_f(X) = \inf_{d>0}\{d : m_f(X, d) = 0\}.$$

Proof. Let $D = \inf_{d>0}\{d : m_f(X, d) = 0\}$. Now take $e > d > d_f(X)$, so that, using (13.2), there is an ϵ_0 such that, for $\epsilon < \epsilon_0$,

$$N(X, \epsilon) < \epsilon^{-d}.$$

From the definition of $m_f(X, d)$,

$$m_f(X, e) \leq \limsup_{\epsilon \to 0} \epsilon^e \epsilon^{-d} = 0,$$

and so $D < e$. Since $e > d_f(X)$ is arbitrary it follows that $D \leq d_f(X)$.

Similarly, suppose that $c < d < d_f(X)$. Then there are infinitely many ϵ_is with

$$\log N(X, \epsilon_i) > d \log(1/\epsilon_i),$$

which is just

$$\epsilon_i^d N(X, \epsilon_i) > 1.$$

Then

$$
\begin{aligned}
m_f(X, d) &\geq \limsup_{i \to \infty} \epsilon_i^c N(X, \epsilon_i) \\
&\geq \limsup_{i \to \infty} \epsilon_i^c \epsilon_i^{-d} \\
&> 1,
\end{aligned}
$$

and so for every $c < d_f(X)$, $m_f(X, c) > 1$, which implies that $D \geq d_f(X)$. ∎

---

* This terminology is somewhat misleading, since unlike the Hausdorff measure, this "fractal measure" is not a bona fide measure. This is one reason why one might prefer the Hausdorff dimension in some situations, even though the fractal dimension has important properties not shared by its weaker counterpart.

The following is an immediate corollary, since clearly

$$\mu(X, d, \epsilon) \leq \epsilon^d N(X, \epsilon).$$

Corollary 13.13. $d_H(X) \leq d_f(X)$.

As a simple application, we have

Corollary 13.14. *If C is the Cantor set, then $d_H(C) = d_f(C) = \log 2/ \log 3$.*

Proof. We saw in Example 13.11 that $d_H(C) \geq \log 2/ \log 3$, and the above corollary shows that $d_H(C) \leq d_f(C) = \log 2/ \log 3$ (Example 13.6). ∎

13.2 Bounding the Attractor Dimension Dynamically

We now proceed to find an analytical method to estimate the fractal dimension of the global attractor. Clearly, if we can bound $d_f(X)$ then Corollary 13.13 immediately gives us the same bound on $d_H(X)$.

The idea is to study the evolution of infinitesimal n-dimensional volumes as they evolve under the flow and to try to find the smallest dimension n at which we can guarantee that all such n-volumes contract asymptotically.

We will consider an abstract problem, written as

$$\frac{du}{dt} = F(u(t)), \qquad u(0) = u_0,$$

with u_0 contained in a Hilbert space H, whose norm we denote by $|\cdot|$. We assume that the equation has unique solutions given by $u(t; u_0) = S(t)u_0$ and a compact global attractor $\mathcal{A}$.

We want to start off with an orthogonal set of displacements near an initial point $u_0 \in \mathcal{A}$, and then watch how the volume they form evolves under the flow. If we have n vectors $\{\xi^{(j)}\}$ $(1 \leq j \leq n)$ then they form a parallelepiped, which we write as

$$\xi^{(1)} \wedge \ldots \wedge \xi^{(n)}.$$

We are interested in the n-volume of this, which we denote as

$$\left| \xi^{(1)} \wedge \ldots \wedge \xi^{(n)} \right|$$

and is given by

$$\left| \xi^{(1)} \wedge \ldots \wedge \xi^{(n)} \right|^2 = \det M\left(\xi^{(1)}, \ldots, \xi^{(n)}\right), \qquad (13.10)$$

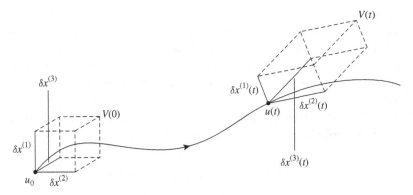

Figure 13.2. Evolution of the infinitesimal volume element $\delta x^{(1)} \wedge \ldots \wedge \delta x^{(n)}$ under the semigroup $S(t)$.

where the matrix M has components

$$M\left(v^{(1)}, \ldots, v^{(n)}\right)_{ij} = \left(v^{(i)}, v^{(j)}\right).$$

This gives the correct volume, since M is the matrix that transforms an orthonormal set of unit vectors into the set $\{\xi^{(j)}\}$ (see Exercise 13.6).

To study the evolution of the volume we have to study the evolution of a set of infinitesimal displacements $\delta x^{(i)}(t)$ about the trajectory $u(t)$ (see Figure 13.2). To be able to consider such displacements, we need to be able to linearise the equation around the trajectory, and for this we need to make some assumption about the smoothness of the flow.

This technical assumption we call "uniform differentiability," and it turns out that we can restrict ourselves to uniform differentiability near trajectories on the attractor.

Definition 13.15. *We say that $S(t)$ is* uniformly differentiable *on $\mathcal{A}$ if for every $u \in \mathcal{A}$ there exists a linear operator $\Lambda(t, u)$, such that, for all $t \geq 0$,*

$$\sup_{u,v\in\mathcal{A};\, 0<|u-v|\leq\epsilon} \frac{|S(t)v - S(t)u - \Lambda(t, u)(v - u)|}{|v - u|} \to 0 \quad as \quad \epsilon \to 0$$

$$and \quad \sup_{u\in\mathcal{A}} \|\Lambda(t, u)\|_{\mathrm{op}} < \infty \quad for\ each \quad t \geq 0. \tag{13.11}$$

This assumption, which is straightforward to check for ordinary differential equations, will often involve technical difficulties in the PDE case; we verify it for the reaction–diffusion equation and the 2D Navier–Stokes equations later on. In fact, we will see that the major work in applying the ideas of this chapter is usually in checking this condition. As with the technicalities of the previous chapter, where we had to use a Galerkin approximation to make sure that our formal estimates really held, this is something that is often relegated to the "we can check that..." status in research papers. Once again, it is worth emphasising in a textbook such as this that these things cannot be assumed without careful proof.

However, now taking this assumption as given, we suppose that $\Lambda(t, u)\xi$ is given by the solution of the linearised equation

$$\frac{dU}{dt} = F'(S(t)u_0)U(t), \qquad U(0) = \xi.$$

We write this as

$$\frac{dU}{dt} = L(t; u_0)U(t), \qquad U(0) = \xi.$$

Then each displacement $\delta x^{(j)}$ evolves according to

$$\frac{d}{dt}\delta x^{(j)} = L(t; u_0)\delta x^{(j)}. \tag{13.12}$$

We need to investigate the evolution of

$$V_n(t) = \left|\delta x^{(1)} \wedge \ldots \wedge \delta x^{(n)}\right|.$$

To do this, consider

$$\frac{d}{dt}\log V_n(t) = \frac{1}{2}\frac{d}{dt}\log V_n^2 = \frac{1}{2}\frac{d}{dt}\log[\det M(t)],$$

where

$$M(t) = M\big(\delta x^{(1)}(t), \ldots, \delta x^{(n)}(t)\big),$$

by (13.10).

Now, $\log[\det M] = \mathrm{Tr}[\log M]$ (see Exercises 13.7 and 13.8), so that

$$\frac{d}{dt}\log V_n(t) = \frac{1}{2}\frac{d}{dt}\mathrm{Tr}[\log M].$$

Furthermore, we can show that

$$\frac{d}{dt}\mathrm{Tr}[\log M] = \mathrm{Tr}\left[M^{-1}\frac{dM}{dt}\right]$$

(see Exercise 13.8 again), and so we have

$$\frac{d}{dt} \log V_n(t) = \tfrac{1}{2}\mathrm{Tr}\left[M^{-1}\frac{dM}{dt}\right]. \tag{13.13}$$

[Notice that the expression $\log V_n(t)$ occurs more naturally here than does $V_n(t)$ itself.] We now have to find a way of evaluating the trace in the above equation.

We do this by introducing a time-dependent set of *orthonormal* vectors $\{\phi^{(j)}(t)\}$ that span the same space as the $\{\delta x^{(j)}(t)\}$. We can define the matrix $m(t)$ given by the components of the $\delta x^{(j)}$ in the $\phi^{(i)}$ directions,

$$m_{ij}(t) = \left(\phi^{(i)}(t), \delta x^{(j)}(t)\right),$$

so that

$$M = m^T m \qquad \text{and} \qquad M^{-1} = m^{-1}(m^T)^{-1}.$$

We can also refer the linearised evolution operator to this basis by setting

$$a_{ij} = \left(\phi^{(i)}, L\phi^{(j)}\right). \tag{13.14}$$

We now write dM/dt in terms of m and a, starting with the equality

$$\begin{aligned}
\frac{dM_{ij}}{dt} &= \left(\frac{d}{dt}\delta x^{(i)}, \delta x^{(j)}\right) + \left(\delta x^{(i)}, \frac{d}{dt}\delta x^{(j)}\right) \\
&= \left(L\delta x^{(i)}, \delta x^{(j)}\right) + \left(\delta x^{(i)}, L\delta x^{(j)}\right),
\end{aligned}$$

using (13.12). Since

$$u = \sum_{j=1}^{n} \phi^{(j)}\left(\phi^{(j)}, u\right) \tag{13.15}$$

if u is an element of the space spanned by the $\{\phi^{(j)}\}$, we can write

$$\begin{aligned}
\frac{dM_{ij}}{dt} &= \sum_{k,l=1}^{n} \left(\delta x^{(i)}, \phi^{(k)}\right)\left[\left(L\phi^{(k)}, \phi^{(l)}\right) + \left(\phi^{(k)}, L\phi^{(l)}\right)\right]\left(\phi^{(l)}, \delta x^{(j)}\right) \\
&= \sum_{k,l=1}^{n} m_{ki}(a_{lk} + a_{kl})m_{lj},
\end{aligned}$$

so that

$$\frac{dM}{dt} = m^T(a^T + a)m.$$

Then from (13.13)

$$2\frac{d}{dt}\log V_n(t) = \mathrm{Tr}\big(m^{-1}(m^T)^{-1}m^T(a^T+a)m\big)$$
$$= \mathrm{Tr}\big(m^{-1}(a^T+a)m\big)$$
$$= \mathrm{Tr}(a^T+a)$$
$$= 2\,\mathrm{Tr}(a),$$

where we have used the cyclic property of the trace and that $\mathrm{Tr}(A^T) = \mathrm{Tr}(A)$.

We now express this equality in terms of a projection operator. We define $P^{(n)}$ as the projection onto the span of the $\{\delta x^{(n)}\}$; since this is the same as the span of the $\{\phi^{(i)}\}$, we have

$$P^{(n)} = \sum_{i=1}^{n} \phi^{(i)}\big(\phi^{(i)}, \cdot\big).$$

Then from (13.14) and using (13.15) again, we get

$$\mathrm{Tr}(a) = \sum_{i=1}^{n} \big(\phi^{(i)}, L\phi^{(i)}\big)$$
$$= \sum_{i,j=1}^{n} \big(\phi^{(j)}, L\phi^{(i)}\big)\big(\phi^{(i)}, \phi^{(j)}\big)$$
$$= \mathrm{Tr}\big[L P^{(n)}\big].$$

Integrating for $V_n(t)$ we obtain

$$V_n(t) = V_n(0)\,\exp\left[\int_0^t \mathrm{Tr}\big(L(s; u_0)P^{(n)}(s)\big)\,ds\right].$$

Therefore, the asymptotic growth rate of this particular initial infinitesimal n-volume is

$$\lim_{t\to\infty} \exp\left[\frac{1}{t}\int_0^t \mathrm{Tr}\big(L(s; u_0)P^{(n)}(s)\big)\,ds\right].$$

We want the maximum possible asymptotic growth rate, taken over all possible initial conditions. Our initial conditions have two parts – the choice of initial point $u_0 \in \mathcal{A}$ and the choice of the initial infinitesimal n-volume. Since the latter is specified by the original set of orthonormal vectors, and so by $P^{(n)}(0)$, we define

$$TR_n(\mathcal{A}) = \sup_{x_0 \in \mathcal{A}} \sup_{P^{(n)}(0)} \limsup_{t \to \infty} \frac{1}{t} \int_0^t \mathrm{Tr}\big(L(s; u_0) P^{(n)}(s)\big)\, ds.$$

If we denote the time average operation by $\langle \cdot \rangle$,

$$\langle f(t) \rangle = \limsup_{t \to \infty} \frac{1}{t} \int_0^t f(s)\, ds, \qquad (13.16)$$

then we can write this as

$$TR_n(\mathcal{A}) = \sup_{x_0 \in \mathcal{A}} \sup_{P^{(n)}(0)} \big\langle \mathrm{Tr}(L(t; u_0) P^{(n)}(t)) \big\rangle.$$

In applications we will seek the smallest n for which the sign of $TR_n(\mathcal{A})$ is negative. If we have a uniform bound on $\mathrm{Tr}\big(L(s; x_0) P^{(n)}(s)\big)$, so that

$$\sup_{x_0 \in \mathcal{A}} \sup_{P^{(n)}(0)} \mathrm{Tr}\big(L(s; u_0) P^{(n)}(s)\big) \leq C$$

then clearly $TR_n(\mathcal{A}) \leq C$. This is often the case and is what we will find below.

If $TR_n(\mathcal{A}) < 0$ then infinitesimal n-volumes decay exponentially. Although it is then reasonable to assume that the attractor contains no n-dimensional subsets, and to expect that $d_f(\mathcal{A}) \leq n$, this is far from straightforward to prove. A proof, due to B. Hunt (personal communication, 1999), is given in Appendix B. See the Notes for more details.

Theorem 13.16. *Suppose that $S(t)$ is uniformly differentiable on $\mathcal{A}$ and that there exists a t_0 such that $\Lambda(t, u_0)$ is compact for all $t \geq t_0$. If $TR_n(\mathcal{A}) < 0$ then $d_f(\mathcal{A}) \leq n$.*

Before we apply this result to give dimension bounds for reaction–diffusion equations and for the Navier–Stokes equations, we will need an auxiliary lemma that gives a lower bound on $\mathrm{Tr}(-\Delta P_n)$, valid for any rank n orthogonal projection.

Lemma 13.17. *Let P_n be a rank n orthogonal projection in $L^2(Q)$, where Q is a periodic domain in $\mathbb{R}^m$. Then*

$$\mathrm{Tr}(-\Delta P_n) \geq C n^{(m+2)/m}. \qquad (13.17)$$

The same result is also valid for any bounded C^2 domain $\Omega \subset \mathbb{R}^m$.

Proof. Write $A = -\Delta$, and denote its orthonormal eigenfunctions as w_j, with corresponding eigenvalues λ_j ordered so that $\lambda_{j+1} \geq \lambda_j$. Now if P_n is the projection onto the space spanned by some orthonormal set of vectors $\{\phi_1, \ldots, \phi_n\}$, then

$$\text{Tr}(AP_n) = \sum_{j=1}^{n} (\phi_j, A\phi_j). \tag{13.18}$$

We show first that

$$\text{Tr}(AP_n) \geq \sum_{j=1}^{n} \lambda_j. \tag{13.19}$$

We write (13.18), expanding the ϕ_j in terms of the eigenbasis $\{w_j\}$, and obtain

$$\text{Tr}(AP_n) = \sum_{j=1}^{n} \sum_{k=1}^{\infty} \lambda_k |(\phi_j, w_k)|^2$$

$$= \sum_{k=1}^{\infty} \lambda_k \left(\sum_{j=1}^{n} |(\phi_j, w_k)|^2 \right).$$

Now, since $|\phi_j| = 1$ we have

$$\sum_{j=1}^{n} \sum_{k=1}^{\infty} |(w_k, \phi_j)|^2 = n,$$

and since $\{\phi_j\}$ are orthonormal but do not span H we have

$$\sum_{j=1}^{n} |(w_k, \phi_j)|^2 \leq 1.$$

Expression (13.19) now follows.

The explicit bound in (13.17) follows easily, using the property of the eigenvalues of the Laplacian (see Exercise 13.9),

$$cj^{2/m} \leq \lambda_j \leq Cj^{2/m},$$

since then

$$\sum_{j=1}^{n} \lambda_j \geq c \sum_{j=1}^{n} j^{2/m} \geq cn^{(2/m)+1} = cn^{(m+2)/m}. \qquad \blacksquare$$

13.3 Example I: The Reaction–Diffusion Equation

First we apply the result to the scalar reaction–diffusion equation from Chapter 8. Note that in the one-dimensional scalar case treated at the end of Chapter 11 we already have in (11.29) a heuristic dimension bound

$$d_f(X) \sim \lambda^{1/2}. \tag{13.20}$$

This will be a useful test of our rigorous estimates.

13.3.1 Uniform Differentiability

The first step towards proving dimension bounds using the trace formula is to verify the differentiability condition (13.11); to do this we usually need some regularity properties of the attractor. In this case we will need the L^∞ bound proved in Theorem 11.6.

Proposition 13.18. *Provided that f is a C^2 function and satisfies the conditions (8.14) and (8.15) imposed before, the flow generated by the reaction–diffusion equation*

$$\frac{du}{dt} = \Delta u + \lambda f(u) \tag{13.21}$$

is uniformly differentiable on $\mathcal{A}$, with $\Lambda(t; u_0)\xi$ the solution of

$$\frac{dU}{dt} = \Delta U + \lambda f'(u(t))U, \qquad U(0) = \xi. \tag{13.22}$$

Furthermore, $\Lambda(t; u_0)$ is compact for all $t > 0$.

Proof. Let $u(t)$ and $v(t)$ be the solutions of (13.21) with initial conditions u_0 and v_0, respectively ($u_0, v_0 \in \mathcal{A}$), and let $U(t)$ be the solution of (13.22) with initial condition $v_0 - u_0$. The error in this linear approximation is given by

$$\theta(t) = v(t) - u(t) - U(t).$$

It follows that θ satisfies the equation

$$\frac{d}{dt}\theta = \Delta\theta + \lambda f'(u)\theta + \lambda g, \qquad \theta(0) = 0, \tag{13.23}$$

where

$$g = f(v) - f(u) - f'(u)(v - u).$$

Since f is C^2 it follows from Taylor's theorem that

$$|g(x)| \leq \tfrac{1}{2}|f''(c)||v(x) - u(x)|^2$$

for some c on the line segment joining $u(x)$ to $v(x)$. Since both u and v lie in $\mathcal{A}$, they are bounded in L^∞, and so

$$|g(x)| \leq C|u(x) - v(x)|^2 \tag{13.24}$$

for some constant C.

Now, using Theorem 5.31, choose a q such that if p is the conjugate exponent to q, then $H^1(\Omega) \subset L^p(\Omega)$. It follows from (13.24), if we write $g(t) = g(u(x, t))$, that

$$\|g(t)\|_{L^q}^q \leq C \int_\Omega |u(t) - v(t)|^{2q} \, dx$$

$$\leq C \int_\Omega |u(t) - v(t)|^{2q-2+\epsilon} |u - v|^{2-\epsilon} \, dx$$

$$\leq C|u(t) - v(t)|^{2-\epsilon},$$

using Hölder's inequality and the L^∞ bound on u and v. So now we have

$$\|g(t)\|_{L^q} \leq C|u(t) - v(t)|^{(2-\epsilon)/q},$$

and if we now choose $\epsilon = 2 - q(1+\delta)$, for some $\delta \in (0, (2-q)/q)$, we obtain

$$\|g(t)\|_{L^q} \leq C|u(t) - v(t)|^{1+\delta}. \tag{13.25}$$

It follows from the uniqueness result (8.24) that in fact

$$|u(t) - v(t)| \leq e^{lt}|u_0 - v_0|,$$

and so (13.25) can be written as

$$\|g(t)\|_{L^q} \leq C e^{(1+\delta)lt}|u_0 - v_0|^{1+\delta}.$$

Taking the inner product of (13.23) with θ then yields

$$\tfrac{1}{2}\frac{d}{dt}|\theta|^2 + \|\theta\|^2 \leq \lambda l|\theta|^2 + \lambda\|g(t)\|_{L^q}|\theta|_{L^p}$$

$$\leq \lambda l|\theta|^2 + C\lambda e^{(1+\delta)lt}|u_0 - v_0|^{1+\delta}\|\theta\|$$

$$\leq \lambda l|\theta|^2 + C e^{2(1+\delta)lt}|u_0 - v_0|^{2(1+\delta)} + \tfrac{1}{2}\|\theta\|^2,$$

so that, neglecting the $\|\theta\|^2$ terms, we have

$$\frac{1}{2}\frac{d}{dt}|\theta|^2 \le \lambda l|\theta|^2 + Ce^{2(1+\delta)lt}|u_0 - v_0|^{2(1+\delta)},$$

Using Gronwall's inequality yields

$$|\theta(t)|^2 \le k(t)|u_0 - v_0|^{2(1+\delta)},$$

so that (13.11) holds – the flow is differentiable, and the derivative is indeed given by the solution of the linearised equation (13.22).

The compactness of $\Lambda(t; u_0)$ follows by obtaining bounds on the solution of (13.22) in $L^2(\Omega)$ and $H_0^1(\Omega)$. Taking the inner product of (13.22) with U we obtain

$$\frac{1}{2}\frac{d}{dt}|U|^2 + \|U\|^2 = \lambda \int_\Omega f'(u)U^2 \, dx$$

$$\le l|U|^2,$$

using $f'(s) \le l$ (8.15). Therefore

$$\frac{1}{2}\frac{d}{dt}|U|^2 \le l|U|^2,$$

which shows that any set of ξ that is bounded in L^2 remains bounded under the evolution of (13.22). Also, for any $t > 0$ we can integrate

$$\frac{1}{2}\frac{d}{dt}|U|^2 + \|U\|^2 \le l|U|^2$$

between $t/2$ and t to bound the integral of $\|U\|^2$:

$$\int_{t/2}^t \|U(s)\|^2 \, ds \le l \int_{t/2}^t |U(s)|^2 \, ds + \frac{1}{2}|U(t/2)|. \qquad (13.26)$$

Taking the inner product of (13.22) with AU gives

$$\frac{1}{2}\frac{d}{dt}\|U\|^2 + |AU|^2 = \lambda \int_\Omega f'(u)U(x)[-\Delta U(x)] \, dx$$

$$\le \lambda C \int_\Omega |U(x)|| - \Delta U(x)| \, dx$$

$$\le \lambda C|U||AU|,$$

since the attractor is bounded in $L^\infty(\Omega)$, and so $|f'(u(x))| \le C$. Rearranging this gives

$$\frac{d}{dt}\|U\|^2 + |AU|^2 \le C|U|^2.$$

Dropping the term in $|AU|^2$ and integrating from s to t, with $t/2 \le s < t$, gives

$$\|U(t)\|^2 \le C \int_s^t |U(\tau)|^2 \, d\tau + \|U(s)\|^2,$$

and then integrating with respect to s from $t/2$ to t we have

$$\|U(t)\|^2 \le \tfrac{1}{2} C t \int_{t/2}^t |U(s)|^2 \, ds + \int_{t/2}^t \|U(s)\|^2 \, ds,$$

which is finite because of (13.26). Thus a bounded set of ξ in L^2 becomes a bounded set in H_0^1. It follows that $\Lambda(t; u_0)$ is compact for any $t > 0$. ∎

13.3.2 A Bound on the Attractor Dimension

With the differentiability property ensured we can apply the result of Theorem 13.16.

Theorem 13.19. *The dimension of the global attractor $\mathcal{A}$ for*

$$u_t - \Delta u = \lambda f(u), \qquad u|_{\partial\Omega} = 0, \qquad \Omega \subset \mathbb{R}^m$$

is bounded by $d_f(\mathcal{A}) \le c\lambda^{m/2}$, for some $c > 0$.

Proof. We need an estimate on

$$\langle \mathrm{Tr} L(t) P_n \rangle,$$

where L is the linearised evolution operator at time t and $\langle \cdot \rangle$ is the maximum time average from (13.16). Since the linearised equation is

$$v_t = \Delta v + \lambda f'(u) v,$$

then, as $f' \le l$ (8.15),

$$\langle \mathrm{Tr} L(t) P_n \rangle = -\langle \mathrm{Tr}(-\Delta P_n) \rangle + \lambda \left\langle \int_\Omega \sum_{j=1}^n \phi_j f'(u) \phi_j \right\rangle$$

$$\le -\langle \mathrm{Tr}(-\Delta P_n) \rangle + l\lambda \left\langle \sum_{j=1}^n \int_\Omega |\phi_j|^2 \, dx \right\rangle$$

$$\le -\langle \mathrm{Tr}(-\Delta P_n) \rangle + n l \lambda.$$

A lower bound on $\mathrm{Tr}(-\Delta P_n)$ is given in Lemma 13.17, so that

$$\langle \mathrm{Tr} L(t) P_n \rangle \leq -cn^{(m+2)/m} + nl\lambda.$$

Thus $\langle \mathrm{Tr} L(t) P_n \rangle$ is certainly negative once

$$n > \left(\frac{l\lambda}{c}\right)^{m/2},$$

and thus, using Theorem 13.16, one obtains the bound given in the statement of the theorem. ∎

Note that this result agrees with (13.20) in the scalar case $m = 1$.

13.4 Example II: The 2D Navier–Stokes Equations

We now do something similar with the 2D Navier–Stokes equations. We will not make any attempt to prove the best possible bound for the attractor dimension here. Sharpening the result we prove in Theorem 13.21 has been a major concern in the mathematical theory of fluid dynamics, and we discuss this more fully in Section 13.5.

13.4.1 Uniform Differentiability

The differentiability property is easier to prove in this case.

Theorem 13.20. *The solutions of the Navier–Stokes equations in two dimensions satisfy (13.11) with $\Lambda(t; u_0)\xi$ the solution of the equation*

$$\frac{dU}{dt} + \nu AU + B(u, U) + B(U, u) = 0, \qquad U(0) = \xi. \tag{13.27}$$

Furthermore, $\Lambda(t; u_0)$ is compact for all $t > 0$.

Proof. As before, we take $u(t)$ and $v(t)$ as solutions of

$$\frac{du}{dt} + \nu Au + B(u, u) = 0$$

with initial conditions u_0 and v_0, respectively ($u_0, v_0 \in \mathcal{A}$), and we consider $U(t)$, the solution of (13.27) with $U(0) = v_0 - u_0$. Then, after some algebra, the equation satisfied by $\theta = v - u - U$ is

$$\frac{d\theta}{dt} + \nu A\theta + B(u, \theta) + B(\theta, u) + B(u - v, u - v) = 0.$$

Writing $w = u - v$ and taking the inner product with θ yields

$$\frac{1}{2}\frac{d}{dt}|\theta|^2 + v\|\theta\|^2 = -b(\theta, u, \theta) - b(w, w, \theta),$$

and so, using (9.26), we have

$$\frac{1}{2}\frac{d}{dt}|\theta|^2 + v\|\theta\|^2 \leq k|\theta|\|\theta\|\|u\| + k|w|\|w\|\|\theta\|.$$

Now, the attractor is bounded in V, so that $\|u\| \leq \rho_V$, and we can use Young's inequality on both terms to give

$$\frac{d}{dt}|\theta|^2 + v\|\theta\|^2 \leq c|\theta|^2 + c|w|^2\|w\|^2.$$

Dropping the $\|\theta\|$ term and using the Gronwall inequality with $\theta(0) = 0$ gives

$$|\theta(t)|^2 \leq k \int_0^t |w(s)|^2\|w(s)\|^2\,ds. \tag{13.28}$$

Now, recall that during the uniqueness proof of Theorem 9.4, we obtained as (9.37) the equation

$$\frac{d}{dt}|w|^2 + v\|w\|^2 \leq \frac{k^2}{v}\|u\|^2|w|^2, \tag{13.29}$$

and from this we deduced the bound (9.38):

$$|w(t)|^2 \leq \exp\left(\int_0^t \frac{k^2}{v}\|u(s)\|^2\,ds\right)|w(0)|^2.$$

Since $\|u\| \leq \rho_V$, this becomes a simple exponential estimate

$$|w(t)|^2 \leq e^{Kt}|w_0|^2$$

(for some K), and we can then multiply (13.29) by $|w(t)|^2$ and integrate between 0 and t to give

$$v\int_0^t |w(s)|^2\|w(s)\|^2\,ds \leq \frac{k^2}{v}\rho_V^2\int_0^t |w(s)|^4\,ds + \frac{1}{2}|w_0|^4,$$

which implies that

$$\int_0^t |w(s)|^2\|w(s)\|^2\,ds \leq Ce^{Kt}|w_0|^4.$$

Feeding this into (13.28) shows that

$$|\theta(t)| \leq C(t)|u_0 - v_0|^2,$$

and so

$$\frac{|v(t) - u(t) - U(t)|}{|v_0 - u_0|} \leq C|v_0 - u_0| \to 0 \qquad \text{as} \qquad v_0 \to u_0,$$

proving the required differentiability.

Exercise 13.10 asks you to check that $\Lambda(t; u_0)$ is compact for all $t > 0$. ∎

13.4.2 A Bound on the Attractor Dimension

With the differentiability ensured we can apply the trace formula to find a bound on the dimension.

Theorem 13.21. *The attractor for the 2D periodic Navier–Stokes equations is finite-dimensional, with*

$$d_f(\mathcal{A}) \leq \alpha \left(\frac{\rho v}{\nu} \right)^2.$$

Proof. The correct form of the linearised equation is given in (13.27), and so

$$L(u)w = \nu Aw - B(w, u) - B(u, w).$$

Thus the time-averaged trace $\langle P_n L(u(t)) \rangle$ is bounded by

$$\langle P_n L(u) \rangle = \left\langle \sum_{j=1}^{n} (L(u)\phi_j, \phi_j) \right\rangle$$

$$= -\left\langle \sum_{j=1}^{n} (-\nu \Delta \phi_j, \phi_j) \right\rangle - \left\langle \sum_{j=1}^{n} b(\phi_j, u, \phi_j) \right\rangle.$$

Using the bound on $b(u, v, w)$ in (9.26) now gives

$$\langle P_n L(u) \rangle \leq -\nu \sum_{j=1}^{n} \langle \|\phi_j\|^2 \rangle + k \left\langle \sum_{j=1}^{n} \|u\| \, \|\phi_j\| \, |\phi_j| \right\rangle$$

$$= -\nu \sum_{j=1}^{n} \langle \|\phi_j\|^2 \rangle + k \left\langle \sum_{j=1}^{n} \|u\| \, \|\phi_j\| \right\rangle$$

$$\leq -\nu \sum_{j=1}^{n} \langle \|\phi_j\|^2 \rangle + k \sum_{j=1}^{n} \langle \|u\| \, \|\phi_j\| \rangle,$$

where we have used the fact that $|\phi_j| = 1$. Now, via Young's inequality on the final term,

$$\langle P_n L(u) \rangle \leq -\nu \sum_{j=1}^{n} \langle \|\phi_j\|^2 \rangle + \sum_{1}^{n} \left\langle \frac{\nu}{2} \|\phi_j\|^2 + \frac{k^2}{2\nu} \|u\|^2 \right\rangle$$

$$= -\frac{\nu}{2} \sum_{j=1}^{n} \langle \|\phi_j\|^2 \rangle + \frac{k^2}{2\nu} \sum_{j=1}^{n} \langle \|u\|^2 \rangle$$

$$= -\frac{\nu}{2} \langle \text{Tr}(-\Delta P_n) \rangle + \frac{k^2 n}{2\nu} \langle \|u\|^2 \rangle.$$

Finally, using Lemma 13.17 with $m = 2$ yields

$$\langle P_n L(u) \rangle \leq -\frac{c\nu}{2} n^2 + \frac{k^2 n}{2\nu} \langle \|u\|^2 \rangle,$$

so that the trace is negative provided that

$$n > \alpha \left(\frac{\langle \|u\|^2 \rangle}{\nu^2} \right).$$

Note that $\langle \|u\|^2 \rangle \leq \rho_V^2$ is finite, since we know from Proposition 12.2 that $\|u\| \leq \rho_V$ on $\mathcal{A}$. Theorem 13.16 now provides the bound on the fractal dimension of $\mathcal{A}$. ∎

13.5 Physical Interpretation of the Attractor Dimension

The existence of finite-dimensional global attractors has been proved for many equations that arise in mathematical physics, and Temam (1988) provides proofs for many of these models.

However, there is little point in proving the existence of these objects for such a plethora of equations if their existence has no implications for the underlying physics. One way of interpreting the physical significance of an attractor is as a means of giving a rigorous notion of the number of independent "degrees of freedom" of the asymptotic dynamics of the system. This is a notion we will try to make precise in the following three chapters, coming closest perhaps with Corollary 16.3, which shows that a finite number of variables (roughly $2d$) will parametrise an attractor with $d_f(\mathcal{A}) = d$.

For now, we assume that the dimension of the attractor is a good indication of the number of degrees of freedom, and we use this to relate $d_f(\mathcal{A})$ to a possible fundamental length scale of the original problem.

Assume that the original equation is posed on some domain $\Omega \subset \mathbb{R}^m$, with volume $|\Omega|$. Suppose that there is a smallest physically relevant length scale l in the problem, the idea being that interactions on scales of less than l do not affect the dynamics (for example, in fluid mechanics the viscosity has a large effect on the very small scales, and we hope that this means that fluctuations on these scales have negligible effects). A heuristic indication of the number of degrees of freedom would then be given by how many "boxes" of side l fit into Ω,

$$n_{\text{heuristic}} \sim |\Omega| l^{-m}.$$

If we assume that $n_{\text{heuristic}}$ is a good estimate of the true number of degrees of freedom, and in turn that this is well estimated by the attractor dimension, we can isolate a length scale, given in terms of $d_f(\mathcal{A})$ by

$$l \sim \left(\frac{|\Omega|}{d_f(\mathcal{A})} \right)^{1/m}. \tag{13.30}$$

Notice that tighter bounds on $d_f(\mathcal{A})$ raise the estimate of the smallest length scale. In particular, much effort has been expended on obtaining optimal bounds for the case of the 2D Navier–Stokes equations. Our bound in Theorem 13.21 is extremely poor, since we made no effort to obtain optimal bounds at any stage. The estimates are usually expressed in terms of the dimensionless Grashof number G, a measure of the strength of the forcing relative to the viscosity,

$$G = \frac{|f|}{\nu^2 \lambda_1}. \tag{13.31}$$

Our estimate was of the order of G^2 as $\nu \to 0$.

However, the best current estimates are

$$d_f(\mathcal{A}) \leq cG^{2/3}(1 + \log G)^{1/3} \tag{13.32}$$

in the case of periodic boundary conditions and

$$d_f(\mathcal{A}) \leq cG$$

in the case of Dirichlet boundary conditions, both due to Constantin *et al.* (1988), with a simpler proof of (13.32) to be found in Doering & Gibbon (1995).

What is remarkable about the bound in (13.32) is that, when our "very heuristic" estimate in (13.30) is used, it corresponds to a length scale l that satisfies

$$\frac{l}{L} \sim G^{-1/3} \tag{13.33}$$

to within logarithmic corrections (L denotes the size of one side of our 2D periodic domain). The length scale in (13.33) is precisely the "Kraichnan length," derived by other (also heuristic) methods as the natural minimum scale in 2D turbulent flows (see Kraichnan, 1967, and Exercise 13.11). This links the rigorous analytical bound on the attractor dimension with an "intuitive" estimate from fluid dynamics.

13.6 Conclusion

We have now seen that the global attractors we found in the previous two chapters are finite-dimensional subsets of L^2, and we have obtained explicit bounds on these dimensions in Theorems 13.19 and 13.21. We spend the rest of this book trying to understand exactly how the existence of a finite-dimensional attractor produces dynamics that are determined by a finite number of degrees of freedom.

Exercises

13.1 Show that if $X \subset \mathbb{R}^m$ then the box-counting dimension defined in the footnote in Section 13.1.1 and the fractal dimension from Definition 13.1 coincide.

13.2 Show that if ϵ_n is a sequence that tends to zero as $n \to \infty$, with $\epsilon_{n+1} \geq \alpha \epsilon_n$ for some α, then

$$d_f(X) = \limsup_{n \to \infty} \frac{\log N(X, \epsilon_n)}{-\log \epsilon_n};$$

that is, one can take the limit over the sequence ϵ_n instead of the continuous $\epsilon \to 0$.

13.3 Let $\{e_n\}$ be an orthonormal basis in a Hilbert space H, and let $H_{\log} = \{e_n / \log n\}_{n=1}^{\infty}$. Show that $d_f(H_{\log}) = \infty$.

13.4 Show that the fractal dimension of the "middle-α" Cantor set, obtained as in Example 13.6, except by removing the middle α from each interval, is $\log 2 / \log \beta$, where $\beta = (1 - \alpha)/2$.

13.5 Show that (13.7),

$$\mathcal{H}^d \left(\bigcup_{k=1}^{\infty} X_k \right) \leq \sum_{k=1}^{\infty} \mathcal{H}^d(X_k),$$

holds.

13.6 Let L be the linear map that takes the basis elements $e^{(k)}$ into a new set $v^{(k)}$. Show that the matrix $M(v^{(1)}, \ldots, v^{(n)})$ used in the derivation of the trace formula,

$$M_{ij} = v^{(i)} \cdot v^{(j)},$$

is in fact $L^T L$. Deduce that the volume V_n of the parallelepiped formed by the vectors $\{v^{(n)}\}$ is indeed given by

$$V_n^2 = \det M,$$

as claimed.

13.7 Show that M as defined above has strictly positive eigenvalues.

13.8 Use the decomposition of M as $M = \sum_{j=1}^n \lambda_j e_j e_j^T$, where $M e_j = \lambda_j e_j$ with $(e_j, e_k) = \delta_{jk}$, to prove that

$$\log[\det M] = \text{Tr}[\log M]$$

and that

$$\frac{d}{dt} \text{Tr}[\log M] = \text{Tr}\left[M^{-1} \frac{dM}{dt}\right]. \qquad (13.34)$$

[When $\det A \neq 0$, $\log A$ is the matrix B such that $e^B = \sum_{j=0}^\infty (B^j)/j! = A$; see Rudin (1991, p. 265).]

13.9 In Lemma 13.17 we used the fact that the eigenvalues λ_j of the Laplacian, arranged in increasing order, satisfy

$$cj^{2/m} \leq \lambda_j \leq Cj^{2/m}.$$

Verify this in the case of periodic boundary conditions on $[0, L]^m$, where the eigenvalues are proportional to the sums of m square integers. For some indications of the proof in the general case see the Notes.

13.10 Show that the operator $\Lambda(t; u_0)$ from Theorem 13.20 (linearisation of the 2D Navier–Stokes equations) is compact for all $t > 0$.

13.11 The Kraichnan length scale L_χ is the only quantity with dimensions of length that can be formed from the viscous enstrophy dissipation χ and the viscosity ν. The quantity χ is defined by

$$\chi = \nu \langle (\Delta u)^2 \rangle = \frac{\nu}{L^2} \limsup_{t \to \infty} \sup_{u_0 \in \mathcal{A}} \frac{1}{t} \int_0^t |Au(s)|^2 \, ds$$

(it is the rate of decrease of $|\nabla u|^2$). Use the inequality (12.6) to show that

$$\chi \leq \nu^3 L^{-6} G^2$$

and hence that L_χ satisfies

$$\frac{L_\chi}{L} \geq G^{-1/3}. \tag{13.35}$$

Compare this with (13.33).

Notes

Because of certain properties to be discussed in Chapter 16, and because Theorem 13.16 allows us to obtain bounds on the fractal dimension of attractors as easily as we could on the Hausdorff dimension, the fractal dimension has become the preferred dimension measure to use in the theory of global attractors. However, no extensive literature exists on fractal dimension, which is treated in passing in Falconer (1990), and in more detail in Eden *et al.* (1994b). Robinson (1995) also treats it in some detail.

The Hausdorff dimension has a much longer history. That $\mathcal{H}^m$ is proportional to Lebesgue measure in $\mathbb{R}^m$ is shown in Falconer (1985). His two books (1985 and 1990) are both good introductions to the geometric aspects of fractal sets in $\mathbb{R}^m$. That $\mathcal{H}^m$ is a measure for all m makes it an appropriate foundation for geometric measure theory (see Rogers, 1970, and Federer, 1991), an extensive and important subject in its own right.

The proof of Propositions 13.2 and 13.9 are taken from Falconer (1990), as is the proof of the Hausdorff dimension of the Cantor set (Example 13.11). The other examples are from Eden *et al.* (1994b), as is the proof of Lemma 13.12.

The use of the trace formula to obtain bounds on the attractor dimension goes back to Constantin & Foias (1985); see also Constantin *et al.* (1985). The theory is discussed in greater generality and more detail than here in the books of Constantin & Foias (1988) and Temam (1988). We follow the simpler presentation in Doering & Gibbon (1995).

The result of Theorem 13.16 is a recent one. The original argument of Constantin & Foias (1985) shows that $d_H \leq n$ but that $d_f \leq 2n$. A proof of Theorem 13.16 in finite-dimensional spaces is given in Hunt (1996), and his proof in the infinite-dimensional case (B. Hunt, personal communication, 1999) is given in Appendix B. Blinchevskaya & Ilyashenko (1999) provide a proof that is valid without the compactness assumption on $\Lambda(t; u_0)$.

The proof of Lemma 13.17 relies on the property of the eigenvalues of the Laplacian

$$cj^{2/m} \leq \lambda_j \leq Cj^{2/m}, \tag{13.36}$$

proved in Exercise 13.9 in the case of periodic boundary conditions. The same result for a general domain follows since one can show (see Davies, 1995) that the eigenvalues depend (in some sense) continuously on the shape of the domain. By putting a collection of rectangular domains "around" and "inside" Ω one obtains (13.36) in this case also.

The proof of the uniform differentiability of the reaction–diffusion equation is taken from Marion (1987) and Debussche (1998), while that of the Navier–Stokes equation is from Temam (1988).

The periodic estimate in (13.32) is sharp, since an example due to Babin & Vishik (1992) shows that there are flows in which (13.32) is also a lower bound on the dimension. The discussion of the relation of these estimates to the classical ideas of the "minimum length scale in the flow" is inspired by Doering & Gibbon (1995). One can also obtain estimates for the attractor of the 3D equation, under the regularity assumptions of Theorem 12.10 (Constantin *et al.*, 1985; Temam, 1988; Gibbon & Titi, 1997).

Finally, we note that there are methods for estimating the attractor dimension which do not rely on differentiability, see Ladyzhenskaya (1982, 1991) and Exercise 14.1.

Part IV
Finite-Dimensional Dynamics

14

The Squeezing Property
Determining Modes

In the previous chapter we proved that the global attractors we had found are in fact finite-dimensional subsets of the infinite-dimensional phase space H. However, this is a property of the *set* $\mathcal{A}$ and a priori says nothing about the dynamics on the attractor. In the next three chapters we will investigate various ways in which we can make precise the statement that the dynamics on the attractor are finite-dimensional.

The main aim is to treat the dynamics of solutions on the attractor, that is, to investigate $S(t)|_{\mathcal{A}}$. Although we are interested in the asymptotic dynamics of all solutions of the original equation, the tracking result of Corollary 10.15 shows that the eventual behaviour of all solutions is determined by those on the attractor, and so by making this restriction we are not sacrificing too much information. Furthermore, we saw in Theorem 10.6 that if the semiflow is injective on $\mathcal{A}$ then in fact $S(t)|_{\mathcal{A}}$ is a dynamical system [$S(t)$ is defined for all $t \in \mathbb{R}$].

This chapter starts from the geometric "squeezing property," which can be proved for a variety of examples, and deduces various consequences. In particular we will show that the attractor is nearly a finite-dimensional submanifold of H and that the behaviour of a finite number of modes in the Fourier expansion "determines" (we will make this precise) the asymptotic behaviour of solutions on $\mathcal{A}$.

14.1 The Squeezing Property

The squeezing property was first introduced in the context of the Navier–Stokes equations, but it holds for many other examples as well. The idea is that there is a natural splitting of the phase space into a finite-dimensional subspace and its infinite-dimensional orthogonal complement, such that the finite-dimensional part of the solution dominates. We write P for the orthogonal projection onto

359

the finite-dimensional subspace, and Q for the projection onto its orthogonal complement. As such, we can split any element of H into two components, $u = Pu + Qu$.

In the examples we will consider, P will be the orthogonal projection onto the first n eigenfunctions w_j (ordered so that $\lambda_{j+1} \geq \lambda_j$) of some linear operator A (usually $-\Delta$),

$$P_n u = \sum_{j=1}^{n} (u, w_j) w_j,$$

and so we refer to the P part of solution as the "low modes" and the Q part of the solution as the "high modes."

The squeezing property treats the difference between two solutions, $w = u - v$, and gives the following alternatives: after both solutions have evolved for a fixed time (we choose $t = 1$), either the high modes are dominated by the low modes, so that the finite-dimensional low mode component is more important; or if not, then at least the solutions are closer together than they were at $t = 0$, which serves to dampen the effect of such "ill-behaved" solutions.

Definition 14.1. *Write* $S = S(1)$. *Then the* squeezing property *holds if, for each* $0 < \delta < 1$, *there exists a finite-rank orthogonal projection* $P(\delta)$, *with orthogonal complement* $Q(\delta)$, *such that for every* $u, v \in \mathcal{A}$ *either*

$$|Q(Su - Sv)| \leq |P(Su - Sv)| \tag{14.1}$$

or, if not, then

$$|Su - Sv| \leq \delta |u - v|. \tag{14.2}$$

In general to decrease δ we need to increase the rank of the orthogonal projection $P(\delta)$ (i.e. to increase the dimension of PH). We will verify that this property holds for our examples in Sections 14.4 and 14.5. For now, we discuss its consequences for the structure of the attractor and the dynamics of the equation. Exercise 14.1 shows that one can use the squeezing property to obtain bounds on the dimension of the attractor.

14.2 An Approximate Manifold Structure for $\mathcal{A}$

One way of interpreting the squeezing property is as follows. If $u, v \in \mathcal{A}$, then since the attractor is invariant we have $u = S(1)u_{-1}$ and $v = S(1)v_{-1}$, for some $u_{-1}, v_{-1} \in \mathcal{A}$. Now suppose that the P components of u and v agree, so that

$Pu = Pv$. Then the squeezing property shows that either $Qu = Qv$ [from (14.1)], in which case $u = v$, or [from (14.2)] that

$$|u - v| \le \delta|u_{-1} - v_{-1}|.$$

Since the attractor lies in a bounded set in H, $\{|u| \le R_H\}$, in either case we have

$$|u - v| \le 2\delta R_H \quad \text{if} \quad Pu = Pv.$$

The finite-dimensional projection onto PH serves to identify points on the attractor to within an error of $2\delta R_H$. By taking δ smaller, and hence n larger, we can increase the accuracy of this approximation.

With a little further work we can use this approach to say something about the structure of the attractor. In fact, $\mathcal{A}$ must lie within a small neighbourhood $(4\delta R_H)$ of an n-dimensional Lipschitz manifold in H.

Proposition 14.2. *If the squeezing property holds then there exists a Lipschitz function* $\Phi : PH \to QH$,

$$|\Phi(p) - \Phi(\overline{p})| \le |p - \overline{p}| \quad \text{for all} \quad p, \overline{p} \in PH,$$

such that $\mathcal{A}$ lies within a $4\delta R_H$ neighbourhood of the graph of Φ,

$$\mathcal{G}[\Phi] = \{u \in H : u = p + \Phi(p),\ p \in PH\}.$$

See Figure 14.1.

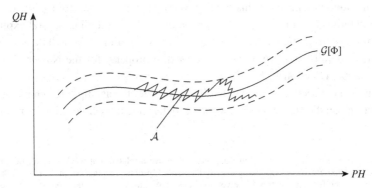

Figure 14.1. Highly schematic illustration of Proposition 14.2.

Proof. Consider a subset X of $\mathcal{A}$ that is maximal for the relation

$$|Q(u-v)| \le |P(u-v)| \qquad \text{for all} \qquad u, v \in X. \tag{14.3}$$

Then, for every $u \in X$, we can define uniquely $\phi(Pu) = Qu$ such that $Pu + \phi(Pu) = u$ [since if $Pu = Pv$ (14.1) shows that $Qu = Qv$]. Furthermore, it follows from (14.3) that

$$|\phi(p_1) - \phi(p_2)| \le |p_1 - p_2| \qquad \text{for all} \qquad p_1, p_2 \in PX.$$

Assume that we can extend ϕ to a function $\Phi : PH \to QH$ (defined on the whole of PH) that satisfies the same Lipschitz bound.* Now, if $u \in \mathcal{A}$ but $u \notin X$, it follows that

$$|Q(u-v)| \ge |P(u-v)|$$

for some $v \in X$. Thus, if $u = S(1)\overline{u}$ and $v = S(1)\overline{v}$, with $\overline{u}, \overline{v} \in \mathcal{A}$, we must have

$$|Q(u-v)| \le \delta |Q(\overline{u} - \overline{v})| \le 2\delta R_H.$$

Since $|P(u-v)| \le |Q(u-v)|$ it follows that $|u-v| \le 4\delta R_H$, and so

$$\text{dist}(u, \mathcal{G}[\Phi]) \le 4\delta R_H,$$

as claimed. ∎

One idea arising from this result involves using the manifold $\mathcal{G}[\Phi]$ to aid computations, and much effort has been expended on finding a good explicit form for Φ, in particular for the Navier–Stokes equations. We will discuss this in more detail when we prove the squeezing property for the Navier–Stokes equations in Section 14.5.

In the next section we show that if we consider the asymptotic behaviour of solutions on the attractor, the low modes determine the whole behaviour of the solution.

---

* This is certainly not obvious, but it is true. We prove a related, but weaker, result in Chapter 16 (Theorem 16.4). Obtaining exactly the same Lipschitz constant for the extension is far from straightforward: A proof is given in Wells & Williams (1975), and a proof for the case $f : \mathbb{R}^m \to \mathbb{R}^n$ that can be extended to the case $f : \mathbb{R}^m \to H$ can be found in Federer (1991).

14.3 Determining Modes

We now show that the squeezing property implies that the flow on the attractor has a finite number of "determining modes." This concept, introduced by Foias & Prodi (1967), says that if two solutions agree asymptotically in their P projection, then they agree asymptotically in their entirety.

Theorem 14.3 (Determining modes). *Suppose that the semiflow $S(t)$ satisfies the Lipschitz condition*

$$|S(t)u_0 - S(t)v_0| \leq L|u_0 - v_0| \qquad \text{for all} \qquad 0 \leq t \leq 1 \qquad (14.4)$$

and that the squeezing property holds. Then the projection onto PH is "determining," in that for two solutions $u(t)$ and $v(t)$ on the attractor, if

$$|P(u(t) - v(t))| \to 0 \qquad \text{as} \qquad t \to \infty \qquad (14.5)$$

then

$$|u(t) - v(t)| \to 0 \qquad \text{as} \qquad t \to \infty. \qquad (14.6)$$

It is as important to stress what this result does not say as what it does say. If you take two solutions of the full equations and know that the P modes are converging, you can deduce that the full solutions are converging. However, it does not say that the solutions of the P-mode Galerkin truncation determine the solution, nor that knowledge of the P modes at any instant will determine the full solution. [Indeed, we saw above that knowing the P modes will give only approximate information as to the Q component on the attractor, $|Qu - \Phi(Pu)| \leq 4\delta R_H$.]

Proof. As in the squeezing property, we write $S = S(1)$, and then (14.5) implies that

$$|P(S^m u_0 - S^m v_0)| \to 0 \qquad \text{as} \qquad m \to \infty. \qquad (14.7)$$

We want to show that

$$|S^m u_0 - S^m v_0| \to 0 \qquad \text{as} \qquad m \to \infty. \qquad (14.8)$$

This implies (14.6) by the Lipschitz property of solutions in (14.4), since

$$|S(t)u_0 - S(t)v_0| = \left|S(t - [t])S^{[t]}u_0 - S(t - [t])S^{[t]}v_0\right| \leq L\left|S^{[t]}u_0 - S^{[t]}v_0\right|,$$

where $[t]$ is the greatest integer less than or equal to t, and clearly $[t] \to \infty$ as $t \to \infty$.

So we suppose that (14.8) does not hold and deduce a contradiction. In this case, there exists an $\epsilon > 0$ and a sequence $m_j \to \infty$ such that

$$|S^{m_j} u_0 - S^{m_j} v_0| \geq \epsilon. \tag{14.9}$$

Since we have

$$|P(S^{m_j} u_0 - S^{m_j} v_0)| \to 0,$$

we must have, for m_j large enough,

$$|Q(S^{m_j} u_0 - S^{m_j} v_0)| > |P(S^{m_j} u_0 - S^{m_j} v_0)|,$$

so that we can use the squeezing property (14.2) to obtain

$$|S^{m_j} u_0 - S^{m_j} v_0| \leq \delta |S^{m_j-1} u_0 - S^{m_j-1} v_0|.$$

We now consider $S^{m_j-1} u_0$ and $S^{m_j-1} v_0$. Either they satisfy

$$|Q(S^{m_j-1} u_0 - S^{m_j-1} v_0)| \leq |P(S^{m_j-1} u_0 - S^{m_j-1} v_0)|$$

or again we can apply the squeezing (14.2) and consider $S^{m_j-2} u_0$ and $S^{m_j-2} v_0$ in the same way. We continue like this M_j times, either until $M_j = m_j$ or until

$$|Q(S^{m_j-M_j} u_0 - S^{m_j-M_j} v_0)| \leq |P(S^{m_j-M_j} u_0 - S^{m_j-M_j} v_0)|.$$

By then we have, applying the squeezing property M_j times,

$$|S^{m_j} u_0 - S^{m_j} v_0| \leq \sqrt{2}\delta^{M_j} |P(S^{m_j-M_j} u_0 - S^{m_j-M_j} v_0)|. \tag{14.10}$$

We now have two possibilities for the sequence M_j. M_j could be bounded, so that $M_j \leq M$ for some M; then

$$|S^{m_j} u_0 - S^{m_j} v_0| \leq \sqrt{2} \max_{0 \leq k \leq M} |P(S^{m_j-k} u_0 - S^{m_j-k} v_0)|,$$

which, using the convergence of the P modes (14.7), gives

$$|S^{m_j} u_0 - S^{m_j} v_0| \to 0, \tag{14.11}$$

contradicting (14.9). The second possibility is that M_j is unbounded; so there is a subsequence (relabel this M_j again) such that $M_j \to \infty$. But since $S^m u_0 \in \mathcal{A}$

for any m, the sequence $\{S^m u_0\}$ is bounded in H. It follows, using (14.10) along with the fact that $\delta < 1$, that (14.11) holds in this case also.

Thus we obtain the discrete convergence (14.8) and hence (14.6). ∎

For a similar result involving the value of $u(x, t)$ at a finite number of points within the domain, called "determining nodes," see Exercises 14.4–14.7.

14.4 The Squeezing Property for Reaction–Diffusion Equations

Checking that the squeezing property holds can be messy and needs to be tailored to particular examples. We prove this property for our two examples, scalar reaction–diffusion equations and the 2D Navier–Stokes equations. Although these proofs could easily be skipped on a first reading, Section 14.5 also discusses another topic, approximate inertial manifolds (cf. Proposition 14.2) for the Navier–Stokes equations.

The main idea is to consider $w(t) = u(t) - v(t)$ and to make estimates on the equation that bound $d(Pw)/dt$ below and $d(Qw)/dt$ above. With some moderate trickery we can then prove the squeezing property.

It will be important that the global attractor is bounded in $L^\infty(\Omega)$ (Theorem 11.6),

$$\|u\|_\infty \leq M \qquad \text{for all} \qquad u \in \mathcal{A}.$$

This enables us to treat the nonlinear term $F(u)$ from

$$du/dt + Au = F(u)$$

[the form in which we wrote $\partial u/\partial t - \Delta u = f(u(x))$ in Chapter 8] as a Lipschitz function from L^2 into itself, since

$$|F(u) - F(v)|^2 = \int_\Omega \left(\int_{u(x)}^{v(x)} f'(s)\, ds \right)^2 dx$$

$$\leq \int_\Omega C^2 |u(x) - v(x)|^2\, dx$$

$$= C^2 |u - v|^2, \tag{14.12}$$

where

$$C = \sup_{s \leq M} |f'(s)|.$$

Here the linear operator A is the negative Laplacian, and we write its eigenfunctions as w_j, ordered so that the corresponding eigenvalues are increasing with j:

$$Aw_j = \lambda_j w_j, \qquad \lambda_{j+1} \geq \lambda_j. \tag{14.13}$$

We denote the orthogonal projector onto the first n modes as P_n,

$$P_n u = \sum_{j=1}^{n} (u, w_j) w_j, \tag{14.14}$$

and that onto the remaining "high" modes by Q_n,

$$Q_n u = (I - P_n)u = \sum_{j=n+1}^{\infty} (u, w_j) w_j. \tag{14.15}$$

We now show that the squeezing property holds. We will make use of the following two inequalities:

$$\|p\| \leq \lambda_n^{1/2} |p| \tag{14.16}$$

and

$$\|q\| \geq \lambda_{n+1}^{1/2} |q|, \tag{14.17}$$

which follow from the eigenvalue expansion of u and the identity

$$\|u\|^2 = a(u, u) = (Au, u);$$

see Exercise 14.2.

Theorem 14.4. *The squeezing property holds for the scalar reaction–diffusion equation*

$$\frac{\partial u}{\partial t} - \Delta u = f(u), \qquad u|_{\partial\Omega} = 0$$

from Chapter 8 [i.e. we make the same assumptions on f ((8.14) and (8.15)) and Ω ("sufficiently smooth") as there].

Proof. The equation for the difference of two solutions $w(t) = u(t) - v(t)$ is

$$\frac{dw}{dt} + Aw = F(u) - F(v), \tag{14.18}$$

and we will write

$$p = P_n w, \qquad q = Q_n w, \qquad w = p + q.$$

First of all we take the inner product of (14.18) with p and get, since A commutes with P_n,

$$\tfrac{1}{2}\frac{d}{dt}|p|^2 + \|p\|^2 = (F(u) - F(v), p).$$

If we use the Lipschitz property of F (14.12) this becomes

$$\tfrac{1}{2}\frac{d}{dt}|p|^2 + \|p\|^2 \geq -C|w||p|.$$

Using (14.16), and writing $\lambda = \lambda_n$, we then have

$$\tfrac{1}{2}\frac{d}{dt}|p|^2 \geq -\lambda|p|^2 - C|w||p|,$$

or, using Exercise 2.5,

$$\frac{d}{dt_+}|p| \geq -\lambda|p| - C(|p| + |q|). \tag{14.19}$$

Similarly, if we take the inner product with q we obtain, using (14.17),

$$\frac{d}{dt_+}|q| \leq -\lambda|q| + C(|p| + |q|). \tag{14.20}$$

We now choose n large enough that

$$\lambda_n - C > 2C. \tag{14.21}$$

Now, either (14.1) holds, and so there is nothing to prove, or it does not, in which case

$$|Qw(1)| > |Pw(1)|. \tag{14.22}$$

In this case, using (14.21), we see that the inequality

$$(\lambda - C)|Qw(t)| > 2C|Pw(t)| \tag{14.23}$$

holds for $t = 1$. Since $w(t)$ is continuous into H, (14.23) holds in a neighbourhood of $t = 1$. We consider two possibilities, the first being the most straightforward.

If (14.23) holds for all $t \in [\frac{1}{2}, 1]$, then we have, by (14.21),

$$(\lambda - c)|q| - C|p| > \tfrac{1}{2}(\lambda - C)|q| > c|q|$$

for $t \in [\frac{1}{2}, 1]$, and so (14.20) becomes

$$\frac{d}{dt_+}|q| \leq -\lambda C|q|,$$

which gives

$$|q(1)| \leq e^{-\frac{1}{2}\lambda C}\left|q\left(\tfrac{1}{2}\right)\right|.$$

Since (14.22) holds, this implies that

$$|w(1)| \leq \sqrt{2}e^{-\frac{1}{2}\lambda C}\left|q\left(\tfrac{1}{2}\right)\right| \leq \sqrt{2}e^{-\frac{1}{2}\lambda C}\left|w\left(\tfrac{1}{2}\right)\right|,$$

and using the Lipschitz property of solutions (8.27),

$$\left|w\left(\tfrac{1}{2}\right)\right| \leq L\left(\tfrac{1}{2}\right)|w(0)|,$$

we have

$$|w(1)| \leq \sqrt{2}L\left(\tfrac{1}{2}\right)e^{-\frac{1}{2}\lambda_n C}|w(0)|.$$

This gives (14.2), provided that λ_n (i.e. n) is chosen large enough.

If (14.23) does not hold on all of $[\frac{1}{2}, 1]$, then it holds on $[t_0, 1]$, with

$$(\lambda - C)|Qw(t_0)| = 2C|Pw(t_0)|. \tag{14.24}$$

We proceed in this case by a trick. Define

$$\Phi(t) = \Phi(p(t), q(t)) = (|p| + |q|)\exp\left(\frac{\lambda|q|}{C(|p| + |q|)}\right); \tag{14.25}$$

we do this since if the inequalities in (14.19) and (14.20) are replaced by equalities, we have $d\Phi/dt = 0$. Happily if we restore the inequalities, elementary but tedious computations (see Exercise 14.3) show that

$$d\Phi/dt \leq 0 \qquad \text{for all} \qquad t \in [t_0, 1].$$

Thus we know that

$$\Phi(1) \leq \Phi(t_0).$$

However, at $t = 1$ we have (14.23), so that

$$\Phi(1) \geq |q(1)|e^{\lambda/C},$$

and at $t = t_0$ the equality (14.24) holds, which gives

$$2C(|p(t_0)| + |q(t_0)|) = (\lambda + C)|q(t_0)|,$$

and so

$$\Phi(t_0) = \frac{\lambda + C}{2C}|q(t_0)|e^{2\lambda/(\lambda+C)}.$$

It follows that

$$|q(1)| \leq e^{-\lambda/C}\frac{\lambda + C}{2C}e^2|q(t_0)|,$$

and using once more the Lipschitz property (8.27) we obtain

$$|q(1)| \leq e^{-\lambda/C}\frac{\lambda + C}{2C}e^2 L(1)|w(0)|.$$

Since $|p(1)| < |q(1)|$, it certainly follows that

$$|w(1)| \leq 2e^{-\lambda_n/C}\frac{\lambda_n + C}{2C}e^2 L(1)|w(0)|.$$

Therefore if n is chosen large enough we have once again ensured (14.2), and the theorem is proved. ∎

14.5 The 2D Navier–Stokes Equations

We now turn to the 2D Navier–Stokes equations. First we sketch a proof that guarantees that the squeezing property holds, and then we consider an explicit form for the function Φ from Proposition 14.2 that gives some information about the global attractor.

14.5.1 Checking the Squeezing Property

In this section we prove that the squeezing property holds for the 2D Navier–Stokes equations. Recall from Chapter 12 that if $f \in H$ then we have bounds on the attractor in H, V, and $D(A)$. The argument is essentially the same as that for the reaction–diffusion equation, and we merely highlight the differences.

In this case, we take A to be the Stokes operator, and we define P_n and Q_n as in the reaction–diffusion case by (14.14) and (14.15), where we order the eigenfunctions as in (14.13).

Theorem 14.5. *If $f \in H$ then the squeezing property holds for the 2D Navier–Stokes equations.*

Note that the constant C may change from line to line in the proof.

Proof. The equation for the difference $w(t) = u(t) - v(t)$ is

$$\frac{dw}{dt} + \nu Aw + B(u, w) + B(w, v) = 0, \qquad (14.26)$$

and we will write

$$p = P_n w, \qquad q = Q_n w, \qquad w = p + q.$$

First we take the inner product of (14.26) with p, using $b(u, w, p) = b(u, p + q, p) = b(u, q, p)$,

$$\frac{1}{2}\frac{d}{dt}|p|^2 + \nu\|p\|^2 = -b(u, q, p) - b(w, v, p).$$

Using the bounds on b in (9.27) and the existence of an absorbing set in H, V, and $D(A)$ (see Chapter 12), we can obtain

$$\begin{aligned}
\frac{1}{2}\frac{d}{dt}|p|^2 + \nu\|p\|^2 &\geq -k\left(|u|^{1/2}|Au|^{1/2}|q|\|p\| - |w|\|p\||v|^{1/2}|Av|^{1/2}\right) \\
&\geq -C_1\left(|q|\lambda^{1/2}|p| + |w|\lambda^{1/2}|p|\right) \\
&\geq -C\lambda^{1/2}|p|(|p| + |q|),
\end{aligned}$$

where $\lambda = \lambda_n$, so that

$$\begin{aligned}
|p|\frac{d|p|}{dt} &\geq -|p|\left(\nu\lambda|p| + C\lambda^{1/2}|p| + C\lambda^{1/2}|p|\right) \\
&\geq -\lambda^{1/2}|p|\left(\nu\lambda^{1/2} + C|p| + C|q|\right).
\end{aligned}$$

Now take the inner product with q,

$$\begin{aligned}
\frac{1}{2}\frac{d}{dt}|q|^2 + \nu\|q\|^2 &= -b(u, p, q) - b(w, v, q) \\
&\leq k|p|\|q\| + k|w|\|q\| \\
&\leq C(|p| + |q|)\|q\|,
\end{aligned}$$

so that

$$|q|\frac{d|q|}{dt} \le \|q\|\left(-\nu\lambda^{1/2}|q| + C|p| + C|q|\right).$$

Provided that the expression in the parentheses is negative,

$$\left(\nu\lambda^{1/2} - C\right)|q| > C|p|,$$

then we have

$$|q|\frac{d|q|}{dt} \le \lambda^{1/2}|q|\left(-\nu\lambda^{1/2}|q| + C|p| + C|q|\right).$$

We now choose n large enough that

$$\nu\lambda_n^{1/2} - C > 2C.$$

If (14.1) does not hold then we follow the previous approach, seeking the first time at which the inequality

$$\left(\nu\lambda^{1/2} - C\right)|Qw(t)| > 2C|Pw(t)|$$

fails to hold. In this case we take

$$\Phi(t) = \Phi(p(t), q(t)) = (|p| + |q|)\exp\left(\frac{\nu\lambda^{1/2}|q|}{C(|p| + |q|)}\right).$$

The argument is essentially unchanged from here. ∎

14.5.2 Approximate Inertial Manifolds

Recall (Proposition 14.2) that if the squeezing property holds we can show that the attractor lies within a neighbourhood of the graph of some function Φ,

$$\mathrm{dist}(\mathcal{A}, \mathcal{G}[\Phi]) \le \epsilon.$$

The graph of such a function Φ has come to be termed an "approximate inertial manifold" (we will see why in the next chapter).

We can obtain a very simple approximate inertial manifold by setting $\Phi = 0$ (this corresponds numerically to the classical Galerkin approximation; see Exercise 14.8). Now we have

$$|Q_n u| = |A^{-1}AQ_n u| = \|A^{-1}Q_n\|_{\mathrm{op}}|Au|.$$

Using the expansion in terms of eigenfunctions of A (cf. (14.16)),

$$
\begin{aligned}
|A^{-1}Q_n u|^2 &= \left| A^{-1} \sum_{j=n+1}^{\infty} (u, w_j) w_j \right|^2 \\
&= \left| \sum_{j=n+1}^{\infty} \lambda_j^{-1}(u, w_j) w_j \right|^2 \\
&= \sum_{j=n+1}^{\infty} \lambda_j^{-2} |(u, w_j)|^2 \\
&\le \lambda_{n+1}^{-2} |Q_n u|^2 \\
&\le \lambda_{n+1}^{-2} |u|^2,
\end{aligned}
$$

and so

$$
\| A^{-1} Q_n \|_{\mathrm{op}} \le \lambda_{n+1}^{-1}.
$$

Therefore, since we know that

$$
|Au| \le R_D
$$

on the attractor, we have

$$
|Q_n u| \le \lambda_{n+1}^{-1} |Au| \le R_D \lambda_{n+1}^{-1}, \tag{14.27}
$$

so that as we take more and more modes we get a better approximation to $u \in \mathcal{A}$. A nontrivial approximation, where $\Phi \ne 0$, should give an improvement on the convergence as $n \to \infty$ in (14.27).

There are many possible forms for the approximate slaving rule Φ, and we will discuss one that has a nice physical interpretation. The "stationary approximate inertial manifold" is obtained by the following simple heuristic argument, which we will then (to some extent) justify rigorously. From the 2D Navier–Stokes equation,

$$
du/dt + \nu Au + B(u, u) = f, \tag{14.28}
$$

we try to obtain a relationship between $q = Q_n u$ and $p = P_n u$. All steady states of (14.28) satisfy

$$
\nu Au + B(u, u) = f,
$$

and projecting this equation onto the higher modes gives

$$\nu A q = -Q B(u, u) + Q f.$$

We now define the stationary approximate inertial manifold, Φ_s, as the solution of

$$\nu A \Phi_s(p) = -Q B(p + \Phi_s(p), p + \Phi_s(p)) + Q f.$$

One can show that such a solution exists by using the contraction mapping theorem applied to the map T, where

$$[T\Phi](p) = (\nu A Q)^{-1}[f - B(p + \Phi(p), p + \Phi(p))].$$

Since we actually want an explicit form for the approximate inertial manifold, we will not use Φ_s itself, but the first iterate of $\Phi \equiv 0$ under the above contraction map,

$$\Phi_1(p) = (\nu A Q)^{-1}[f - B(p, p)]. \tag{14.29}$$

One can in fact show that the theoretical estimate for the accuracy of Φ_s is no better than that of Φ_1. We will therefore concentrate on the explicit approximate inertial manifold given by (14.29).

To show that Φ_1 is a improvement on $\Phi \equiv 0$, we will appeal to a result due to Foias & Temam (1979) that guarantees that the solutions of the complexified form of the 2D equation,

$$du/d\zeta + \nu A u + B(u, u) = f,$$

are analytic in ζ in a strip of width δ about the real axis, and if $u \in \mathcal{A}$ then

$$\sup_{\{\zeta : |\mathrm{Im}\, \zeta| < \delta\}} |A u(\zeta)| \le M.$$

We can then use the Cauchy integral formula

$$q(t) = \int_\Gamma \frac{q(z)}{z - t_0}\, dz,$$

where Γ is a circle, centred on t_0 of radius $\delta/2$, to obtain an expression for dq/dt,

$$\frac{dq}{dt} = \int_\Gamma \frac{q(z)}{(z - t_0)^2}\, dz,$$

and hence the bound (following (14.27))

$$\left|\frac{dq}{dt}\right| \le \frac{\pi\delta}{(\delta/2)} M\lambda_{n+1}^{-1} = K\lambda_{n+1}^{-1}.$$

Now, we subtract

$$dq/dt + \nu Aq + QB(u,u) = Qf$$

from

$$\nu A\Phi_1(p) + QB(p,p) = Qf$$

to obtain

$$\nu A(q - \Phi_1(p)) = \frac{dq}{dt} + QB(p+q, p+q) - QB(p,p)$$

$$= \frac{dq}{dt} + B(p,q) + B(q,p) + B(q,q).$$

Bounding $\frac{dq}{dt}$ as above and the terms in B by using the inequalities (9.27), we obtain

$$|\nu A(q - \Phi_1(p))| \le C\lambda_{n+1}^{-1} + k\big[|p|^{1/2}|Ap|^{1/2}\|q\|$$
$$+ |q|^{1/2}|Aq|^{1/2}\|p\| + |q|^{1/2}|Aq|^{1/2}\|q\|\big]$$
$$\le K\lambda_{n+1}^{-1/2},$$

since there are absorbing sets in H, V, and $D(A)$. As above, the estimate

$$|A(q - \Phi(p))| \le K\lambda_{n+1}^{-1/2}$$

implies that

$$|q - \Phi(p)| \le k\lambda_{n+1}^{-3/2},$$

an improvement on (14.27) by a factor of $\lambda_{n+1}^{-1/2}$.

14.6 Finite-Dimensional Exponential Attractors

The global attractors whose existence was proved in Chapters 10–12 can attract solutions arbitrarily slowly. To remedy this problem, the notion of an exponential attractor was introduced in 1990 by Eden *et al.* These are positively invariant, finite-dimensional sets, that attract all trajectories exponentially fast.

When the squeezing property holds, it is possible to prove the existence of such an exponential attractor with finite fractal dimension (earning this section its place in the current chapter). Instead of giving the proof of this, which is quite involved, we prove a much simpler result that relies on the fact that any countable set of points has zero Hausdorff dimension, namely the existence of an exponential attractor whose Hausdorff dimension is (essentially) the same as that of the attractor.

The only assumption we will need here is a Lipschitz property of solutions,

$$|S(t)u_0 - S(t')v_0| \le K(T)(|u_0 - v_0| + |t - t'|) \qquad \text{for all} \qquad 0 \le t, t' \le T \tag{14.30}$$

for $u_0, v_0 \in X$. Not only does this contain the Lipschitz property of $S(t)$ for each $t \ge 0$ that we derived for our examples in Chapters 8 (8.25) and 9 (9.39),

$$|S(t)u_0 - S(t)v_0| \le k(T)|u_0 - v_0| \qquad \text{for all} \qquad 0 \le t \le T,$$

but it also tells us that the solutions are Lipschitz in t. This follows easily for both our model equations.

For the reaction–diffusion example,

$$du/dt = -Au + F(u),$$

we know that the global attractor is bounded in $D(A)$ and in $L^\infty(\Omega)$. This implies that the right-hand side is bounded in $H = L^2(\Omega)$. Similarly, the bound in $D(A)$ for the attractor of the Navier–Stokes equations shows that the right-hand side of

$$du/dt = -\nu Au - B(u, u) + f$$

is bounded in H. It follows for both examples that $|du/dt| \le k$ for some k, and so

$$|u(t) - u(t')| \le k|t - t'|.$$

Expression (14.30) then follows with $K(T) = \max(k(T), k)$, via the triangle inequality.

Theorem 14.6 (Exponential attractors). *Suppose that $S(t)$ has a compact absorbing set X, has a global attractor $\mathcal{A}$ with finite Hausdorff dimension $d_H(\mathcal{A}) < \infty$, and satisfies the Lipschitz property (14.30). Then for any $\sigma > 0$ there exists a set $\mathcal{E}$, an "exponential attractor" that is positively invariant,*

$$S(t)\mathcal{E} \subset \mathcal{E} \qquad \text{for all} \qquad t > 0,$$

attracts all solutions exponentially fast,

$$\text{dist}(S(t)u_0, \mathcal{E}) \leq Ce^{-\sigma t},$$

and has finite Hausdorff dimension,

$$d_H(\mathcal{E}) \leq d_H(\mathcal{A}) + 1.$$

Note that the construction can produce a set with an arbitrarily fast rate of attraction σ. The more refined constructions obtain sets with finite fractal dimension but place limits on this rate of attraction.

Proof. We first prove the existence of a set $\mathcal{E}_S$ with $d_H(\mathcal{E}_S) = d_H(\mathcal{A})$, which is an exponential attractor for the time $t_0(X)$ map,

$$S = S(t_0(X)),$$

where $t_0(X)$ is the time it takes for X to absorb itself. Note that we therefore have

$$S(t)X \subset X \qquad \text{for all} \qquad t \geq t_0(X).$$

We then take the image of this set under the flow between $t = 0$ and $t = t_0(X)$ to obtain an exponential attractor $\mathcal{E}$ for the continuous semiflow.
   Set

$$\theta = e^{-\sigma t_0(X)}. \tag{14.31}$$

Note that we have $S(X) \subset X$, and by the characterisation from Exercise 10.5 we have

$$\mathcal{A} = \bigcap_{n \geq 0} S^n X.$$

Since X is compact, we have $X \subseteq B(0, R)$ for some $R > 0$. We now choose a covering of SX by balls of radius θR, centred at a set $E^{(1)}$ of points in SX, that is, we choose

$$E^{(1)} = \{a_{1,j} : j = 1, \ldots, N_1\} \subset SX,$$

such that

$$SX \subseteq \bigcup_{j=1}^{N_1} B(a_{1,j}, \theta R).$$

Since SX is compact, this is a finite set. We now choose a covering by balls of radius $\theta^2 R$ of $S^2 X$, that is, we choose a set of points

$$E^{(2)} = \{a_{2,j} : j = 1, \ldots, N_2\} \subset S^2 X,$$

such that

$$S^2 X \subseteq \bigcup_{j=1}^{N_2} B(a_{2,j}, \theta^2 R).$$

We continue in the same way, defining at each stage a finite set of points $E^{(k)}$,

$$E^{(k)} = \{a_{k,j} : j = 1, \ldots, N_k\} \subset S^k X,$$

such that

$$S^k X \subseteq \bigcup_{j=1}^{N_k} B(a_{k,j}, \theta^k R).$$

Finally, we gather all these points together into the countable set $E^{(\infty)}$ given by

$$E^{(\infty)} = \bigcup_{k=1}^{\infty} E^{(k)}.$$

We claim that

$$\mathcal{E}_0 \equiv \overline{E^{(\infty)}} = E^{(\infty)} \cup \mathcal{A}.$$

Indeed, if $\{a_n\} \in E^{(\infty)}$ and $a_n \to a$, then certainly

$$a_n \in E^{(k_n)}$$

for each n. Either k_n is bounded, say $k_n \leq k$, in which case

$$a_n \in \bigcup_{j=1}^{k} E^{(j)},$$

which is just a finite set of points, so clearly $a \in E^{(j)}$ for some $j \leq k$; or k_n is unbounded, so that for a subsequence (relabeled a_n) we have $k_n \to \infty$, so that

$$a_n = S^{k_n} x_n,$$

with $x_n \in X$ and $k_n \to \infty$, and we must have $a \in \mathcal{A}$.

To make $\mathcal{E}_0$ positively invariant under S, we have to add the images of all the points in $\mathcal{E}_0$ under all iterates of S,

$$\mathcal{E}_S = \mathcal{A} \cup \bigcup_{j=0}^{\infty} S^j \left(E^{(\infty)} \right).$$

Note that the addition to $\mathcal{A}$ is still a countable set of points, and so by (13.7) we have

$$d_H(\mathcal{E}_S) = d_H(\mathcal{A}).$$

We also have

$$\text{dist}(S^k X, \mathcal{E}_S) \le R\theta^k. \tag{14.32}$$

Indeed, take any $x \in X$. Then by construction there exists an $a \in E^{(k)}$ such that

$$|S^k x - a| \le R\theta^k,$$

and since $E^{(k)} \subset \mathcal{E}_S$, (14.32) follows.

To obtain an exponential attractor for the continuous semiflow, we set

$$\mathcal{E} = \bigcup_{0 \le t \le t_0(X)} S(t)\mathcal{E}_S,$$

and noting that

$$S(t) : [0, 1] \times \mathcal{E}_S \to H$$

is Lipschitz by (14.30), we see that

$$d_H(\mathcal{E}) \le d_H([0, 1] \times \mathcal{E}_S) \le d_f([0, 1]) + d_H(\mathcal{A}) \le 1 + d_H(\mathcal{A}),$$

by the properties of d_H given in Proposition 13.9. To show that $\mathcal{E}$ attracts exponentially, write

$$t = kt_0(X) + s, \qquad s \in [0, t_0(X)),$$

and then, for $u \in X$,

$$\begin{aligned}
\text{dist}(S(t)u, \mathcal{E}) &= \inf_{v \in \mathcal{E}} |S(t)u - v| \\
&= \inf_{v \in \mathcal{E}} |S(s)S^k u - v| \\
&\le \inf_{v \in S(s)\mathcal{E}} |S(s)S^k u - v|,
\end{aligned}$$

since $S(s)\mathcal{E} \subset \mathcal{E}$. Now using the Lipschitz property (14.30) and the exponential attraction of $\mathcal{E}_S \subset \mathcal{E}$, we have

$$\begin{aligned}
\text{dist}(S(t)u, \mathcal{E}) &= \inf_{v \in \mathcal{E}} |S(s)S^k u - S(s)v| \\
&\leq \inf_{v \in \mathcal{E}} L(t_0(X))|S^k u - v| \\
&\leq L(t_0(X))R\theta^k.
\end{aligned}$$

Finally, using the definition of θ (14.31), we obtain

$$\begin{aligned}
\text{dist}(S(t)u, \mathcal{E}) &\leq L(t_0(X))e^{-\sigma k t_0(X)} \\
&= L(t_0(X))e^{\sigma s}e^{-\sigma t} \\
&\leq L(t_0(X))\exp(\sigma t_0(X))e^{-\sigma t},
\end{aligned}$$

and the theorem is proved. ∎

14.7 Conclusion

Using the squeezing property we have shown that the attractor lies close to a finite-dimensional manifold given as a graph over PH and that the dynamics "projected" onto PH give a good indication of the asymptotic behaviour of solutions. Furthermore, it is possible to find a large set $\mathcal{E} \supset \mathcal{A}$ that attracts all solutions exponentially fast.

In the next chapter we show that under a strengthened version of the squeezing property we can prove that the attractor is a subset of a finite-dimensional manifold $\mathcal{M}$ that attracts all solutions exponentially fast. As such, the projection of the dynamics onto some finite-dimensional space PH determines exactly the dynamics of the full equation. We call such a manifold $\mathcal{M}$ an "inertial manifold."

Exercises

14.1 Suppose that for $\delta = 1/8$ the squeezing property holds for some projection P of rank n_0. Then it is possible to show that if $a \in \mathcal{A}$, the set

$$Z = S(B(a, r) \cap \mathcal{A})$$

can be covered by K_0 balls of radius $r/2$, where

$$K_0 \leq \alpha^{n_0},$$

for some constant α independent of r. Deduce that the fractal dimension of the global attractor must be bounded by

$$\left(\frac{\log \alpha}{\log 2}\right) n_0. \tag{14.33}$$

This result has the advantage over the methods of Chapter 13 that it does not require the semiflow to be uniformly differentiable (Definition 13.15). [Hint: If r is chosen such that $\mathcal{A} \subset B(0, r)$, show that $N(\mathcal{A}, 2^{-k}r) \leq K_0^k$.]

14.2 By writing

$$u = \sum_{j=1}^{\infty} (u, w_j) w_j,$$

where $\{w_j\}$ are the eigenvalues of $A = -\Delta$ ordered as in (14.13), show that (14.16) and (14.17) hold. Deduce that

$$|Ap| \leq \lambda_n |p| \qquad \text{for all} \qquad u \in H$$

and

$$|Aq| \geq \lambda_{n+1} |q| \qquad \text{for all} \qquad u \in D(A).$$

14.3 Verify that if

$$\Phi(t) = (a + b) \exp\left(\frac{\lambda b}{C(a + b)}\right)$$

[this is essentially (14.25)],

$$\frac{da}{dt} \geq -(\lambda + C)a - Cb,$$

$$\frac{db}{dt} \leq -(\lambda - C)b + Ca$$

[these are essentially (14.19) and (14.20)], and

$$(\lambda - C)b > 2Ca \tag{14.34}$$

[which is (14.23)] then $d\Phi/dt \leq 0$ (these are the "elementary but tedious computations" referred to in Section 14.1).

The following four exercises provide a proof, in the case of the 2D Navier–Stokes equations, that one can replace the finite-dimensional projection P in Theorem 14.3 with the value of u at a finite number of points throughout the domain Q. The notes discuss further generalisations.

14.4 Use the Fourier expansion to show that if $u \in H^2_p(Q)$, where $Q = [0, L]^2$, then

$$|u(x) - u(y)| \leq c\|u\|_{H^2}|x - y|^{1/2}.$$

(You may assume that

$$\left|e^{2\pi i k \cdot x/L} - e^{2\pi i k \cdot y/L}\right| \leq C|k|^{\lambda}|x - y|^{1/2}. \tag{14.35}$$

You can prove this easily by considering the two cases $|k||x - y| \leq 1$ and $|k||x - y| > 1$ separately.)

14.5 Consider a finite set $\mathcal{N}$ of points $x_j \in Q$, and set

$$d(\mathcal{N}) = \sup_{x \in Q} \min_j |x - x_j|.$$

Use the result of the previous exercise and (9.13) to deduce that for $u \in D(A)$ we have

$$\|u\|_\infty \leq \left[\sup_{x_j \in \mathcal{N}} |u(x_j)|\right] + cd(\mathcal{N})^{1/2}|Au|. \tag{14.36}$$

14.6 Show that if $X(t)$ is bounded, $b(t) \to 0$ as $t \to \infty$, and

$$\frac{dX}{dt} + aX \leq b(t),$$

then $X(t) \to 0$ as $t \to \infty$.

14.7 Suppose that u and v are two solutions lying on the global attractor of the 2D Navier–Stokes equation

$$du/dt + \nu Au + B(u, u) = f$$

(periodic boundary conditions), and write $w = u - v$. Define

$$\eta(w) = \sup_{x_j \in \mathcal{N}} |w(x_j)|$$

($\mathcal{N}$ as in Exercise 14.5) and suppose that $\eta(w(t)) \to 0$ as $t \to \infty$. Starting from the identity

$$\frac{1}{2}\frac{d}{dt}\|w\|^2 + \nu|Aw|^2 = b(w, w, Au),$$

use (9.25) and Poincaré's inequality to show that

$$\frac{d}{dt}\|w\|^2 + \mu\|w\|^2 \leq C\eta(w) \qquad \text{for some} \qquad \mu > 0,$$

if $d(\mathcal{N})$ is small enough. Deduce, using the result of Exercise 14.6, that in this case

$$\|u(t) - v(t)\| \to 0 \quad \text{as} \quad t \to \infty. \tag{14.37}$$

We say that $\mathcal{N}$ is a set of "determining nodes." [Recall that $\mathcal{A}$ is bounded in H, V, and $D(A)$, so you can replace $|u|$, $\|u\|$, $|Au|$, etc. with constants if necessary.]

14.8   In previous chapters we used the Galerkin truncation to prove existence of solutions of various PDEs, but it can also be used as the basis of a numerical scheme. Suppose that we want to approximate the solution of

$$du/dt + Au = F(u), \qquad u(0) = u_0 \tag{14.38}$$

where we will assume for simplicity that $F : H \to H$ is bounded and Lipschitz continuous,

$$|F(u)| \le C_0 \quad \text{and} \quad |F(u) - F(v)| \le C_1|u - v|.$$

The numerical version of the Galerkin method is to approximate the solution of (14.38) by the solution of the finite-dimensional truncation

$$dp_n/dt + Ap_n = P_n F(p_n), \qquad p_n(0) = P_n u(0). \tag{14.39}$$

(i)   Show that if $u(t)$ is the solution of (14.38) then

$$|Q_n u(t)| \le \frac{C_0}{\lambda_{n+1}} + |Q_n u(0)|. \tag{14.40}$$

(ii)   Use (14.40) to show that if $u(t)$ solves (14.38) then the difference between $P_n u(t)$ and the solution $p_n(t)$ of (14.39) is bounded according to

$$|P_n u(t) - p_n(t)| \le C_1^{-1}\left[\frac{C_0}{\lambda_{n+1}} + |Q_n u(0)|\right] e^{C_1 t},$$

and hence

$$|u(t) - p_n(t)| \to 0$$

uniformly on $[0, T]$ as $n \to \infty$. This shows that the solutions of (14.39) converge to those of (14.38) as $n \to \infty$.

Notes

The squeezing property was introduced by Foias & Temam in 1979 in the context of the Navier–Stokes equations. The proof of the existence of determining modes given above is nonstandard but emphasises the role of the squeezing property in this particular kind of "finite-dimensional" dynamics. The standard method consists of making estimates on the P and Q components of the equation for the difference of two solutions, and this has led to the best estimates on the number of modes required. For the 2D Navier–Stokes equations with periodic boundary conditions the best current estimate is due to Jones & Titi (1993), who show that one can take $n \leq cG^{1/2}$, where G is the Grashof number defined in (13.31). A similar theory for certain stochastic equations has been developed by Flandoli & Langa (1998) and Kuksin & Shirikyan (2000).

The idea of determining modes has been extended in other directions [Cockburn *et al.* (1997) give a unifying framework for all these "determining things."] For example, Foias & Temam (1984) introduced the notion of determining nodes (see Exercises 14.4–14.7). Here you take a finite set of points $\{x_j\}$ in $Q = [0, L]^2$ and consider the values of the solution only at these "nodes." They showed, for the 2D Navier–Stokes equations, that if these points are close enough together and

$$\max_j |u(x_j) - v(x_j)| \to 0 \qquad \text{as} \qquad t \to \infty$$

then

$$\sup_{x \in Q} |u(x) - v(x)| \to 0 \qquad \text{as} \qquad t \to \infty.$$

As with the determining modes, the estimates on the maximum separation have been improved in a series of papers, and currently the best estimate (which is naturally a length scale) is $\delta \leq cG^{-1/2}$, due to Jones & Titi (1993). See also Section 16.1.1 for a related result.

The proof of the squeezing property for the Navier–Stokes equation given here is taken from Temam (1985); the reaction–diffusion proof is adapted from this. Proofs for many other examples are given in Eden *et al.* (1994b).

Approximate inertial manifolds were introduced as a separate object of study in Foias *et al.* (1988c) and in Foias *et al.* (1989), and there is now a large literature on approximate inertial manifolds and the numerical methods associated with them. The stationary approximate inertial manifold and the rigorous contraction mapping analysis discussed in this section are due to Titi (1990); the approximation in (14.29) turns out to be precisely that considered in Foias *et al.* (1988c). The best theoretical result (Debussche & Temam, 1994) is that one can find a family of (calculable) approximate inertial manifolds of exponential order, so that, if $\mathcal{G}[\Phi_m]$ has dimension m, then

$$|q - \Phi_m(p)| \leq Ce^{-k\lambda_{m+1}^\alpha},$$

for some $C, k, \alpha > 0$.

Discussion of the effectiveness of the approximate inertial manifold in computations is given in Marion & Temam (1989), Jauberteau *et al.* (1990), Heywood & Rannacher (1993), Jones *et al.* (1995), and Temam (1995). Results are at present still inconclusive. For recent references, and a more promising method, see García-Archilla *et al.* (1998, 1999).

Exponential attractors were introduced in 1990 by Eden *et al.*, and in fact these postdate inertial manifolds, the subject of the next chapter. Indeed, when they were first introduced they were termed "inertial sets." The construction given in this chapter is very simple minded, and much more sophisticated techniques are available: a fairly up-to-date summary is given in the monograph by Eden *et al.* (1994b). If one makes use of the squeezing property it is possible to produce an exponential attractor with finite fractal dimension, and with yet more work one can obtain an exponential attractor whose dimension agrees with that of the global attractor estimated using the techniques of Chapter 13 [this appears in the article by Eden *et al.* (1994a)]. Finally, a nice discussion of the relevance of exponential attractors to fluid turbulence is given in Eden *et al.* (1993).

15

The Strong Squeezing Property
Inertial Manifolds

In this chapter we introduce the "strong squeezing property" and use it to prove the existence of inertial manifolds. When an inertial manifold exists we can obtain a finite-dimensional system of ordinary differential equations (ODEs) that determines the dynamics on the attractor.

The essential idea is to find an invariant manifold that contains the attractor (cf. Proposition 14.2). Since this manifold has a smooth structure that is not given by the existence of an attractor alone, it becomes easy to write down a system of ODEs that determines the attractor dynamics.

In this chapter we will give a construction of inertial manifolds that works for equations on a Hilbert space H, in the form

$$du/dt + Au = F(u), \qquad (15.1)$$

where A is a positive linear operator (the negative Laplacian, for example) and F satisfies certain conditions, which we will discuss below (Section 15.3). We will show that our general result implies the existence of inertial manifolds for reaction–diffusion equations in one dimension and in certain rectangular domains in two dimensions. We will then discuss how these results can be extended by using different methods (in fact the theory applies to reaction–diffusion equations on any rectangular domain in two dimensions and to cubic ones in three dimensions) and give an indication of why the method cannot presently be applied to the Navier–Stokes equations.

15.1 Inertial Manifolds and "Slaving"

The standard construction of inertial manifolds is based on the Fourier expansion of the solution $u(t)$ in terms of the eigenfunctions of A,

$$u(t) = \sum_{j=1}^{\infty} \big(u(t), w_j\big) w_j = \sum_{j=1}^{\infty} c_j(t) w_j.$$

We would like to express the high Fourier coefficients ($j > n$ for some sufficiently large n) in terms of the lower Fourier coefficients ($j \leq n$), but we allow for an error, so that

$$c_j(t) = \phi_j(c_1, \ldots, c_n) + \text{error} \qquad \text{for all} \qquad j > n. \qquad (15.2)$$

This is commonly referred to as a "slaving rule," since the higher modes are governed by or "slaved to" the lower modes (see Haken (1978) for example). Although we cannot expect that the relationship in (15.2) holds exactly for all solutions, we will insist that the error decays exponentially to zero in time. Also, we would like to be sure that once equality holds in (15.2), it continues to hold.

We can write (15.2) more concisely in terms of the projection P_n onto the first n eigenfunctions of A (in their usual order, such that $\lambda_{j+1} \geq \lambda_j$) and its orthogonal complement Q_n,

$$P_n u = \sum_{j=1}^{n} (u, w_j) w_j \qquad \text{and} \qquad Q_n u = \sum_{j=n+1}^{\infty} (u, w_j) w_j.$$

The projection P_n has a finite-dimensional range spanned by the $\{w_j\}_{j=1}^{n}$, but Q_n has infinite-dimensional range. If we write $p = P_n u$ and $q = Q_n u$, (15.2) can be rewritten as

$$q = \phi(p) + \text{error}, \qquad (15.3)$$

where $\phi : P_n H \to Q_n H$.

The graph of ϕ,

$$\mathcal{G}[\phi] \equiv \{u : u = p + \phi(p), p \in P_n H\},$$

defines an n-dimensional manifold $\mathcal{M}$ in the phase space H. Insisting that an exact equality in (15.3) remains an equality means that we need this manifold to be positively invariant. Requiring the error to decay exponentially to zero means that the manifold should be exponentially attracting: if $u(t)$ is a solution of (15.1) and we write $p(t) = P_n u(t)$ and $q(t) = Q_n u(t)$, we need

$$|q(t) - \phi(p(t))| \leq C(|u_0|)e^{-kt}. \qquad (15.4)$$

(We have used $|\cdot|$ for the norm in H.) This is illustrated in Figure 15.1.

Ideally, we would like to be able to write down a finite set of ODEs that gives the dynamics on $\mathcal{M}$. If we project Equation (15.1) onto $P_n H$ we obtain

$$dp/dt + Ap = P_n F(u),$$

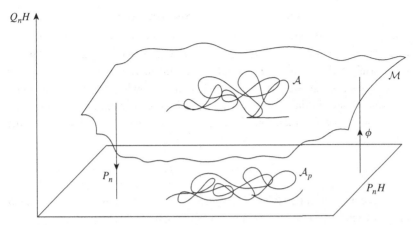

Figure 15.1. Via the inertial manifold $\mathcal{M}$, the inertial from gives rise to a set of ODEs on $P_n H \simeq \mathbb{R}^n$ with a global attractor $\mathcal{A}_p = P_n \mathcal{A}$.

since P_n commutes with A. If $u \in \mathcal{M}$, then we know that $u = p + \phi(p)$, and so in fact we obtain an n-dimensional system of ODEs for p,

$$dp/dt + Ap = P_n F(p + \phi(p)). \tag{15.5}$$

This set of ODEs is called the "inertial form." Since the global attractor is a subset of $\mathcal{M}$ and $\mathcal{M}$ is positively invariant, the system (15.5) must have a global attractor $\mathcal{A}_p$ that is just the projection of $\mathcal{A}$ onto $P_n H$, $\mathcal{A}_p = P_n \mathcal{A}$. The dynamics on $\mathcal{A}$ are entirely determined by the dynamics on $\mathcal{A}_p$, and so the asymptotic behaviour of our original PDE is now completely determined by the dynamics of the finite-dimensional system (15.5); see Figure 15.1.

If we want to ensure that the inertial form has unique solutions we need the nonlinear right-hand side to be a Lipschitz function of p (cf. Theorem 2.3). Even if we know that F is Lipschitz (in some appropriate way, which we will ensure below), we will need ϕ to be Lipschitz too. So we build this into our definition, requiring ϕ to satisfy

$$|\phi(p_1) - \phi(p_2)| \le L|p_1 - p_2|.$$

Definition 15.1. *An inertial manifold $\mathcal{M}$ is a finite-dimensional Lipschitz manifold, which is positively invariant and attracts all trajectories exponentially,*

$$\mathrm{dist}(S(t)u_0, \mathcal{M}) \le C(|u_0|)e^{-kt} \qquad \textit{for all} \qquad u_0 \in H. \tag{15.6}$$

That the convergences given in the definition here and in the discussion above (15.4) are equivalent is shown in Exercise 15.1.

15.2 A Geometric Existence Proof

There are many different methods in the literature for proving the existence
of an inertial manifold. We will give a geometric proof, based on the "graph
transform" method. The idea is simple. Since the inertial manifold is attracting
(if it exists), we should be able to find it by following the evolution under
the flow of any initial manifold $\mathcal{G}[\psi]$. It is most convenient to follow the
evolution of the flat manifold $P_n H$, so we investigate the behaviour of the
sets

$$\mathcal{M}_t = S(t)[P_n H].$$

Under a condition called the *strong squeezing property* we will be able to show
that for each t the new set is still given as a graph,

$$\mathcal{M}_t = \mathcal{G}[\phi_t],$$

and that the functions ϕ_t converge as $t \to \infty$. The method is illustrated by
some computations in Figure 15.2.

15.2.1 The Strong Squeezing Property

We now introduce a stronger version of the squeezing property, which will
enable us to prove the existence of inertial manifolds. First, we motivate the
definition.

If u and $\bar{u}$ lie on a manifold given as a Lipschitz graph with Lipschitz constant
1, then

$$|Q_n u - Q_n \bar{u}| \le |P_n u - P_n \bar{u}|. \tag{15.7}$$

Since we wish to ensure that manifolds given as Lipschitz graphs remain Lipschitz graphs under the evolution and we would like to ensure that the Lipschitz
constant does not increase under the flow (so that the limit is Lipschitz), we
require that the condition (15.7) be preserved so that it also holds with u and $\bar{u}$
replaced by $S(t)u$ and $S(t)\bar{u}$ for any $t \ge 0$.

Because we would like trajectories to satisfy (15.7), we will also require that
the distance between two trajectories that do not satisfy (15.7) decay to zero.
The combination of these two requirements leads us to consider the strong
squeezing property.

In all that follows, we will write $P = P_n$, $Q = Q_n$, and always $p = P_n u$,
$\bar{p} = P_n \bar{u}$, $\tilde{p} = P_n \tilde{u}$, $q = Q_n u$, etc.

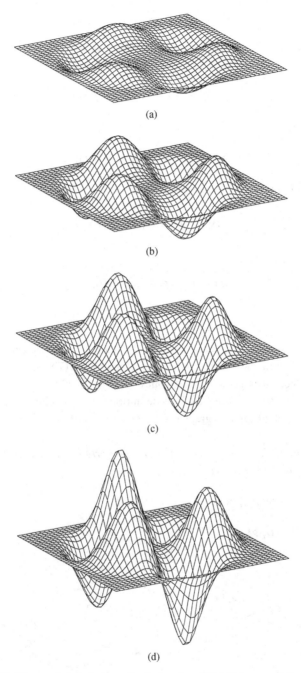

(a)

(b)

(c)

(d)

Figure 15.2. The graph transform method, showing $S(t)PH$ for $t = 2, 5, 10$, and 25. The final set ($t = 25$) is close to the inertial manifold illustrated on the front cover of the book ($t = 100$).

389

Definition 15.2. *The* strong squeezing property *holds if, for any two solutions* $u(t)$ *and* $\bar{u}(t)$, *we have (i) the cone invariance property:*

$$|q(0) - \bar{q}(0)| \le |p(0) - \bar{p}(0)|$$

implies that

$$|q(t) - \bar{q}(t)| \le |p(t) - \bar{p}(t)| \qquad (15.8)$$

for all $t > 0$ *and (ii) the decay property: if, for some* $t > 0$,

$$|q(t) - \bar{q}(t)| \ge |p(t) - \bar{p}(t)|$$

then

$$|q(t) - \bar{q}(t)| \le |q(0) - \bar{q}(0)|e^{-kt}. \qquad (15.9)$$

See Figure 15.3.

(The strong squeezing property can also be defined with a general Lipschitz constant L taking the place of the 1, so that $|\phi(p) - \phi(\bar{p})| \le L|p - \bar{p}|$, etc., but this affects the argument only by introducing extra algebra.)

An immediate consequence of this definition is that the global attractor is in fact a subset of a Lipschitz graph (cf. Proposition 14.2).

Proposition 15.3. *If the strong squeezing property holds then there exists a Lipschitz function* $\Phi : P_n H \to Q_n H$,

$$|\Phi(p_1) - \Phi(p_2)| \le |p_1 - p_2| \qquad \text{for all} \qquad p_1, p_2 \in P_n H, \qquad (15.10)$$

such that $\mathcal{A} \subset \mathcal{G}[\Phi]$.

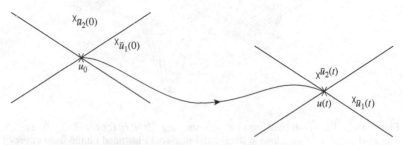

Figure 15.3. The strong squeezing property. The behaviour of $\bar{u}_1(t)$ and $\bar{u}_2(t)$ is determined relative to the cone $|Q\bar{u} - Qu| \le |P\bar{u} - Pu|$.

Proof. Take $u, v \in \mathcal{A}$; then the claim is that $|Qu - Qv| \leq |Pu - Pv|$. If not, then we fall into (15.9) of the strong squeezing property. Since $\mathcal{A}$ is invariant we know that for each t there exist $u_t, v_t \in \mathcal{A}$ such that $u = S(t)u_t$ and $v = S(t)v_t$, and so (15.9) gives

$$|u - v| \leq e^{-kt}|u_t - v_t|.$$

Since $\mathcal{A}$ is bounded ($|u| \leq R_H$ for all $u \in \mathcal{A}$) we have

$$|u - v| \leq 2R_H e^{-kt}.$$

Since this is valid for all t, we have $u = v$. It follows that we can define a Lipschitz function Φ on $P\mathcal{A}$ by $\Phi(Pu) = Qu$ for every $u \in \mathcal{A}$. As in Proposition 14.2, we can now extend Φ to a Lipschitz function as in (15.10). ∎

In fact we can show that there is such a Lipschitz manifold that is positively invariant and attracts solutions exponentially fast: this is an inertial manifold.

15.2.2 The Existence Proof

We need to make one additional assumption for our proof to work.

Definition 15.4. *We will say that the equation has been "appropriately prepared" if*

(i) there exists an absorbing ball in H, $B(0, \rho)$, such that $B(0, \rho) \cap PH$ is positively invariant and
(ii) for every $t \geq 0$,

$$PS(t)[PH] = PH,$$

that is, for every $p \in PH$ there exists a $p_0 \in PH$ such that

$$p = P(S(t)p_0).$$

We now show that the strong squeezing property implies the existence of an inertial manifold.

Theorem 15.5. *If the strong squeezing property (Definition 15.2) holds for an appropriately prepared equation, then this equation has an inertial manifold $\mathcal{M}$, given as the graph of a Lipschitz function $\phi : P_n H \to Q_n H$,*

$$|\Phi(p) - \phi(\overline{p})| \leq |p - \overline{p}| \qquad \text{for all} \qquad p, \overline{p} \in P_n H.$$

The manifold attracts exponentially, at the rate k from (15.9) in the strong squeezing property,

$$\text{dist}(u(t), \mathcal{M}) \leq C(|u_0|)e^{-kt}.$$

Proof. We set $\phi_0 = 0$ and $\mathcal{M}_0 = P_n H = \mathcal{G}[\phi_0]$, and we follow the evolution of $\mathcal{M}_0$ under the flow as outlined above. We write

$$\mathcal{M}_t = S(t)\mathcal{M}_0 = \{S(t)u_0 : u_0 \in \mathcal{M}_0\},$$

and we want first to show that

$$\mathcal{M}_t = \mathcal{G}[\phi_t] \tag{15.11}$$

for some ϕ_t that has Lipschitz constant at most 1. Since any two points u and $\overline{u}$ in $\mathcal{M}_0$ have $q = \overline{q} = 0$, we certainly have $|q - \overline{q}| \leq |p - \overline{p}|$. The cone invariance part of the strong squeezing property (15.8) then shows that

$$|q(t) - \overline{q}(t)| \leq |p(t) - \overline{p}(t)|.$$

Thus $|q_1 - q_2| \leq |p_1 - p_2|$ for any two points u_1 and u_2 in $\mathcal{M}_t$. Thus for every point $p \in P\mathcal{M}_t$, there is a unique $\phi_t(p)$ such that $p + \phi_t(p) \in \mathcal{M}_t$. Equation (15.11) now follows, since part (ii) of the "appropriately prepared" assumption ensures that there are no "holes" in $\mathcal{M}_t$, and so $P\mathcal{M}_t = PH$.

Thus we have to investigate the behaviour of the functions ϕ_t (from (15.11)) as $t \to \infty$. Consider the two points $u = p + \phi_t(p)$ and $\overline{u} = p + \phi_\tau(p)$ with $\tau > t$. Since u lies on $\mathcal{M}_t$, it is equal to $S(t)u_0$ for some $u_0 \in PH$. Since $\overline{u}$ lies on $\mathcal{M}_\tau$, it is equal to $S(\tau)\overline{u}_0$ for some $\overline{u}_0 \in PH$. Now, if $\phi_t(p) \neq \phi_\tau(p)$ we have

$$|QS(t)u_0 - QS(\tau)\overline{u}_0| = |\phi_t(p) - \phi_\tau(p)| > 0 = |PS(t)u_0 - PS(\tau)\overline{u}_0|,$$

and so using the decay property (15.9) we have

$$|\phi_t(p) - \phi_\tau(p)| \leq |Qu_0 - QS(\tau - t)\overline{u}_0|e^{-kt}$$
$$\leq |QS(\tau - t)\overline{u}_0|e^{-kt}.$$

Now, $u_0 \in PH$, and so by the preparation assumption part (i), $S(t)u_0 \in B(0, \rho) \cup PH$ for all $t \geq 0$. It follows that $|QS(\tau - t)\overline{u}_0| \leq \rho$, and we obtain

$$\|\phi_t - \phi_\tau\|_\infty \leq \rho e^{-kt}, \qquad \tau > t. \tag{15.12}$$

This shows that the sequence $\{\phi_n\}$ is Cauchy, and so it converges to some Lipschitz function ϕ. Taking limits in (15.12) then shows that ϕ_t converges exponentially fast to ϕ,

$$\|\phi_t - \phi\|_\infty \le \rho e^{-kt}.$$

The graph of this function ϕ we will call $\mathcal{M}$.

The invariance of ϕ follows from the uniform convergence of ϕ_t to ϕ. Suppose that we have an initial condition $u_0 \in \mathcal{M}$. Then $u_0 = p + \phi(p)$ and evolves under $S(\tau)$ into $S(\tau)u_0$. Now consider the approximations to u_0, $u_0^{(t)} \in \mathcal{M}_t$, so that

$$u_0^{(t)} = p + \phi_t(p).$$

We know that $S(\tau)u_0^{(t)} \in \mathcal{M}_{t+\tau}$. We now let $t \to \infty$, and by the continuity of solutions in terms of the initial conditions we have that

$$S(\tau)u_0^{(t)} \to S(\tau)u_0,$$

but the left-hand side is equal to

$$P\left[S(\tau)u_0^{(t)}\right] + \phi_{t+\tau}\left(P\left[S(\tau)u_0^{(t)}\right]\right),$$

and so its limit is

$$P[S(\tau)u_0] + \phi\big(P[S(\tau)u_0]\big)$$

by the uniform convergence of ϕ_t to ϕ. Thus $S(\tau)u \in \mathcal{M}$ and $\mathcal{M}$ is invariant.

To show that the manifold is exponentially attracting, we follow an argument similar to that used above to show the convergence of the functions ϕ_t. Consider an initial condition $u_0 \in B(0, \rho)$ and its image after a time t, $S(t)u_0 = u = p + q$. Consider also the point $\overline{u} \in \mathcal{M}$ given by $\overline{u} = p + \phi(p)$. Then

$$|Qu - Q\overline{u}| > 0 = |Pu - P\overline{u}|,$$

so the decay property (15.9) shows that

$$\begin{aligned}|u - \overline{u}| &= |q - \overline{q}| \\ &\le |Qu_0 - \phi(z)|e^{-kt},\end{aligned}$$

where $S(t)[z + \phi(z)] = \overline{u}$. Then

$$\begin{aligned}\operatorname{dist}(S(t)u_0, \mathcal{M}) &\le |u - \overline{u}| \\ &\le (\rho + \|\phi\|_\infty)e^{-kt}.\end{aligned}$$

For a general initial condition u_0 not necessarily in $B(0, \rho)$ there is a time $t_0(Y)$, uniform for u_0 in a bounded set $Y \in H$, such that $S(t)u_0 \in B(0, \rho)$. Thus

$$\begin{aligned} \text{dist}(S(t)u_0, \mathcal{M}) &= \text{dist}(S(t - t_0)[S(t_0)u_0], \mathcal{M}) \\ &\leq (\rho + \|\phi\|_\infty)e^{-k(t-t_0(Y))}, \\ &\leq C(Y)e^{-kt}, \end{aligned}$$

with the constant $C(Y)$ uniform on bounded sets Y. ∎

Note that the above proof does not depend in any way on the choice of phase space H. Since the conditions are geometric, the proof works in any phase space provided that all distances are measured in the norm of that space. This comment will prove useful later when we discuss problems in proving the existence of inertial manifolds for the Navier–Stokes equations.

15.3 Finding Conditions for the Strong Squeezing Property

We now turn to a general result that gives a condition for the strong squeezing property to hold. We will deal with the equation

$$du/dt + Au = F(u),$$

making the assumption that $F(u)$ is a globally Lipschitz function from H into itself,

$$|F(u) - F(v)| \leq C_1|u - v|, \qquad u, v \in H. \tag{15.13}$$

We will be able to use regularity results for the attractor to "prepare" reaction–diffusion equations to ensure that (15.13) holds.

The condition given in the following theorem, which is based on the eigenvalues of A, is known as the "spectral gap condition."

Proposition 15.6. *If there exists an n such that the eigenvalues λ_n and λ_{n+1} satisfy*

$$\lambda_{n+1} - \lambda_n > 4C_1 \tag{15.14}$$

then the strong squeezing property holds, with the k in (15.9) bounded below according to $k \geq \lambda_n + 2C_1$.

Proof. The idea is simply to let w be the difference of two solutions $w = u - \overline{u}$ and to consider what happens on the boundary of the "cone":

$$\left\{ (u, \overline{u}) : |Q(u - \overline{u})| \le |P(u - \overline{u})| \right\}.$$

The first step is to show that pairs of trajectories cannot leave this cone, and so we must show that

$$\frac{d}{dt}(|Qw| - |Pw|)$$

is negative when $|Qw| = |Pw|$. We do this by bounding $d|Qw|/dt$ above and $d|Pw|/dt$ below.

The difference w of two solutions satisfies the equation

$$dw/dt + Aw = F(u) - F(\overline{u}). \tag{15.15}$$

Set $p = Pw$ and $q = Qw$. We can write (15.15) as two coupled equations for p and q. Since P and Q commute with A we can apply both operators to (15.15) to obtain

$$\begin{aligned} dp/dt + Ap &= PF(u) - PF(\overline{u}), \\ dq/dt + Aq &= QF(u) - QF(\overline{u}). \end{aligned} \tag{15.16}$$

It is straightforward to obtain a lower bound on $d|Pw|/dt$ by taking the scalar product of the P component of (15.16) with p to give [using (14.17)]

$$\begin{aligned} \tfrac{1}{2}\frac{d}{dt}|p|^2 &= -\|p\|^2 - \left(PF(u) - PF(\overline{u}), p \right) \\ &\ge -\lambda_n |p|^2 - C_1 |w| |p|, \end{aligned}$$

or when $|q(0)| = |p(0)|$,

$$\left(\frac{d}{dt_+} |p| \right)_{t=0} \ge -(\lambda_n + 2C_1)|q|.$$

We now take the inner product of the Q equation in (15.16) with q to obtain

$$\tfrac{1}{2}\frac{d}{dt}|q|^2 + \|q\|^2 = \left(QF(u) - QF(\overline{u}), q \right),$$

and so [using (14.18)] we get

$$\tfrac{1}{2}\frac{d}{dt}|q|^2 \le -\lambda_{n+1}|q|^2 + C_1 |q| |w|. \tag{15.17}$$

When $|p(0)| = |q(0)|$ this becomes

$$\left(\frac{1}{2}\frac{d}{dt}|q|^2\right)_{t=0} \leq -\lambda_{n+1}|q|^2 + 2C_1|q|^2.$$

Therefore,

$$\left(\frac{d}{dt_+}|q|\right)_{t=0} \leq -(\lambda_{n+1} - 2C_1)|q|.$$

Then at $t = 0$

$$\frac{d}{dt_+}(|q| - |p|)_{t=0} \leq -(\lambda_{n+1} - \lambda_n - 4C_1)|q(0)|,$$

which is negative provided that (15.14) holds, and so we obtain the cone invariance property.

The exponential squeezing outside the cone (15.8) follows directly from the Q component of the equation with $|q| \geq |p|$; in that case (15.17) becomes

$$\frac{1}{2}\frac{d}{dt}|q|^2 \leq -\lambda_{n+1}|q|^2 + 2C_1|q|^2,$$

and the exponential decay in (15.9) follows with $k = \lambda_{n+1} - 2C_1$ using the Gronwall lemma (Lemma 2.8). The lower bound on k follows from the spectral gap condition. ∎

15.4 Inertial Manifolds for Reaction–Diffusion Equations

We now want to apply the abstract theorem above to the scalar reaction–diffusion equation

$$\frac{\partial u}{\partial t} - \Delta u = f(u), \qquad u|_{\partial\Omega} = 0.$$

The first stage is to prepare the equation "appropriately."

15.4.1 Preparing the Equation

The preparation comes in two parts. First we adjust $f(s)$ to give a new function $g(s)$ that is globally Lipschitz. For this we use the L^∞ bound on the attractor. Then we truncate the resulting functional $G[u]$ to give a new nonlinearity $F[u]$ that is zero outside some ball $B(0, \rho)$. The construction yields an equation that is "appropriately prepared" and a nonlinear term that is globally Lipschitz from L^2 into itself.

Although we will have changed the original equation, we will be careful to leave everything unchanged on the set

$$\{u \in H : |u| \leq R, \|u\|_\infty \leq M\},$$

where R and M are the bounds on $\mathcal{A}$ in $L^2(\Omega)$ and $L^\infty(\Omega)$, respectively (from Proposition 11.1 and Theorem 11.6). If we adjust the nonlinearity outside this set we do not affect the dynamics on $\mathcal{A}$.

For the first step we use the L^∞ bound and define

$$g(s) = \theta_M(s)f(s),$$

where $\theta_R(s) = \theta(s/R)$ and θ is a C^∞ cutoff function, $\theta : \mathbb{R} \mapsto [0, 1]$ such that

$$\theta(r) = \begin{cases} 1, & |r| \leq 1, \\ 0, & |r| \geq 2, \end{cases}$$

with $|\theta'(r)| \leq 2$. The function $g(s)$ is now a globally Lipschitz function, so we have, for some C,

$$|g(s) - g(r)| \leq C|s - r|.$$

This allows us to prove the global Lipschitz bound (15.13) we need for Proposition 15.6. Writing $G[u](x) = g(u(x))$ we have

$$|G[u] - G[v]|^2 \leq \int_\Omega |g(u(x)) - g(v(x))|^2 \, dx$$

$$\leq C^2 \int_\Omega |u(x) - v(x)|^2 \, dx$$

$$= C^2 |u - v|^2.$$

Compare this with the general result (Proposition 8.6) we used before (to prove continuity of the solution semigroup in H_0^1), where we could show only that F was locally Lipschitz from H^1 into L^2 (in the case $m = 1$ or $m = 2$).

To ensure that the equation is appropriately prepared, we now replace $G(u)$ with

$$F(u) = \theta_R(|u|)G(u).$$

If we set $\rho = 2R$, then $F(u) = 0$ outside $B(0, \rho)$. It then follows that $B(0, \rho)$ is absorbing, since outside $B(0, \rho)$ the equation becomes

$$\dot{u} = -Au, \tag{15.18}$$

whose solutions decay exponentially. $B(0, \rho)$ is also positively invariant, and if $u_0 \notin B(0, \rho)$ with $u_0 \in P_n H$ then (15.18) implies that $u(t) \in P_n H$ until $u(t)$ enters the positively invariant set $B(0, \rho)$, which gives property (i).

For property (ii), note that the form of the equation outside $B(0, \rho)$, (15.18), implies that

$$S(t)\left\{u \in PH : |u| \leq 2\rho e^{\lambda_n t}\right\} \supset \left\{u \in PH : \rho \leq |u| \leq 2\rho\right\}$$

and

$$S(t)\left\{u \in PH : |u| \geq \rho e^{\lambda_n t}\right\} \supset \left\{u \in PH : |u| \geq \rho\right\}.$$

It follows, since both P and $S(t)$ are continuous, that $P\mathcal{M}_t = PS(t)\mathcal{M}_0$ is closed in PH.

Now, a simple variant of the analysis in Chapter 11 (Theorem 11.11) shows that $S(t)$ is injective on PH:

$$S(t)p_1 = S(t)p_2 \quad \Rightarrow \quad p_1 = p_2 \quad \text{for all} \quad t \geq 0, \ p_1, p_2 \in PH.$$

Thus we can define a continuous map $S(-t) : \mathcal{M}_t \to \mathcal{M}_0$ (using Exercise 10.8 on each bounded subset of M_0 in turn). It follows that the map $S(-t)\phi_t : P\mathcal{M}_t \to PH$ is continuous, and so $P\mathcal{M}_t$ is also open in PH. Since $P\mathcal{M}_t \neq \emptyset$ we must have $P\mathcal{M}_t = PH$, as required.

We can also show that this preparation has not sacrificed the global Lipschitz property of G.

Lemma 15.7. *$F(u)$ is bounded and globally Lipschitz from H into H.*

Proof. F is globally bounded, since

$$\|F\|_\infty \leq C_0 = \sup_{|u| \leq \rho} |G(u)| \leq |G(0)| + \rho C.$$

To show that F is globally Lipschitz we consider three cases.

(i) First, if $|u_1|, |u_2| \geq \rho$ then $F(u_1) = F(u_2) = 0$, so the Lipschitz property is clear.

(ii) If $|u_1| \geq \rho \geq |u_2|$ then $\theta_R(|u_1|) = 0$, and so

$$|F(u_1) - F(u_2)| = \left|\theta_R(|u_1|)G(u_2) - \theta_R(|u_2|)G(u_2)\right|$$

$$\leq \frac{C_0}{\rho}|u_1 - u_2|.$$

(iii) Finally, if $|u_1|, |u_2| \leq \rho$, then

$$|F(u_1) - F(u_2)| \leq \big|\theta_R(|u_1|) - \theta_R(|u_2|)\big||G(u_1)|$$
$$+ \theta_R(|u_2|)|G(u_1) - G(u_2)|$$
$$\leq \left(\frac{C_0}{\rho} + C\right)|u_1 - u_2|,$$

and we are done. ∎

15.4.2 Checking the Spectral Gap Condition

We need to satisfy the spectral gap condition (15.14), which is

$$\lambda_{n+1} - \lambda_n > 4C_1.$$

To look for large gaps in the spectrum we clearly need to have a good idea of the behaviour of the spectrum itself. We will consider two cases: first, general domains $\Omega \subset \mathbb{R}^m$, where one can show that

$$\lambda_n \sim Cn^{2/m}, \tag{15.19}$$

and second for box domains $\prod_{j=1}^m [0, L_j]$, where the eigenvalues of the Laplacian are proportional to weighted sums of squares of integers,

$$\lambda_{(n_1,\dots,n_m)} \propto \sum_{j=1}^m \frac{n_j^2}{L_j^2}.$$

[Note that (15.19) still holds, but we have more control over the eigenvalues since we can write them explicitly.]

In a one-dimension domain, where $\lambda_n \propto n^2/L^2$, the gap between consecutive eigenvalues satisfies

$$\lambda_{n+1} - \lambda_n \propto (2n + 1)/L^2,$$

and so the gap condition is essentially, for large n, $2n + 1 > L^2 C_1$. This can easily be satisfied by choosing n large enough. It follows that a one-dimensional reaction–diffusion equation always has an inertial manifold.

For two-dimensional domains it is not so clear that the spectrum has the required gaps. Indeed, on a general two-dimensional domain (15.19) tells us only that $\lambda_n \sim n$. Large gaps look unlikely, and it is not known whether or not they exist for such general domains.

However, for the rectangular domain $[0, L_1] \times [0, L_2]$, the eigenvalues are appropriately weighted sums of squares. Although they still satisfy $\lambda_n \sim n$, we can appeal to a result from number theory due to Richards (1982), which guarantees that there are arbitrarily large gaps in the weighted sums of two squares provided that L_1/L_2 is rational. This implies that $\lambda_{n+1} - \lambda_n$ can be big; to achieve a gap of size h, the result requires that $n \sim e^{kh}$ for some constant k, giving an estimate of the dimension of the inertial manifold that is exponential in C_1 (the required gap is $4C_1$).

We now consider briefly how these dimension estimates compare with those for the global attractor we obtained in Chapter 13. Suppose that we introduce a parameter λ into the equation

$$du/dt + Au = \lambda F(u),$$

as we did when we were considering the structure of the attractor in Chapter 11 and its dimension in Chapter 13. The effect of this on the gap condition would be to replace C_1 with λC_1, and therefore the dependence of the dimension* of the inertial manifold on λ is

$$n \sim \begin{cases} \lambda, & m = 1, \\ e^{k\lambda}, & m = 2. \end{cases}$$

We already know that the dimension of the attractor is much less than this for large λ, since we found in Chapter 13 that $d_f(\mathcal{A}) \sim \lambda^{m/2}$. We will come back to this in the next chapter.

Let us now return to our discussion of spectral gaps as we increase the dimension. In three dimensions, results from number theory show that the only gaps between the sums of three integers are 0, 1, 2, and 3; and any integer can be represented as the sum of four squares [see Hardy & Wright (1960) and Exercise 15.4]. We therefore cannot prove the strong squeezing property in these cases by using Proposition 15.6. However, it is possible to get around this problem and improve the results slightly.

15.4.3 Extensions to Other Domains and Higher Dimensions

What happens when we are in a 2D rectangular domain $[0, L_1] \times [0, L_2]$ and L_1 and L_2 are not rationally related? Mallet-Paret & Sell (1988), using methods dramatically different from those of Proposition 15.6, showed directly that the

---

* This underestimates the dimension if we introduce the λ dependence in the original equation,

$$u_t - \Delta u = \lambda f(u),$$

since then the bounds on $\mathcal{A}$ in L^2 and L^∞ that we used in our truncation also depend on λ. Since the Lipschitz constant of $F(u)$ depends on these bounds, C_1 also increases as λ increases.

strong squeezing property holds in this case. Their analysis also works for the cube $[0, L]^3$, extending the results into this one special 3D domain. In any higher dimensions results are unlikely. Indeed, Mallet-Paret *et al.* (1993) have shown that there is no inertial manifold for "hyper-cubic" domains in $m = 4$ in the case of Neumann boundary conditions ($\nabla u \cdot n = 0$).

15.5 More General Conditions for the Strong Squeezing Property

For many equations it is not possible to prepare them so that they can be put in the form of (15.1) with F Lipschitz from L^2 into L^2. However, it is often possible to adjust the equation so that F is Lipschitz from one fractional power space $D(A^\alpha)$ into another, $D(A^\beta)$, with $\beta < \alpha$, so that

$$|A^\beta(F(u) - F(v))| \leq C_1 |A^\alpha(u - v)| \qquad \text{for all} \qquad u, v \in D(A^\alpha). \quad (15.20)$$

(Fractional power spaces were defined in Section 3.10.) The difference between α and β, $\gamma = \alpha - \beta$, measures the "loss of smoothness" involved in the mapping $u \mapsto F(u)$.

We say that the strong squeezing property holds in a space X if all the norms in Definition 15.2 are norms in X. Under the assumption (15.20) it is possible to adapt the argument of Proposition 15.6 to guarantee that the strong squeezing property holds in $D(A^\alpha)$. However, we need to strengthen the conditions, requiring that both

$$\lambda_{n+1}^{1-\gamma} > 2C_1$$

(which can be satisfied easily by choosing n large enough) and the strengthened gap condition,

$$\lambda_{n+1} - \lambda_n > 2C_1\left(\lambda_n^\gamma + \lambda_{n+1}^\gamma\right), \quad (15.21)$$

hold. The new gap condition requires not only the occurrence of large gaps, but means that they have to occur soon enough in the spectrum to outweigh the growth on the right-hand side.

15.5.1 Inertial Manifolds and the Navier–Stokes Equations

Problems with spectral gaps have obstructed the proof of existence of inertial manifolds for the Navier–Stokes equations. This is a particularly important open problem, since the question "Is turbulence a finite-dimensional phenomenon?" is of great interest.

For the Navier-Stokes equations the nonlinear term $B(u, u) = u \cdot \nabla u$ contains a derivative, and we cannot expect it to be Lipschitz from $\mathbb{L}^2$ into $\mathbb{L}^2$. However, we can show that $B(u, u)$ is locally Lipschitz from $\mathbb{H}^1$ into $\mathbb{L}^2$. Writing $F(u) = B(u, u) + f$ and $w = u - v$, we have

$$F(u) - F(v) = B(u, u) - B(v, v) = B(w, u) + B(v, w).$$

It follows from the inequalities for b (9.27) that

$$|B(w, u)| + |B(v, w)| \le k\left[|w|^{1/2}\|w\|^{1/2}\|u\|^{1/2}|Au|^{1/2} + |v|^{1/2}|Av|^{1/2}\|w\|\right],$$

and so on the intersection of the absorbing sets in $D(A)$ and V (and so certainly on the attractor)

$$|F(u) - F(v)| \le C_1\|u - v\|. \tag{15.22}$$

We will now see that this Lipschitz property will not be of any use to us, even if we can prepare the equation "appropriately" and preserve it. Since $\mathbb{H}^1_p(Q) = D(A^{1/2})$, we need to take $\alpha = \frac{1}{2}$ and $\beta = 0$ [i.e. $\gamma = \frac{1}{2}$ in (15.21)]. Thus we must satisfy the stronger gap condition

$$\lambda_{n+1} - \lambda_n > 2C_1\left(\lambda_n^{1/2} + \lambda_{n+1}^{1/2}\right). \tag{15.23}$$

We mentioned above that on a general domain the eigenvalues of the Laplacian (and so of the Stokes operator) satisfy $\lambda_n \sim Cn$, and so we cannot expect the condition to hold in general, since this would require $C > 4C_1 n^{1/2}$. Even on a square domain $[0, L]^2$,

$$\{4\pi^2 n^2 / L^2, \ n \in \mathbb{Z}\}$$

is a subset of the spectrum of A. Thus if the gap condition holds for the equation with viscosity ν,

$$\nu\lambda_{n+1} - \nu\lambda_n > 2C_1\left(\nu^{1/2}\lambda_{n+1}^{1/2} + \nu^{1/2}\lambda_n^{1/2}\right)$$

for some n, then we must have $\pi/L > \nu^{-1/2}C_1$, which cannot be satisfied for small ν. Since small ν is the parameter range in which the flow becomes complicated and turbulent, a result valid only for large ν is of no account.

The final chapter (Exercises 17.12 onward) treats an example in which the general form of the spectral gap condition *can* be applied to deduce the existence of an inertial manifold.

15.6 Conclusion

We have shown that inertial manifolds exist for scalar reaction–diffusion equations on one-dimensional domains, two-dimensional rectangular domains, and the cube in three dimensions. We do not even have an existence result for general two-dimensional domains. Furthermore, the dimension estimates obtained are much larger than that for the global attractor of the system.

The existence or otherwise of inertial manifolds for the most important example, the Navier–Stokes equations, is still entirely open. We therefore need to develop other methods if we want to investigate finite-dimensional behaviour in fluid mechanical systems or in reaction–diffusion equations in arbitrary bounded domains in dimensions higher than one.

In the next chapter we adopt a more direct approach, starting from the existence of a global attractor.

Exercises

15.1 Show that the characterisations of "exponential convergence" towards $\mathcal{M}$ given in Equations (15.4) and (15.6) are equivalent. [Hint: Consider $\mathrm{dist}(u, \mathcal{M})^2 = \inf_{p \in PH}(|Pu-p|^2 + |Qu-\phi(p)|^2)$ and $|Qu-\phi(Pu)|^2$.]

15.2 Suppose that F is a Lipschitz function from H into H,

$$|F(u) - F(v)| \le C|u - v|.$$

Show that if the strong squeezing property holds then there is a Lipschitz ODE on $P_n H \simeq \mathbb{R}^n$ for which $P_n \mathcal{A}$ is an invariant set. [Use Proposition 15.3 and the ideas behind the inertial form (15.5).]

15.3 Suppose that the strong squeezing property holds for an equation that has been "appropriately prepared," and let

$$\Gamma = \partial P_n B(0, \rho).$$

Show that, for each $t_0 < \infty$,

$$\Sigma_{t_0} = \bigcup_{0 \le t \le t_0} S(t)\Gamma$$

is a subset of a Lipschitz graph over $P_n H$ with constant at most 1. Define

$$\Sigma = \bigcup_{t \ge 0} S(t)\Gamma$$

and

$$\mathcal{M} = \overline{\Sigma} \cup \{u \in P_n H : |u| \geq \rho\}.$$

Assuming that $P_n \overline{\Sigma} = P_n B(0, \rho)$ (this is true provided that n is large enough) show that $\mathcal{M}$ is an invariant Lipschitz manifold that contains $\mathcal{A}$. (This is the basis of the geometric, or "Cauchy," method of construction; see Notes.)

15.4   Express 135, 136, 137, and 138 as sums of at most four squares.

15.5   Suppose that

$$du/dt + Au = F(u) \tag{15.24}$$

has been "appropriately prepared" and has an inertial manifold given as the graph of a Lipschitz function $\Phi : P_n H \to Q_n H$ with Lipschitz constant 1 (as in Theorem 15.5).

(i)   Show that if $u(t)$ is any solution of (15.24) then $p(t) = P_n u(t)$ solves

$$dp/dt + Ap = F(p + \Phi(p)) + h(t), \tag{15.25}$$

where $|h(t)| \leq Ce^{-kt}$.

(ii) Suppose that if $h(t)$ satisfies $|h(t)| \leq Ce^{-kt}$, then for each solution $p(t)$ of (15.25) there is a corresponding solution $\overline{p}(t)$ of

$$d\overline{p}/dt + A\overline{p} = F(\overline{p} + \Phi(\overline{p}))$$

such that

$$|p(t) - \overline{p}(t)| \leq De^{-kt}.$$

Show that if $u(t)$ solves (15.24) then there exists a solution $\overline{u}(t)$ of (15.24), which lies on the inertial manifold, such that

$$|u(t) - \overline{u}(t)| \leq Me^{-kt}. \tag{15.26}$$

(This property, known as *asymptotic completeness*, is stronger than just exponential attraction to $\mathcal{M}$; see Notes.)

Notes

The subject of inertial manifolds has some overlap with the theory of normally hyperbolic invariant manifolds, extensively developed for the case of ODEs; see for example Hale (1969), Hirsch *et al.* (1977), Fenichel (1971), and Wiggins (1994). Direct precursors of the theory of inertial manifolds are contained in Mañé (1977), Mora (1983), Henry (1984), and Kamaev (1984).

Inertial manifolds themselves were introduced in a short paper by Foias *et al.* in 1985. From the start the hope has been to apply the ideas to the Navier–Stokes equations, but this has so far been frustrated. There are many methods for proving the existence of inertial manifolds. The one here is based on Mallet-Paret & Sell (1988) and Robinson (1993). The most powerful results, due to Chow *et al.* (1992), follow from a contraction mapping argument, such as was used in the first detailed paper on the subject by Foias *et al.* (1988a).

Other possible methods of proof are the geometric "Cauchy" method of Constantin *et al.* (1989) and Constantin (1989), reformulated in terms of the strong squeezing property and significantly simplified in Robinson (1995), and the method of "elliptic regularisation" in Fabes *et al.* (1991) and Debussche (1990). Mention should also be made of Ninomiya (1992) and Rodriguez Bernal (1990); the latter provides a proof valid in general Banach spaces.

For general equations there appears to be no way around the spectral gap condition, although this was circumvented for some reaction–diffusion equations, as discussed in Section 15.4.3, by Mallet-Paret & Sell (1988), and there are examples of particular equations in which it is not necessary (e.g., Bloch & Titi, 1990).

The number theoretic results on sums of squares can be found in Hardy & Wright (1960). The result of Richards (1982) can also be found in Mallet-Paret & Sell (1988).

Asymptotic completeness was discussed in Foias *et al.* (1989) and in detail, following on from the result of Exercise 15.5, in Robinson (1996).

16

A Direct Approach

Although we have obtained finite-dimensional global attractors in Chapter 13 and investigated various approaches towards showing that the dynamics of the equation is therefore finite-dimensional, all these methods have been somewhat indirect. The most successful, the theory of inertial manifolds as described in the previous chapter, relied on finding a smooth manifold containing the global attractor in order to construct the "inertial form" (15.5), a set of ODEs reproducing the dynamics on $\mathcal{A}$.

In this chapter we explore the possibility of working directly from the existence of the attractor to construct such a finite-dimensional system. Since we would like the results obtained to hold in the greatest possible generality, we will discuss this question for the attractor of an abstract evolution equation on a Hilbert space H, with semigroup $S(t)$ that is injective on $\mathcal{A}$.

An ideal result would be the following.

Conjecture 16.1 (Utopian Theorem). *For some k, comparable with $d_f(\mathcal{A})$, there exists a map $\varphi : H \to \mathbb{R}^k$ that is injective on $\mathcal{A}$ and a smooth ordinary differential equation (ODE) on $\mathbb{R}^k$, with corresponding solution operator $T(t)$ and global attractor X, such that the dynamics on $\mathcal{A}$ and X are conjugate under φ,*

$$ T(t)|_X = \varphi \circ S(t) \circ \varphi^{-1}. $$

Currently this theorem appears out of reach, but we will give two partial results in this direction. One of these weakens the requirement that the ODE have unique solutions and the other reproduces only the time T-map on the attractor $(S(T)|_\mathcal{A})$.

The first step is to show that it is possible to parametrise a finite-dimensional set with a finite number of coordinates.

406

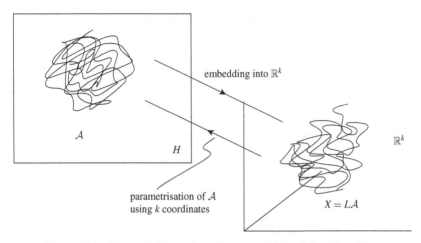

embedding into $\mathbb{R}^k$

$\mathcal{A}$

H

$\mathbb{R}^k$

parametrisation of $\mathcal{A}$
using k coordinates

$X = L\mathcal{A}$

Figure 16.1. Schematic illustration of Theorem 16.2 and Corollary 16.3.

16.1 Parametrising the Attractor

We will show that, if k is large enough (roughly twice the dimension of the attractor), then the attractor can be embedded into $\mathbb{R}^k$. This result can be considered in two ways (see Figure 16.1). First, the idea is that we can take the attractor "out" of the infinite-dimensional space and map it, using some linear map L, homeomorphically onto a subset of $\mathbb{R}^k$. This makes sense of the idea that $\mathcal{A}$ is a finite-dimensional set. Second, and perhaps more importantly, it follows that L^{-1} provides a way of parametrising the attractor using a finite set of coordinates. We will make use of both interpretations in what follows.

Theorem 16.2. *Let X be a compact subset of H, with $d_f(X) < d$, d an integer, and let $k \geq 2d + 1$. Then if L_0 is a bounded linear map into $\mathbb{R}^k$, for any $\epsilon > 0$ there exists another bounded linear map into $\mathbb{R}^k$, $L = L(\epsilon)$, such that L is injective on X and*

$$\|L - L_0\|_{\mathrm{op}} \leq \epsilon.$$

Proof. We will use the Baire category theorem (Theorem 3.5) applied to an appropriate countable intersection.

First, we set

$$Y = \{v - w : v, w \in X\}.$$

Since Y is the image of $X \times X$ under the Lipschitz map $(v, w) \mapsto v - w$ it follows from the result of Proposition 13.2 that $d_f(Y) \leq 2d_f(X) < 2d$. We now define

$$A_r = \{v - w : v, w \in X, \text{with } |v - w| \geq 1/r\}$$

and

$$A_{r,j,n} = \left\{ u \in A_r : |(e_j, u)| \geq \frac{1}{n} \right\}.$$

Note that $A_{r,j,n}$ is compact, and since $0 \notin A_r$, we have

$$A_r = \bigcup_{j=1}^{\infty} \bigcup_{n=1}^{\infty} A_{r,j,n}.$$

We now define

$$\mathbb{L}_{r,j,n} = \{L \in L(H, \mathbb{R}^k) : L^{-1}(0) \cap A_{r,j,n} = \emptyset\}.$$

We first show that

$$\bigcap_{r=1}^{\infty} \bigcap_{j=1}^{\infty} \bigcap_{n=1}^{\infty} \mathbb{L}_{r,j,n} \tag{16.1}$$

consists precisely of those linear maps that are injective on X. First, observe that

$$\bigcap_{j=1}^{\infty} \bigcap_{n=1}^{\infty} \mathbb{L}_{r,j,n}$$

consists of those linear maps for which $L^{-1}(0) \cap A_r = \emptyset$, that is, for which

$$\text{diam}(L^{-1}(x) \cap X) < 1/r \qquad \text{for all} \qquad x \in \mathbb{R}^k.$$

It is therefore clear that (16.1), the intersection of these over all r, consists of maps that are injective on X.

If we can show that each $\mathbb{L}_{r,j,n}$ is open and dense in $L(H, \mathbb{R}^k)$ then we can apply Baire's theorem. First, we show that each $\mathbb{L}_{r,j,n}$ is open. To do this, we show that if $L \in \mathbb{L}_{r,j,n}$ then there exists an ϵ such that $\|\tilde{L} - L\| \leq \epsilon$ implies that $\tilde{L} \in \mathbb{L}_{r,j,n}$. This amounts to showing that if $Lx \neq 0$ for all $x \in A_{r,j,n}$, then $\tilde{L}x \neq 0$ also. Since $A_{r,j,n}$ is compact, it follows that

$$\min_{x \in A_{r,j,n}} |Lx| = \eta > 0 \qquad \text{and} \qquad \max_{x \in A_{r,j,n}} |x| \leq R,$$

for some $0 < \eta \le R$. We now write

$$
\begin{aligned}
|\tilde{L}x| &= |Lx - (L - \tilde{L})x| \\
&\ge |Lx| - |(L - \tilde{L})x| \\
&\ge \eta - \|L - \tilde{L}\|_{op} R.
\end{aligned}
$$

So if we choose $\epsilon < \eta/2R$ we can ensure that $|\tilde{L}x| \ne 0$ as required.

To show that $\mathbb{L}_{r,j,n}$ is dense, we will make use of the Hausdorff dimension. We choose $L_0 \in L(H, \mathbb{R}^k)$ and some $\epsilon > 0$. We let ϕ be the map from $\mathbb{R}^k$ into the unit sphere $S(0, 1)$ defined by

$$
\phi(x) = \begin{cases} x/|x|, & x \ne 0, \\ p, & x = 0, \end{cases}
$$

where p is some point in $S(0, 1)$. Now, ϕ is a Lipschitz map on $\mathcal{O}_\rho \equiv \{x : |x| \ge \rho\}$ for any $\rho > 0$. For any set W we can write

$$
\phi(W) = \left\{ \bigcup_{k=1}^{\infty} \phi(W \cap \mathcal{O}_{1/k}) \right\} \underbrace{\bigcup p}_{\text{if } 0 \in W}.
$$

Since $d_H(p) = 0$, by the countable additivity property of the Hausdorff dimension we have

$$
d_H(\phi(W)) \le \sup_k d_H\big(\phi(W \cap \mathcal{O}_{1/k})\big),
$$

and since ϕ is Lipschitz on each $\mathcal{O}_{1/k}$ this yields

$$
d_H(\phi(W)) \le d_H(W).
$$

It follows in particular that

$$
d_H(\phi(L_0 A_r)) \le d_H(L_0 A_r) \le d_H(A_r) \le d_f(A_r) < 2d.
$$

Since the dimension of $S(0, 1)$ is $k - 1 = 2d$, there must exist a $z \in S(0, 1)$ such that $z \notin \phi(L_0 A_r)$.

Now set

$$
L = L_0 + \epsilon z e_j^*,
$$

where e_j^* is the linear functional that takes $u \mapsto (u, e_j)$. Clearly $L \in L(H, \mathbb{R}^k)$, and

$$
\|L - L_0\| \le \epsilon.
$$

Of course, it remains to show that in fact $L \in \mathbb{L}_{r,j,n}$. If not, then there must exist a $u \in A_{r,j,n}$ with $Lu = 0$, i.e.

$$L_0 u = -(e_j, u)\epsilon z.$$

Since $u \in A_{r,j,n}$ implies that $|(e_j, u)| \geq 1/r > 0$, we can write z in terms of u:

$$z = -((e_j, u)\epsilon)^{-1} L_0 u.$$

Using the definition of ϕ, we have

$$z = \phi(z) = \phi(L_0 u) \in \phi(L_0 A_r),$$

which is a contradiction. It follows that $\mathbb{L}_{r,j,n}$ must be dense.

To conclude, we have shown that, for each r, j, and n, $\mathbb{L}_{r,j,n}$ is open and dense in $L(H, \mathbb{R}^k)$. It follows from the Baire category theorem (Theorem 3.5) that their countable intersection (16.1) is open and dense in $L(H, \mathbb{R}^k)$ also. Since this intersection consists precisely of those linear maps that are injective on X, the result follows. ∎

It follows from Exercise 10.8 that L^{-1} is continuous when restricted to LX, since LX is compact. This gives a continuous parametrisation of the attractor using a finite number of coordinates.

Corollary 16.3. *Let X be a compact subset of H, with $d_f(X) < d$, d an integer, and take $k \geq 2d + 1$. Then there exists a continuous parametrisation of X using k coordinates.*

By using different methods this corollary can be strengthened considerably. The most powerful current result is due to Hunt & Kaloshin (1999), who guarantee further that the inverse of L is in fact Hölder continuous,

$$|L^{-1}x - L^{-1}y| \leq C|x - y|^\theta, \qquad x, y \in LX, \qquad (16.2)$$

and give sharp bounds on θ. See Exercise 16.2 for more details.

In fact one can also prove Theorem 16.2 and the stronger version in (16.2) for finite-dimensional orthogonal projections P as well as general linear maps. What this result says is that the attractor lies in a manifold that is the graph of some Hölder continuous function ϕ over PH,

$$\mathcal{A} \subset \mathcal{G}[\phi] = \{u = p + \phi(p) : p \in PH\}.$$

This is the general version of Proposition 15.3 (which needed the strong squeezing property). Note that in this case the function ϕ need not be Lipschitz and the projection P has nothing to do with the "natural" finite-dimensional projections P_n we have used in previous chapters.

16.1.1 Experimental Measurements as Parameters

One problem with embedding results such as Theorem 16.2 and the corresponding parametrisation (as in Corollary 16.3) is that the parameters do not correspond to any natural physical quantities. We mention here a recent result due to Friz & Robinson (2000) that gives one physically relevant consequence of the existence of a global attractor. In this situation each "parameter" corresponds to an experimental measurement. For the 2D Navier–Stokes equations on the periodic box $Q = [0, L]^2$, if the forcing term $f(x)$ is real analytic and $k > 8d_f(\mathcal{A})$, then the map

$$u \mapsto \big(u(x_1), \ldots, u(x_k)\big)$$

is one-to-one between the attractor and its image for almost every choice $\{x_1, \ldots, x_k\}$ of k points in Q.

This shows that a sufficient number of instantaneous point measurements will determine the velocity field *everywhere* within the domain Q. Similar results are also valid in the case of Dirichlet boundary conditions (Friz *et al.* 2000).

16.2 An Extension Theorem

In what follows we will need the following extension theorem. It guarantees that a continuous function defined on closed subsets of $\mathbb{R}^k$ has a continuous extension to the whole of $\mathbb{R}^k$, with essentially the same modulus of continuity (cf. footnote in Section 14.2). In particular, Lipschitz (Hölder) functions have Lipschitz (Hölder) extensions (set $\omega(r) = Cr^\theta$).

Theorem 16.4. *Let X be a compact subset of $\mathbb{R}^m$, and let f be a continuous function from X into $\mathbb{R}^k$ such that*

$$|f(x) - f(y)| \leq \omega(|x - y|), \tag{16.3}$$

where ω, the modulus of continuity of f, is convex:

$$\omega(r + s) \leq \omega(r) + \omega(s). \tag{16.4}$$

Then f has a continuous extension $F : \mathbb{R}^m \to \mathbb{R}^k$ that satisfies

$$|F(x) - F(y)| \leq \sqrt{k}\,\omega(|x - y|). \tag{16.5}$$

Exercise 16.1 shows that one can always find such an ω for a continuous function defined on a compact set. Note also that the result of the theorem can be extended to cover a function defined on a closed subset of $\mathbb{R}^m$, provided that f satisfies (16.3) and is globally bounded ($\|f\|_\infty < \infty$).

Proof. It follows from (16.3) that

$$|f_j(x) - f_j(x)| \le \omega(|x - y|) \tag{16.6}$$

for each component of f. We extend each component in turn preserving this property and then combine them to give an extension of f itself. Set

$$F_j(y) = \sup_{x \in X}[f_j(x) - \omega(|x - y|)].$$

First, note that if $x, y \in X$ then

$$F_j(x) - F_j(y) + \omega(|x - y|) \le |f_j(x) - f_j(y)| - \omega(|x - y|) \le 0,$$

and so $F_j(y) = f_j(y)$. To show (16.5), we use (16.4),

$$\begin{aligned} F_j(y) - F_j(z) &= \sup_{x \in X}[f_j(x) - \omega(|x - y|)] - \sup_{w \in X}[f_j(w) - \omega(|w - z|)] \\ &\le \sup_{x \in X}[f_j(x) - \omega(|x - y|) - f_j(x) + \omega(|x - z|)] \\ &\le \sup_{x \in X}[\omega(|x - z|) - \omega(|x - y|)] \\ &\le \omega(|z - y|), \end{aligned}$$

giving (16.6) for each F_j. Combining these inequalities yields (16.5) for F. ∎

16.3 Embedding the Dynamics Without Uniqueness

We will show that there is a finite set of ODEs, with dimension comparable with that of the global attractor, that reproduces its dynamics. However, these ODEs do not have unique solutions, so we cannot really speak about the "dynamical system" they generate, nor the corresponding attractor (as required by our Utopian Theorem).

Theorem 16.5. *Let $S(t)$ be the semigroup generated by the PDE*

$$du/dt = F(u), \tag{16.7}$$

where $F(u)$ is (Hölder) continuous from $\mathcal{A}$ into H (with Hölder exponent α). Suppose that $S(t)$ has a global attractor $\mathcal{A}$, with $d_f(\mathcal{A}) < d$. Then, for any

$k \geq 2d + 1$, there exists a system of ODEs in $\mathbb{R}^k$,

$$dx/dt = f(x), \qquad (16.8)$$

where $f : \mathbb{R}^k \to \mathbb{R}^k$ is (Hölder) continuous, and a bounded linear map $L : H \to \mathbb{R}^k$, which is injective on $\mathcal{A}$, such that for every solution $u(t)$ of (16.7) with $u(t) \in \mathcal{A}$ there is a solution $x(t)$ of (16.8) such that

$$u(t) = L^{-1}[x(t)]. \qquad (16.9)$$

Note the assumption that $F : \mathcal{A} \to H$ is a (Hölder) continuous function. It is certainly not reasonable to expect that F is continuous from the whole of H into H, since in most of our examples $F(u)$ contains the term Au, which is unbounded on H. However, we can often use the increased regularity of solutions on $\mathcal{A}$ to prove the required continuity, and we do this for our two examples after the proof of the theorem.

Proof. Theorem 16.2 guarantees that there exists a bounded linear map L from H into $\mathbb{R}^k$, which is injective on $\mathcal{A}$ and has a continuous inverse on $L\mathcal{A}$. Now consider the ODE for $x \in L\mathcal{A}$ obtained from the equation on $\mathcal{A}$,

$$\dot{x} = LF(L^{-1}x).$$

The function $\tilde{f} : L\mathcal{A} \to \mathbb{R}^k$ given by

$$\tilde{f}(x) = LF(L^{-1}x)$$

is certainly continuous. If we have assumed Hölder continuity of F, then $\tilde{f}$ is also Hölder, since we can use (16.2) to write

$$
\begin{aligned}
|\tilde{f}(x) - \tilde{f}(y)| &= |LF(L^{-1}x) - LF(L^{-1}y)| \\
&\leq \|L\|_{\mathrm{op}} |F(L^{-1}x) - F(L^{-1}y)| \\
&\leq K\|L\|_{\mathrm{op}} |L^{-1}x - L^{-1}y|^{\alpha} \\
&\leq CK\|L\|_{\mathrm{op}} |x - y|^{\theta\alpha}. \qquad (16.10)
\end{aligned}
$$

One can then use Theorem 16.4 to extend $\tilde{f}$ to a function $f : \mathbb{R}^k \to \mathbb{R}^k$ that is (Hölder) continuous and bounded. Then we have a system of ODEs

$$\dot{x} = f(x), \qquad x \in \mathbb{R}^k, \qquad (16.11)$$

with a (Hölder) continuous right-hand side and solutions that exist for all time, since $f(x)$ is globally bounded.

However, the solutions of (16.11) *may not be unique*, as f is only (Hölder) continuous, not Lipschitz. Nevertheless, by construction, we have ensured that one of the solutions of (16.8) through $x_0 = Lu_0$ with $u_0 \in \mathcal{A}$ will be precisely

$$x(t) = Lu(t).$$

This guarantees (16.9). ∎

Note that we have that $L\mathcal{A}$ is "weakly invariant," in the sense that for any initial condition $x_0 \in L\mathcal{A}$ we cannot guarantee that $x(t)$ remains in $L\mathcal{A}$, but we do know that there is at least one solution $x(t; x_0)$ that remains in $L\mathcal{A}$ [see Bhatia & Szegö (1967) for some discussion of such sets].

16.3.1 Continuity of F on $\mathcal{A}$ for the Scalar Reaction–Diffusion Equation

To show that the "right-hand side"

$$F(u) = \Delta u + f(u) = -Au + f(u)$$

is continuous we need to use some regularity of solutions on the attractor. We will assume that f is a C^∞ function and that Ω is a bounded C^∞ domain, and then we can appeal to Theorem 11.8, which guarantees that $\mathcal{A}$ is bounded in $H^k(\Omega)$ for every $k \geq 0$. In particular, $\mathcal{A}$ is bounded in $H^3(\Omega)$, and so in $D(A^{3/2})$.

We can now use the interpolation inequality for fractional powers proved in Lemma 3.27,

$$|Aw| \leq |w|^{1/3} |A^{3/2}w|^{2/3},$$

to show that the map $u \mapsto Au$ is Hölder on the attractor, since

$$|Au - Av| \leq |u - v|^{1/3} |A^{3/2}(u - v)|^{2/3} \leq C|u - v|^{1/3}, \qquad (16.12)$$

as $\mathcal{A}$ is bounded in $D(A^{3/2})$.

Also, we saw before in Equation (11.22) that

$$|f(u) - f(v)| \leq \|f'\|_\infty |u - v|,$$

so that f is continuous from $L^2(\Omega)$ into $L^2(\Omega)$. It follows that

$$-Au + f(u)$$

is Hölder continuous from $\mathcal{A}$ into $L^2(\Omega)$, and Theorem 16.5 applies.

16.3.2 Continuity of F on 𝒜 for the 2D Navier–Stokes Equations

To show that the nonlinear term

$$F(u) = -\nu A u - B(u, u) + f$$

from the Navier–Stokes equations is continuous we once again need to use some regularity of solutions on the attractor. We will appeal to the regularity theorem (Theorem 12.7), which shows that if we take $f \in \dot{C}_p^\infty(Q)$ [and in fact $f \in V$ is sufficient] then $\mathcal{A}$ is bounded in $\mathbb{H}^3$ and so in $D(A^{3/2})$ [using the result of (9.13) to move between $D(A^{k/2})$ and $\mathbb{H}^k$].

The Hölder continuity of A follows as above in (16.12), and we saw in the last chapter (15.22) that $B(u, u)$ is Lipschitz from V into H,

$$|B(u, u) - B(v, v)| \leq C_1 \|u - v\|.$$

Clearly this implies that

$$|B(u, u) - B(v, v)| \leq C_1 \lambda_1^{-1/2} |Au - Av|,$$

and so we obtain via (16.12) that $B(u, u)$ is also Hölder continuous. It follows that F itself is Hölder continuous from $\mathcal{A}$ into H:

$$|F(u) - F(v)| \leq K |u - v|^{1/3}, \qquad u, v \in \mathcal{A}.$$

Once again, Theorem 16.5 applies.

16.4 A Discrete-Time Utopian Theorem

Although our Utopian Theorem requires a set of ODEs on $\mathbb{R}^k$ with X as an attractor, any such construction is made extremely difficult by the fact that $f(x) = LF(L^{-1}x)$ is only Hölder continuous on X [see (16.10)].

Instead of struggling against this lack of uniqueness in ODEs, we will look for a map on $\mathbb{R}^k$ that reproduces the time T-map on the attractor and consider the discrete semidynamical system that comes from iterating $S(T)$. Since any numerical algorithm produces only the values of the solutions at discrete-time steps, this approach is more akin to numerics than a continuous system would be. We discuss this in more detail at the end of this section.

We will show that there is an iterated map on $\mathbb{R}^k$ that comes "within ϵ" of our Utopian Theorem in a discrete-time formulation. The result relies heavily on the topological properties of the global attractor. A detailed discussion of these is beyond the scope of this textbook, but we outline the main ideas below.

16.4.1 The Topology of Global Attractors

The topological structure of a global attractor (at least in simple cases) can be deduced from our construction in Chapter 10. There, we saw that (Exercise 10.5)

$$\mathcal{A} = \bigcap_{n=1}^{\infty} S(nT)B$$

for any absorbing set B and any $T > 0$. Since $\mathcal{A}$ is bounded, $B(0, R)$ is an absorbing set for some R. If we take T to be the time $B(0, R)$ takes to absorb $B(0, 2R)$, then we can write

$$\mathcal{A} = \bigcap_{n=1}^{\infty} C_n, \qquad\qquad (16.13)$$

with $C_n = S(nT)B(0, R)$ a sequence of sets satisfying

$$C_{n+1} \subset \text{int}(C_n), \qquad\qquad (16.14)$$

and we know that for every $\epsilon > 0$ there exists an n_0 such that

$$C_{n_0} \subset N(\mathcal{A}, \epsilon). \qquad\qquad (16.15)$$

If the flow has the injectivity property then in addition for each n there is a homeomorphism h_n that maps

$$(C_n, \partial C_n) \qquad \text{onto} \qquad (B(0, 1), \partial B(0, 1)). \qquad\qquad (16.16)$$

If $\mathcal{A}$ satisfies (16.13)–(16.16) we say that $\mathcal{A}$ is *strongly cellular*. With more powerful results from infinite-dimensional topology it is possible to show that a global attractor is always strongly cellular (even if we do not have the injectivity property).

The useful thing about this topological property is that it is invariant under linear embeddings of the set into finite-dimensional spaces. We state without proof the following topological theorem.

Theorem 16.6. *Suppose that $\mathcal{A}$ is strongly cellular and that $L : H \to \mathbb{R}^k$ is linear and one-to-one between $\mathcal{A}$ and its image $X = L\mathcal{A}$. Then X is a strongly cellular subset of $\mathbb{R}^k$. In particular there exists a homeomorphism $h : \mathbb{R}^k \setminus \{0\} \to \mathbb{R}^k \setminus X$ such that, for each $n \in \mathbb{N}$, h maps*

$$(\mathbb{R}^k \setminus B(0, 2^{-n}), \partial B(0, 2^{-n})) \qquad \text{onto} \qquad (\mathbb{R}^k \setminus C_n, \partial C_n). \qquad\qquad (16.17)$$

We now use this to show that if $X \subset \mathbb{R}^k$ is strongly cellular then there is a dynamical system that has X as an attractor. Note that this theorem in fact shows that strong cellularity fully characterises the topology of global attractors in dynamical systems.*

Theorem 16.7. *If $X \subset \mathbb{R}^k$ is strongly cellular then there exists a dynamical system $\varphi(t; x)$ on $\mathbb{R}^k$ for which X is an attractor consisting entirely of fixed points.*

Proof. We will take $a_n = 2^{-n}$, write $S = \partial B(0, 1)$, and use the strong cellularity of X in $\mathbb{R}^k$ to deduce the existence of a homeomorphism as in (16.17).

Now, using the uniform continuity of h on the closed set $[a_{n+3}, a_n]S$ it follows that for any $n \in N$ there exists a b_n such that

$$|h(\lambda s) - h(\mu s)| \le a_n \qquad (16.18)$$

whenever $\lambda, \mu \in [a_{n+3}, a_n]$ and $|\lambda - \mu| < b_n$. Clearly we can assume without loss of generality that $b_{n+1} \le b_n$ and that $b_n \le a_n$.

Now define $T_0 = 0$ and

$$T_n = \sum_{j=1}^{n} \frac{1}{b_n}.$$

Since b_n is decreasing we must have $T_n \to \infty$ as $n \to \infty$. Define a continuous scalar function $m : \mathbb{R} \to (0, 1]$ by

$$m(t) = \begin{cases} 1 - t, & t \le 0, \\ a_n \left[1 - \frac{1}{2}(t - T_n) b_n \right], & T_n < t \le T_{n+1}. \end{cases}$$

Observe that $m(0) = 1$ and $m(T_n) = a_n$. The idea is that $m(t) \to 0$ as $t \to \infty$ in such a way that it slows down enough to compensate for the larger changes in $h(s)$ as $s \to 0$.

We use $m(t)$ to define a dynamical system $\mu(t; x)$ on $\mathbb{R}^k \setminus \{0\}$,

$$\mu(t; x) = m(t + s)x \qquad \text{when} \qquad y = m(s)x, \qquad \text{for some} \qquad x \in S.$$

---

* There is a subtlety here. In infinite-dimensional spaces, global attractors of semiflows are strongly cellular, whereas in finite-dimensional spaces this is unknown. The result is that strong cellularity fully characterises the topology of attractors of semiflows in infinite-dimensional spaces and of flows in finite-dimensional spaces.

Using μ, we define a dynamical system $\varphi(t; x)$ on $\mathbb{R}^k$ by

$$\varphi(t; x) = \begin{cases} h \circ \mu\big(t; h^{-1}(x)\big), & x \in \mathbb{R}^k \setminus X, \\ x, & x \in X. \end{cases}$$

The function φ is clearly continuous away from X, and since μ is a dynamical system, φ satisfies the uniqueness property:

$$\begin{aligned} \varphi\big(t; \pi(s; x)\big) &= h \circ \mu\big(t; h^{-1}(h \circ \mu(s; h^{-1}(x)))\big) \\ &= h \circ \mu\big(t; \mu(s; h^{-1}(x))\big) \\ &= h \circ \mu\big(t + s; h^{-1}(x)\big) \\ &= \varphi(t + s; x). \end{aligned}$$

It remains to check that $\varphi(t; x)$ is continuous for every $(t; z)$ with $z \in X$. Assume that $x \in C_{n+1} \setminus C_{n+2}$. Then $h^{-1}(x) = m(s)\xi$ for some $\xi \in S$, and

$$m(s) \in [a_{n+2}, a_{n+1}].$$

If $|t| < 1/a_n$ then $|t| < 1/b_n$, and so it follows that we have

$$m(t + s) \in [a_{n+3}, a_n]. \tag{16.19}$$

Since $|t| < 1/a_n$, it follows from the definition of $m(t)$ that we must have

$$|m(t + s) - m(t)| < b_n. \tag{16.20}$$

Since

$$\mu(t; h^{-1}(x)) = \mu(t; m(s)\xi) = m(t + s)\xi$$

it follows from (16.19), (16.20), and (16.18) that

$$\begin{aligned} |\varphi(t; x) - x| &= |h \circ \mu(t; h^{-1}(x)) - h(h^{-1}(x))| \\ &\leq |h(m(t + s)\xi) - h(m(s)\xi)| \\ &\leq a_n, \end{aligned}$$

whenever $x \in C_{n+1} \setminus C_{n+2}$ and $|t| < 1/a_n$. Since a_n is decreasing we obtain

$$|\varphi(t; x) - z| \leq |\varphi(t; x) - x| + |x - z| \leq a_k + |y - z|$$

whenever $z \in X$, $x \in C_{k+1}$, and $|t| < 1/a_k$.

That X is an attractor is clear by construction. ∎

We will use the following immediate corollary, obtained by setting $\Phi(x) = \varphi(1, x)$.

Corollary 16.8. *If X is a strongly cellular subset of $\mathbb{R}^k$ then there exists a map $\Phi : \mathbb{R}^k \mapsto \mathbb{R}^k$ for which X is the attractor, and $\Phi(x) = x$ for all $x \in X$.*

16.4.2 The "within ϵ" Discrete Utopian Theorem

The following theorem guarantees the existence of a discrete dynamical system on $\mathbb{R}^k$ that reproduces the time T-map on $\mathcal{A}$ and has a global attractor arbitrarily close to X.

Theorem 16.9. *Let $\mathcal{A}$ have finite fractal dimension, $d_f(\mathcal{A}) < d$. Then given $T > 0$ and $\epsilon > 0$, for any $k \geq 2d + 1$ there exists a bounded linear map $L : \mathcal{A} \to \mathbb{R}^k$, injective on $\mathcal{A}$, and a continuous map f from $\mathbb{R}^k$ into itself, such that if we write $X = L\mathcal{A}$, then*

(i) the dynamics on $\mathcal{A}$ and X are conjugate under L:

$$f|_X = L \circ S(T) \circ L^{-1},$$

and

(ii) the global attractor for the discrete dynamical system generated by f, X_f, satisfies

$$X_f \subset N(X, \epsilon),$$

and has X as an invariant subset.

For a schematic illustration of this result, see Figure 16.2. Note that part (ii) says that

$$\text{dist}_{\mathcal{H}}(X, X_f) \leq \epsilon,$$

and so X_f is very nearly (within ϵ) X, as is required in the full version of the Utopian Theorem.

Proof. First use Theorem 16.2 to find a bounded linear map $L : H \to \mathbb{R}^k$ that is injective on $\mathcal{A}$, and set $X = P\mathcal{A}$.

Let Φ be the map on $\mathbb{R}^k$ from Corollary 16.8. Then, since X is a global attractor for the dynamical system generated by Φ, it follows that for the given ϵ there exists a δ such that

$$x \in N(X, \delta) \quad \Rightarrow \quad \Phi^j(x) \in N(X, \epsilon) \quad \text{for all} \quad j \geq 0. \quad (16.21)$$

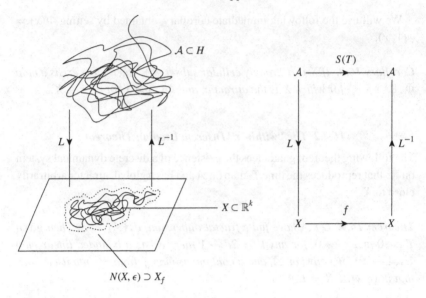

Figure 16.2. Schematic illustration of the "within ϵ" discrete-time Utopian Theorem.

Using the continuity of $S(t)$ and the compactness of $\mathcal{A}$, choose an M such that the map $\tilde{g}(x)$ given by

$$\tilde{g}(x) = L \circ S(T/M) \circ L^{-1} \qquad (16.22)$$

satisfies

$$|\tilde{g}(x) - x| \le \delta/4 \qquad \text{for all} \quad x \in X.$$

Now extend the function

$$p_1(x) = \begin{cases} \tilde{g}(x) - x, & x \in X, \\ 0, & x \notin N(X, \delta/2) \end{cases}$$

to a function $p_2(x)$ defined on the whole of $\mathbb{R}^k$, using* Theorem 16.4. Defining

---

* In fact we use the modification of Theorem 16.4 discussed briefly after its statement, since $p_1(x)$ is defined on a set that is not compact. One way to avoid this would be to find an extension $\tilde{p}_2$ of the function

$$\tilde{p}_1(x) = \begin{cases} \tilde{g}(x) - x, & x \in X, \\ 0, & x \in \partial N(X, \delta/2), \end{cases}$$

using Theorem 16.4 as stated, and then set

$$p_2(x) = \begin{cases} \tilde{p}_2(x), & x \in N(X, \delta/2), \\ 0, & x \notin N(X, \delta/2). \end{cases}$$

$p(x)$ by

$$p(x) = p_2(x)\,\theta\left(\frac{|p_2(x)|}{\delta/4}\right),$$

where $\theta : [0, \infty) \to [0, 1]$ is continuous and satisfies

$$\theta(r) = \begin{cases} 1, & 0 \le r \le 1, \\ 0, & r \ge 2, \end{cases}$$

we obtain an extension $p(x)$ of $p_1(x)$, with $|p(x)| \le \delta/2$ for all x.

Now let $g(x) = x + p(x)$ to obtain an extension of the function $\tilde{g}(x)$ defined in (16.22) above, and consider $h(x) = g^M(x)$. Then $h(x)$ extends

$$\begin{cases} L \circ S(T) \circ L^{-1}x, & x \in X, \\ x, & x \notin N(X, \delta/2) \end{cases}$$

and has the property that

$$x \in N(X, \delta/2) \quad \Rightarrow \quad h(x) \in N(X, \delta).$$

Finally, define f as the composition of h with Φ, which is again a map from $\mathbb{R}^k$ to itself,

$$f = \Phi \circ h.$$

Then, as Φ is the identity on X,

$$f|_X = L \circ S(T) \circ L^{-1},$$

and, since h is the identity on $\mathbb{R}^k \setminus N(X, \delta/2)$,

$$f|_{\mathbb{R}^k \setminus N(X, \delta/2)} = \Phi.$$

This ensures that $N(X, \epsilon)$ is an absorbing set for the dynamical system $\{f^n\}$, since $f = \Phi$ until the trajectory enters $N(X, \delta/2)$. From then, the iterates under $f \ne \Phi$ either remain in $N(X, \delta/2)$ or leave into $N(X, \delta)$. There we have $f = \Phi$ once again, so that the iterates under f remain in $N(X, \epsilon)$ by (16.21).

Thus the system has a global attractor X_f, which must lie within $N(X, \epsilon)$. That X is invariant and that $X \subset X_f$ are obvious. ∎

The result of this theorem provides some optimism that in time it may be possible to develop numerical schemes that take advantage of the finite-dimensional nature of the system. It is notoriously difficult to compute the solutions of PDEs, and efficient codes for the Navier–Stokes equations, for example, are extremely valuable. Since most numerical methods correspond to an iterated map on some finite-dimensional phase space, we could interpret the result of this theorem as proving the existence, at least in theory, of an efficient method based on the intrinsic dynamics of the equation.

16.5 Conclusion

The results of this chapter give the best indication so far of the sense in which finite-dimensional attractors imply that the underlying system has a finite number of degrees of freedom.

Not only have we proved the embedding Theorem 16.2, which shows that the global attractor can be parametrised by a finite set of coordinates, we have also proved the discrete-time Utopian Theorem (Theorem 16.9). This says, in effect, that even though the models we have studied are infinite-dimensional, their asymptotic dynamics are no more complicated than those of an iterated map on a finite-dimensional space. In particular, this means that numerical investigations, by their very nature finite-dimensional, can be expected to provide insight into the behaviour of these extremely complicated systems.

The main focus of research into global attractors over the past decade and a half has been to extend the theory to an ever increasing array of models. The two examples that we have treated in the previous chapters have been supplemented by a host of others, and new equations are succumbing all the time. A large, though by no means exhaustive, set of these is covered in Temam (1988) and more in the second edition of the same book (Temam, 1997).

However, the theory is now clearly valid in a great many situations, and there is a need to concentrate more specifically on the direct physical implications of the existence of a finite-dimensional global attractor. The results of this chapter (in particular those of Section 16.1.1) offer a first step in this direction, but there is much more to be done if we are to make the most of the dynamical systems approach to these infinite-dimensional problems.

Exercises

16.1   Suppose that X is a compact subset of $\mathbb{R}^m$ and f is a continuous function from X into $\mathbb{R}^k$. Show that it is possible to find a modulus of continuity for f as in (16.3), $\omega : \mathbb{R}^+ \to \mathbb{R}^+$, such that

$$\omega(r + s) \le \omega(r) + \omega(s).$$

[Hint: Set

$$\omega(r) = \sup_{\{x,y \in X \,:\, |x-y| \le r\}} |f(x) - f(y)|,$$

and show that ω is well defined and has the required convexity property.]

16.2 The result of Hunt & Kaloshin (16.2) shows that linear maps whose inverse satisfies

$$|L^{-1}x - L^{-1}y| \le C|x - y|^\theta, \qquad x, y \in LX$$

form a dense subset of $\mathcal{L}(H, \mathbb{R}^k)$ for any θ in the range

$$0 < \theta < \frac{k - 2d_f(X)}{k(1 + \tau(X)/2)}.$$

Here $\tau(X)$ is the "thickness" of X, a measure of how well the set can be approximated by finite-dimensional linear subspaces. If $d(X, \epsilon)$ is the minimum dimension of any finite-dimensional subspace, V, such that every point of X lies within ϵ of V, then the "thickness" $\tau(X)$ is defined by

$$\tau(X) = \limsup_{\epsilon \to 0} \frac{\log d(X, \epsilon)}{\log(1/\epsilon)}.$$

(i) Show that $\tau(X) \le d_f(X)$.
(ii) Find an example of a set X with $\tau(X) = 0$ but $d_f(X) = n$.

16.3 Let A be the negative Laplacian on Ω. Then its eigenfunctions form an orthonormal basis for $L^2(\Omega)$, and its eigenvalues (ordered to increase with n) satisfy $\lambda_n \sim n^{2/m}$ for n large. By considering the linear subspace consisting of the first n eigenfunctions of A, show that, if X is bounded in $D(A^r)$, then the quantity $d(X, \epsilon)$ (from the previous exercise) is bounded by

$$d(X, \epsilon) \le C\epsilon^{-m/2r}.$$

Deduce that

$$\tau(X) \le \frac{m}{2r}. \tag{16.23}$$

If X is bounded in $D(A^r)$ for any r deduce that the Hölder exponent of L^{-1} can be made as close to 1 as possible provided that k is large enough. [See Friz & Robinson (1999).]

16.4 Suppose that $A : \mathcal{A} \to H$ is Lipschitz continuous. By writing $w = u - v$ in two pieces, one consisting of the first n modes and one of the remaining modes (an infinite number) show that in this case $\mathcal{A}$ must lie in a Lipschitz manifold given as a graph over $P_n H$. (Hint: Use the result of Exercise 14.2.)

16.5 Under the same assumptions as those in the previous exercise, we can write down a Lipschitz ODE on $\mathbb{R}^n \simeq P_n H$ for which $P_n \mathcal{A}$ is an invariant set (cf. Exercise 15.2). Denote this by

$$\dot{x} = f(x).$$

Günther (1995) has proved that there exists a Lipschitz ODE on $\mathbb{R}^n$,

$$\dot{x} = g(x),$$

for which $X = P_n \mathcal{A}$ is the global attractor, with $g|_X = 0$. Prove a "within ϵ" continuous-time Utopian Theorem in this case. [Hint: Consider an equation of the form $F(x) + g(x)$, where $F(x) = f(x)$ in a neighbourhood of X.]

Notes

The proof of Theorem 16.2 is based on that of Mañé (1981), with some simplifications due to the fact that we are considering the problem in a Hilbert space setting. Many papers have treated the problem since, all concentrating on the case of orthogonal projections until the result given in (16.2), due to Hunt & Kaloshin (1999) [this is also valid for orthogonal projections; see Friz & Robinson (1999)]. Eden *et al.* (1994b) gave a constructive proof in a Hilbert space and showed that the inverse is Hölder continuous if $X \subset \mathbb{R}^N$; Ben-Artzi *et al.* (1993) gave sharp bounds on this Hölder inverse (for $X \subset \mathbb{R}^N$); Sauer *et al.* (1993) replaced density by a probabilistic version ("almost all" in some sense); Foias & Olson (1996) gave the first proof that the inverse is Hölder continuous when X is a subset of an infinite-dimensional space.

The proof of the extension theorem here is due to McShane (1934); a proof that works when $f : \mathbb{R}^k \to H$ is given in Stein (1970). As mentioned before, it is possible to extend the function keeping the same Lipschitz constant, but the proof is much more involved [see Wells & Williams (1975); one can also adapt the proof of Kirszbraun's theorem in Federer (1991)].

Eden *et al.* (1994b) provide another construction aimed at satisfying the demands of the Utopian theorem. Let L be a bounded linear map from H into $\mathbb{R}^k$ that is one-to-one on $\mathcal{A}$, and set $X = L\mathcal{A}$. They define a function $\nu : \mathbb{R}^k \mapsto X$ by $\nu(x) = y$, where y is chosen to be one of those points in X such that $|x - y| = \text{dist}(x, X)$. Although this function v is not continuous, they show that solutions of the ODE

$$\dot{x} = f(v(x)) + \sigma(v(x) - x) \tag{16.24}$$

are attracted to X and reproduce the dynamics on $\mathcal{A}$. However, the right-hand side of (16.24) is not even continuous, so there is no hope that the solutions can be unique.

The discussion of the topology of global attractors, given in more detail in Robinson (1999b), combines several results, motivated by the treatment in Garay (1991). That strong cellularity is preserved under linear embeddings (as in Theorem 16.6) follows from results of Geoghegan & Summerhill (1973) and Borsuk (1969, 1970). The existence of the appropriate homeomorphism is proved in McCoy (1973). The construction of the dynamical system in Theorem 16.7 is adapted from Garay (1991); see also Günther & Segal (1993). Note that Theorem 16.7 does not claim that there is a smooth system of ordinary differential equations with X as an attractor, but this has also been proved (Günther, 1995).

Finally, the proof of the discrete-time Utopian Theorem (Theorem 16.9) is taken from Robinson (1999b).

17

The Kuramoto–Sivashinsky Equation

This final chapter is devoted to a series of exercises that provide an analysis of the scalar Kuramoto–Sivashinsky equation (KSE),

$$u_t + u_{xxxx} + u_{xx} + uu_x = 0, \qquad u(x,t) = u(x+L,t). \tag{17.1}$$

This equation, introduced independently by Kuramoto & Tsuzuki (1976) and Sivashinsky (1977), is a model of instabilities in various physical situations, including flame fronts and the flow of thin liquid films on inclined planes. It has since become a popular model, since it is nontrivial but nonetheless amenable to analysis.

The exercises below treat existence, uniqueness, existence of a global attractor $\mathcal{A}$, uniform differentiability on $\mathcal{A}$, a proof that the attractor is finite-dimensional, and the existence of inertial manifolds. All function spaces are assumed to have domain $(0, L)$, and we will write D for both d/dx and $\partial/\partial x$.

17.1 Preliminaries

Along with the Poincaré inequality (Lemma 5.40)

$$|u| \le C_p |Du| \qquad \text{for all} \qquad u \in \dot{H}_p^1 \tag{17.2}$$

we will use the result of the following exercise repeatedly.

Exercise 17.1. *If $u \in \dot{C}_p^2$, integrate the identity*

$$D(uDu) = uD^2u + (Du)^2 \tag{17.3}$$

to deduce that

$$|Du|^2 \le |u||D^2u|. \tag{17.4}$$

Show that this also holds if $u \in \dot{H}_p^2$.

The dominant linear part of the equation is the one-dimensional version of the"bi-Laplacian" Δ^2, D^4. We make sure that this operator has an orthonormal set of eigenvalues.

Exercise 17.2. *By first taking the inner product of the equation*

$$D^4 u = f \tag{17.5}$$

with a suitable smooth function, write (17.5) in its weak form: given $f \in H^{-2}$, find $u \in \dot{H}_p^2$ such that

$$a(u, v) = \langle f, v \rangle \qquad \text{for all} \qquad v \in \dot{H}_p^2, \tag{17.6}$$

where $a(u, v) : \dot{H}_p^2 \times \dot{H}_p^2 \to \mathbb{R}$ is an appropriate bilinear form.

Exercise 17.3. *Show that (17.6) has a unique solution. Defining a linear operator $A : \dot{H}_p^2 \to H^{-2}$ by*

$$\langle Au, v \rangle = a(u, v) \qquad \text{for all} \qquad u, v \in \dot{H}_p^2,$$

show that A^{-1} is compact and symmetric and deduce that A has a set of orthonormal eigenfunctions that form a basis of $\dot{L}^2$.

We now want to write down the weak form of the KSE itself. If we take the inner product with a function in $\dot{C}_p^2$, integrate by parts, and take limits, we obtain

$$\frac{d}{dt}(u, v) + a(u, v) + (D^2 u, v) + \int_0^L u(Du)v \, dx = 0 \qquad \text{for all} \qquad v \in \dot{H}_p^2. \tag{17.7}$$

By analogy with the Navier–Stokes equations we define a trilinear form $b : \dot{H}_p^2 \times \dot{H}_p^2 \times \dot{H}_p^2 \to \mathbb{R}$ by

$$b(u, v, w) = \int_0^L u(Dv)w \, dx.$$

Exercise 17.4. *Show that*

$$b(u, u, u) = 0, \tag{17.8}$$

$$b(u, v, w) + b(v, w, u) + b(w, u, v) = 0, \tag{17.9}$$

and that

$$|b(u, v, w)| \leq \begin{cases} |u||D^2 v||w|, & u \in \dot{L}^2, v \in \dot{H}_p^2, w \in \dot{L}^2, \\ |Du||Dv||w|, & u, v \in \dot{H}_p^1, w \in \dot{L}^2. \end{cases} \tag{17.10}$$

We also define the bilinear form $B(u, v) : \dot{H}_p^2 \times \dot{H}_p^2 \to H^{-2}$ by

$$\langle B(u, v), w \rangle = b(u, v, w) \qquad \text{for all} \qquad u, v, w \in \dot{H}_p^2.$$

In terms of A and B we can write the KSE as

$$\frac{du}{dt} + Au + D^2 u + B(u, u) = 0. \tag{17.11}$$

17.2 Existence and Uniqueness of Solutions

We now want to use the eigenfunctions of the bi-Laplacian operator from Exercise 17.2 to prove the existence of solutions via the Galerkin method.

Exercise 17.5. *With*

$$u_n = \sum_{j=1}^{n} u_{nj} w_j$$

consider the equation for u_n given by

$$\frac{du_n}{dt} + Au_n + D^2 u_n + P_n B(u_n, u_n) = 0. \tag{17.12}$$

Take the inner product of this equation with u_n to show that

$$
\begin{array}{lll}
u_n & \text{is uniformly bounded in} & L^\infty(0, T; \dot{L}^2), \\
u_n & \text{is uniformly bounded in} & L^2(0, T; \dot{H}_p^2), \qquad \text{and} \qquad (17.13) \\
du_n/dt & \text{is uniformly bounded in} & L^2(0, T; H^{-2}).
\end{array}
$$

[Hint: You will need to use (17.4).]

Exercise 17.6. *Use the bounds in (17.13) along with the compactness theorem (Theorem 8.1) to extract a subsequence of the $\{u_n\}$ that converges to a weak solution u of (17.7) and show that*

$$u \in C^0([0, T]; \dot{L}^2) \cap L^2(0, T; \dot{H}_p^2). \tag{17.14}$$

[Hint: Theorem 8.1 can be applied to show that some subsequence converges strongly in $L^2(0, T; \dot{H}_p^1)$.]

Exercise 17.7. *Show that*

$$\frac{d}{dt}|w|^2 + |D^2 w|^2 \le C(1 + |D^2 u| + |v|^2)|w|^2,$$

and use the fact that any two solutions u and v of (17.7) are elements of
$L^2(0, T; \dot{H}_p^2)$ *(17.14) to deduce that the solutions satisfy*

$$|w(t)|^2 \leq |w(0)|^2 \exp \left(C \int_0^t 1 + |D^2 u(s)| + |v(s)|^2 \, ds \right), \qquad (17.15)$$

which gives uniqueness.

17.3 Absorbing Sets and the Global Attractor

With existence and uniqueness ensured, we now turn to finding absorbing sets in L^2 and $\dot{H}_p^1$ to give a global attractor. This is not going to be an entirely straightforward calculation, since taking the inner product of (17.11) with u and integrating yield

$$\tfrac{1}{2} \frac{d}{dt} |u|^2 + |D^2 u|^2 = |Du|^2, \qquad (17.16)$$

and we have a "forcing" term on the right-hand side that is not always dominated by the dissipative term $|D^2 u|^2$. Indeed, if we use (17.4) and Young's inequality then we end up with

$$\frac{d}{dt} |u|^2 + |D^2 u|^2 \leq 2|u|^2,$$

from which we can deduce only that $|u(t)|$ grows no worse than exponentially,

$$|u(t)|^2 \leq e^t |u(0)|^2.$$

We therefore have to proceed in a more roundabout way.

We will restrict all that follows to the case of odd solutions, that is, those $u(x, t)$ for which

$$u(x, t) = -u(-x, t).$$

This is at least consistent, since it follows from the equation that if the initial condition is odd then the solution is odd for all time.

The key is to consider not the variable u but a transformed variable $v = u - \phi$, where we choose the "gauge function" $\phi \in \dot{C}_p^\infty$ in an appropriate way. With this substitution (17.1) becomes

$$\frac{\partial v}{\partial t} + D^4 v + D^2 v + \phi Dv + v D\phi + v Dv = g(\phi), \qquad (17.17)$$

where

$$g(\phi) = -D^4\phi - D^2\phi - \phi D\phi.$$

Note that $g \in \dot{L}^2$.

Multiplying (17.17) by v and integrating over $(0, L)$ we obtain

$$\tfrac{1}{2}\frac{d}{dt}|v|^2 + |D^2v|^2 - |Dv|^2 + \tfrac{1}{2}\int_0^L v^2 D\phi\, dx = (g, v).$$

To make use of the potentially helpful terms on the left-hand side,

$$|D^2v|^2 + \tfrac{1}{2}\int_0^L v^2 D\phi\, dx,$$

Nicolaenko *et al.* (1985) proved that for any $\alpha > 0$ there exists an odd function $\phi \in \dot{C}_p^\infty$ such that

$$|D^2v|^2 + \int_0^L v^2 D\phi\, dx \geq \alpha|u|^2$$

for all odd functions $u \in \dot{H}_p^2$. [An alternative proof can be found in Temam (1988).]

Exercise 17.8. *Use the above result for an appropriate value of α to show that*

$$\frac{d}{dt}|v|^2 + \tfrac{1}{2}|D^2v|^2 + 2|v|^2 \leq \tfrac{1}{2}|g|^2.$$

Deduce that there is a $t_0(|u_0|)$ and constant R and M such that

$$|u(t)| \leq R \quad \text{and} \quad \int_t^{t+1} |D^2u(s)|^2\, ds \leq M \quad \text{for all} \quad t \geq t_0(|u_0|).$$

$$(17.18)$$

It is comparatively easy to find an absorbing set in $\dot{H}_p^1$.

Exercise 17.9. *Take the inner product of (17.11) with $-D^2u$, and use (17.10) to obtain*

$$\frac{d}{dt}|Du|^2 \leq |D^2u|^2 + |D^2u|^2|Du|.$$

Use (17.18) and the uniform Gronwall lemma of Exercise 11.2 to obtain an absorbing set in $\dot{H}_p^1$, and deduce the existence of a global attractor $\mathcal{A}$ in L^2.

17.4 The Attractor is Finite-Dimensional

To show that the attractor is finite-dimensional we first have to check that the semiflow is uniformly differentiable on the attractor $\mathcal{A}$. The analysis is very similar to that for the 2D Navier–Stokes equations in Theorem 13.19.

Exercise 17.10. *Show that the semigroup obtained in Exercise 17.6 is uniformly differentiable on $\mathcal{A}$ in the sense of Definition 13.15, with $\Lambda(t; u_0)\xi$ given by the solution $U(t)$ of the equation*

$$U_t + U_{xxxx} + U_{xx} + Uu_x + uU_x = 0, \qquad U(0) = \xi. \qquad (17.19)$$

Exercise 17.11. *Show that $\Lambda(t; u_0)$ is compact for all $t > 0$.*

We now want to show that the attractor has finite fractal dimension. We know that $d_f(\mathcal{A}) < n$ provided that

$$\langle \operatorname{Tr} P_n L(u(t)) \rangle < 0.$$

If $\{\phi_j\}_{j=1}^n$ are a set of n orthonormal functions in $\dot{L}^2$ then we need to bound

$$\sum_{j=1}^n (L\phi_j, \phi_j).$$

Since

$$(L\phi_j, \phi_j) = -|D^2\phi_j|^2 + |D\phi_j|^2 - \int_0^L u(D\phi_j)\phi_j \, dx - \int_0^L \phi_j^2 Dv \, dx$$

$$= -|D^2\phi_j| + |D\phi_j|^2 - \tfrac{1}{2}\int_0^L \phi_j^2 Dv \, dx,$$

after an integration by parts, we need to consider

$$\sum_{j=1}^n (L\phi_j, \phi_j) = -\sum_{j=1}^n |D^2\phi_j| + \sum_{j=1}^n |D\phi_j|^2 - \tfrac{1}{2}\sum_{j=1}^n \int_0^L \phi_j^2 Du \, dx.$$

Exercise 17.12. *Show that*

$$\sum_{j=1}^n |D\phi_j|^2 \le n + \tfrac{1}{4}\sum_{j=1}^n |D^2\phi_j|^2$$

and that

$$\int_0^L \phi_j^2 Du \, ds \le C|\phi_j|^2 + \tfrac{1}{4}|D^2\phi_j|^2.$$

[For this second estimate use $H^1 \subset L^4$ and (17.4).] Deduce that

$$\sum_{j=1}^{n} (L\phi_j, \phi_j) \leq -\tfrac{1}{2} \sum_{j=1}^{n} |D^2\phi_j|^2 + Mn$$

for some constant M and hence, using the argument of Lemma 13.17, deduce that $d_f(\mathcal{A}) < \infty$.

17.5 Inertial Manifolds

Finally, we use the generalised spectral gap condition (15.21) to check that the strong squeezing property holds. We let A be the linear operator associated with D^4, as in Exercise 17.2, and analyse the remaining terms in the equation,

$$N(u) = D^2 u + u Du.$$

First we show that if $u \in L^2$ then $N(u) \in D(A^{-1/2})$ [the dual of $D(A^{1/2})$].

Exercise 17.13. *Show that*

$$|(N(u), v)| \leq C(1 + |u|)|u||A^{1/2}v|$$

for all $v \in D(A^{1/2})$.

Since $D(A^{-1/2})$ is the dual of $D(A^{1/2})$, we can rewrite this as

$$|A^{-1/2}N(u)| \leq C(1 + |u|)|u|.$$

We now show in addition that $N(u)$ is locally Lipschitz from L^2 into $D(A^{-1/2})$.

Exercise 17.14. *Show that*

$$|(N(u) - N(v), w)| \leq C(1 + |u + v|)|u - v||A^{1/2}w|$$

for all $w \in D(A^{1/2})$.

This gives

$$|A^{-1/2}(N(u) - N(v))| \leq C(1 + |u + v|)|u - v|.$$

One can now prepare the equation, following (essentially) the argument in Lemma 15.7 to show that the function

$$F(u) = \theta_R(|u|)N(u)$$

(with θ_R defined as in Section 15.4.1) is globally Lipschitz from L^2 into $D(A^{-1/2})$ and zero outside the ball $B(0, 2R)$.

Since the eigenvalues λ_n of A satisfy $\lambda_n \sim n^4$, it is easy to satisfy the spectral gap condition (15.21) and to deduce the existence of an inertial manifold for the KSE.

Notes

Almost all of the analysis needed for these exercises, or something very similar, can be found in Temam (1988).

It is possible to show the existence of an absorbing set in $\dot{L}^2$ for arbitrary initial data, but the various arguments are significantly more involved; see Il'yahsenko (1992), Collet *et al.* (1993), or Goodman (1994). In particular the approach of Collet *et al.* is a generalisation of that given here. Foias *et al.* (1988b) prove directly that the strong squeezing property holds for the KSE and hence that it has an inertial manifold. For some numerical studies of the Kuramoto–Sivashinsky equation see Hyman & Nicolaenko (1986), Hyman *et al.* (1986), Kevrekidis *et al.* (1990), and Jolly *et al.* (1990).

Appendix A

Sobolev Spaces of Periodic Functions

In this brief appendix we give simple proofs of the Sobolev embedding theorem (Theorem 5.31) and the Rellich–Kondrachov compactness theorem (Theorem 5.32) in the setting of spaces of periodic functions. The arguments use the Fourier series expansion rather than the calculus methods of Chapter 5.

To prevent too clumsy a notation we will drop the "p" subscripts in what follows and assume that all sums over k are for $k \in \mathbb{Z}^m$.

A.1 The Sobolev Embedding Theorem – H^s, C^r, and L^p

In this section we investigate how the spaces $H^s(Q)$, $C^r(\overline{Q})$, and $L^p(Q)$ are related. We first find conditions to ensure that a function in $H^s(Q)$ is in fact continuous and then investigate its integrability properties.

A.1.1 Conditions for $H^s(Q) \subset C^0(\overline{Q})$

We will show that if $s > m/2$ then $H^s(Q) \subset C^0(\overline{Q})$, i.e. that if a function is in H^s then it is in fact continuous. Note that the caveats given by Proposition 5.22 apply equally in this case.

Theorem A.1. *If $u \in H^s(Q)$ with $s > m/2$, then $u \in C^0(\overline{Q})$ and*

$$\|u\|_\infty \leq C_s \|u\|_{H^s}.$$

Proof. Write

$$u = \sum_k c_k e^{2\pi i k \cdot x / L},$$

and then

$$\|u\|_\infty \le \sum_k |c_k|$$

$$\le \sum_k \frac{1}{(1 + |k|^{2s})^{1/2}} (1 + |k|^{2s})^{1/2} |c_k|$$

$$\le \left(\sum_k (1 + |k|^{2s}) |c_k|^2 \right)^{1/2} \sum_k \frac{1}{(1 + |k|^{2s})}.$$

Now,

$$\sum_k \frac{1}{(1 + |k|^{2s})} = C_s < \infty$$

provided that $s > m/2$. Therefore

$$\|u\|_\infty \le C_s \|u\|_{H^s},$$

and the absolute convergence of the coefficients yields the uniform convergence of the Fourier series and hence continuity of u. ∎

A straightforward corollary is:

Corollary A.2. *If $s > m/2 + j$ in Theorem A.1, then $u \in C^j(\overline{Q})$ and*

$$\|u\|_{C^j} \le C \|u\|_{H^s}.$$

A.1.2 Integrability Properties of Functions in H^s

To obtain bounds on u in various L^p spaces given $u \in H^s(Q)$ we will need to know that if $\mathbf{c} = \{c_k\} \in l^p$ (the sequence space introduced in Section 1.4.5), then

$$u = \sum_k c_k e^{ik \cdot x} \in L^q$$

(p and q conjugate) with

$$\|u\|_{L^q} \le \alpha \|\mathbf{c}\|_{l^p}. \tag{A.1}$$

The proof is not straightforward and relies on complex variable methods [cf. Rudin (1974), Theorem 12.11].

Theorem A.3. *If $u \in H^s(Q)$ with $s < m/2$, then $u \in L^p(Q)$ with*

$$p \in \left[2, \frac{m}{(m/2) - k} \right).$$

If $s = m/2$ then the embedding holds for all $p < \infty$ (but not for $p = \infty$).

We show the result for the half-open interval, but in fact it holds for the closed interval (this requires the calculus methods used in Chapter 5).

Proof. It is immediate that $H^s(Q) \subset L^2(Q)$ with $|u| \leq \|u\|_{H^s}$, so we prove the embedding only for $p > 2$. Then, using (A.1), we have

$$
\|u\|_{L^p} \leq \alpha \left(\sum |c_k|^q \right)^{1/q}
$$

$$
\leq \alpha \left(\sum (1 + |k|^{2s})^{q/2} (1 + |k|^{2s})^{-q/2} |c_k|^q \right)^{1/q}
$$

$$
\leq \alpha \left[\left(\sum (1 + |k|^{2s}) |c_k|^2 \right)^{\frac{q}{2}} \left(\sum (1 + |k|^{2s})^{-q/(2-q)} \right)^{\frac{2-q}{2}} \right]^{1/q}
$$

$$
\leq \alpha \|u\|_{H^s} \left(\sum (1 + |k|^{2s})^{-q/(2-q)} \right)^{(2-q)/2q}.
$$

The sum in this final expression is finite provided that $qs/(2 - q) > m/2$, that is, provided that $q > 2m/(2s + m)$. Since $p^{-1} + q^{-1} = 1$, this requires

$$
p < \frac{m}{\frac{m}{2} - s},
$$

as in the statement of the theorem. ∎

A.2 Rellich–Kondrachov Compactness Theorem

Theorem A.4. $H^1(Q)$ *is compactly embedded in* $L^2(Q)$.

The proof is taken from Temam (1985).

Proof. Consider a sequence $\{u_n\}$,

$$
u_n = \sum_k c_{nk} e^{2\pi i k \cdot x / L},
$$

bounded in H^1. Then, for some M,

$$
\sum_k (1 + |k|^2) |c_{nk}|^2 \leq M. \tag{A.2}
$$

In particular, each Fourier component is uniformly bounded in n, so one can extract a subsequence $u_{n_{1j}}$ such that $c_{n_{1j}1}$ converges. From this take a subsequence $u_{n_{2j}}$ such that $c_{n_{2j}1}$ and $c_{n_{2j}2}$ converge. Continue in this way with subsequences $u_{n_{lj}}$ such that $c_{n_{lj}1}, \ldots, c_{n_{lj}l}$ all converge. Now take the diagonal sequence $\tilde{u}_j = u_{n_{jj}}$ and write c_{jk} for the corresponding Fourier coefficients. Then, for each k, c_{jk} converges to a limit c_k^*. From (A.2) we know that

$$
\sum_k (1 + |k|^2) |c_k^*|^2 \leq M.
$$

Thus

$$\sum_k (1 + |k|^2)|c_{jk} - c_k^*|^2 \le 2M.$$

Now,

$$|\tilde{u}_j - u^*|^2 = \sum_k |c_{jk} - c_k^*|^2$$

$$= \sum_{|k| \le K} |c_{jk} - c_k^*|^2 + \frac{1}{K^2} \sum_{|k| \ge K} |c_{jk} - c_k^*|^2 |k|^2$$

$$\le \sum_{|k| \le K} |c_{jk} - c_k^*|^2 + \frac{2M}{K^2}.$$

Given $\epsilon > 0$, choose K large enough that the second term is $\le \epsilon/2$, and choose j large enough that the first term is too. Then $u_j \to u^*$ in L^2. ∎

Corollary A.5. $H^{s+1}(Q)$ *is compactly embedded in* $H^s(Q)$.

Appendix B

Bounding the Fractal Dimension Using the Decay of Volume Elements

This appendix provides a proof of Theorem 13.16, due to Hunt.* [See Blinchevskaya & Ilyashenko (1999) for a similar result valid under less restrictive assumptions.]

The essential idea of the proof is to take a covering of $\mathcal{A}$ by $N(\mathcal{A}, \epsilon)$ balls of radius ϵ, and then to apply the time T-map of the flow (for some suitable T) to obtain another covering. With some appropriate bounds one can iterate this idea to show that $N(\mathcal{A}, \epsilon)$ obeys some scaling law, which gives the dimension estimate.

We will consider the action of the map $F = S(T)$ on X. How would we measure the effect of each iteration of F on infinitesimal n-volumes? Clearly we need to know about DF at each point on X. Now, the image of a ball of initial conditions under a linear map L is an ellipse, and the semi-axes of this ellipse describe the change in the volume of this ball. In this appendix we treat the case when $DS(T)$ is compact, but the argument can be extended to more general situations using the methods in Temam (1988) or Blinchevskaya & Ilyashenko (1999).

Lemma B.1. *If L is a compact linear map from H into itself, then $L(B)$, the image of the unit ball, is an ellipse, with the length of the semi-axes given by the eigenvalues of $(L^T L)^{1/2}$.*

Proof. $L^T L$ is a positive compact self-adjoint operator, as is its square root $(L^T L)^{1/2}$. Thus there is an orthonormal basis $\{e_i\}$ of H that consists of eigenvectors of $(L^T L)^{1/2}$, with corresponding eigenvalues $\alpha_i(L)$, which we order so that

$$\alpha_1(L) \geq \alpha_2(L) \geq \cdots \geq 0, \qquad (L^T L)^{1/2} e_i = \alpha_i e_i.$$

Since

$$(Le_i, Le_j) = (L^T Le_i, e_j) = \alpha_i^2(e_i, e_j) = \alpha_i^2 \delta_{ij},$$

the vectors $\{Le_i\}$ are orthogonal and $|Le_j| = \alpha_j$. So $\{Le_j/\alpha_j\}$ form another orthonormal basis for H. If

$$u = \sum_{j=1}^{\infty} c_j e_j$$

---

* B. R. Hunt, Department of Mathematics and Institute for Physical Science and Technology, University of Maryland, College Park, MD 20742, USA. I would like to thank him for permission to reproduce the argument of his proof here. Any errors that may have crept in are entirely my own fault.

439

then

$$Lu = \sum_{j=1}^{\infty} c_j (Le_j) = \sum_{j=1}^{\infty} c_j \alpha_j \frac{Le_j}{\alpha_j},$$

and the unit ball $\sum_{j=1}^{\infty} |c_j|^2 \leq 1$ is mapped into the ellipse

$$\left\{ u = \sum_{j=1}^{\infty} k_j e_j : \sum_j \frac{|k_j|^2}{\alpha_j^2} \leq 1 \right\}. \qquad \blacksquare$$

The lengths of the semi-axes of $L(B)$ can be used to give an estimate of the effect of L on n-dimensional volumes. Line displacements can be magnified by a maximum factor of α_1, areas by a factor of at most $\alpha_1\alpha_2$, etc. Therefore we can define a maximum factor for expansion of n-volumes, ω_n, by

$$\omega_n(L) = \alpha_1(L) \cdots \alpha_n(L), \qquad (B.1)$$

and if $\omega_n(L) < 1$ then "L contracts in dimension n."
     If we define

$$\omega_n(F, x) = \omega_n(DF(x)),$$

we can state the result precisely.

Theorem B.2 (B. Hunt, personal communication, 1999). *Let X be a compact set, and let F be a C^1 map defined on a neighbourhood of X whose derivative DF is compact. If for some integer d*

$$\omega_d(F, x) \leq 1 \qquad \text{for all} \quad x \in X,$$

(where in the case of equality d is the largest integer with this property), we have $d_f(X) \leq d_L(F, X)$.

The first part of the proof shows that image of a ball of points can be covered by a certain number of balls of a different radius. Note that the singular values of an operator L are the eigenvalues of $(L^T L)^{1/2}$ (cf. Lemma B.1).

Lemma B.3. *Let G be a C^1 map defined on a neighbourhood of a compact set X, and let $\alpha_1(G, x) \geq \alpha_2(G, x) \geq \alpha_3(G, x) \ldots \geq \alpha_n(G, x) \geq \ldots$ be the singular values of $DG(x)$. Assume that there exists an integer d and constants $\rho \geq \sigma > 0$ such that for all $x \in X$,*

(i) $\alpha_{d+1}(G, x) \leq \rho/2$ and

(ii) $\omega_j(G, x) \leq (\sigma/2)^{j-d}$ for $1 \leq i \leq d$.

Let $\beta(x) = \max(2\alpha_{d+1}(G, x), \sigma) \in [\sigma, \rho]$. Then there are constants $c > 1$ and $\epsilon_0 > 0$ such that for all $x \in X$ and $0 < \epsilon < \epsilon_0$, the set $G(B(x; \epsilon))$ can be covered by at most

$$c\beta(x)^{-d} \qquad (B.2)$$

balls of radius $\beta(x)\epsilon$.

Proof. Let $\beta = \beta(x) = \max(2\alpha_{d+1}(G, x), \sigma) \in [\sigma, \rho]$. If $\beta > \sigma$ take $j = d$, whereas if $\beta = \sigma$ take j to be the smallest nonnegative integer such that $2\alpha_{j+1} < \beta$. We have

$j \leq d$ and

$$\omega_j \leq (\beta/2)^{j-d}, \qquad (B.3)$$

from (ii).

First we study the image of the unit ball under the linearisation of G; we then use this to treat the case of $B(x, \epsilon)$ under G for small enough ϵ.

Let $E = DG(x)[B(0, 1)]$, so that (by Lemma B.1) the lengths of the semi-axes of E are precisely $a_j = a_j(G, x)$ for $j = 1, \ldots$. Set $\delta = \sigma/4$, and let E' be the δ-neighbourhood of E, $E' = N(E, \delta)$. Since $a_{j+1} \leq \beta/2$, every point of E' lies within $(\beta/2) + \delta$ of $E' \cap \Pi$, where Π is the hyperplane spanned by the first j semi-axes of E. So if $E' \cap \Pi$ is covered by a collection of balls of radius $\beta/2 - \delta$ with centres in Π, then E' can be covered by balls with the same centres but with radius β.

To estimate this number, consider a covering of $E' \cap \Pi$ with a grid of j-dimensional boxes with sides of length $(\beta - 2\delta)/\sqrt{j}$. Then the number is balls is bounded by

$$\left(\frac{(2a_1 + 2\delta)}{(\beta - 2\delta)/\sqrt{j}} + 1 \right) \left(\frac{(2a_2 + 2\delta)}{(\beta - 2\delta)/\sqrt{j}} + 1 \right) \cdots \left(\frac{(2a_j + 2\delta)}{(\beta - 2\delta)/\sqrt{j}} + 1 \right). \qquad (B.4)$$

Now, $\beta \geq \sigma = 4\delta$, so $\beta - 2\delta \geq \frac{1}{3}(\beta + 2\delta)$ and therefore

$$\frac{(2a_l + 2\delta)}{(\beta - 2\delta)/\sqrt{j}} + 1 \leq \frac{3\sqrt{j}(2a_l + 2\delta)}{\beta + 2\delta} + 1.$$

If $1 \leq l \leq j$ then $2a_l \geq 2a_j \geq \beta$, and so this is bounded by

$$\frac{3\sqrt{j} \cdot 2a_l}{\beta} + 1 \leq \frac{(3\sqrt{j} + 1)(2a_l)}{\beta}.$$

Therefore the number of balls in (B.4) is bounded by

$$\frac{(6\sqrt{j} + 2)a_1}{\beta} \frac{(6\sqrt{j} + 2)a_2}{\beta} \cdots \frac{(6\sqrt{j} + 2)a_j}{\beta}$$

$$= (6\sqrt{j} + 2)^j \frac{\omega_j}{\beta^j} \leq c\beta^{-d},$$

with $c = (6\sqrt{d} + 2)^d$.

The image of $B(x, \epsilon)$ under the linearisation of G is

$$G(x) + DG(x)[B(x, \epsilon)] = G(x) + \epsilon E.$$

Now, DG is uniformly continuous on X, since G is C^1 and X is compact. Thus for ϵ_0 small enough we can guarantee that the linear approximation is within $\delta\epsilon$ of the true value,

$$|G(x) + DG(x)u - G(u)| \leq \delta\epsilon, \qquad u \in B(x; \epsilon),$$

for all $x \in X$ whenever $\epsilon < \epsilon_0$. For these values of ϵ, we have that

$$G(B(x; \epsilon)) \subset G(x) + \epsilon E + \epsilon\delta B(0, 1) = G(x) + \epsilon E',$$

and so $G(B(x; \epsilon))$ can be covered by $c\beta^{-d}$ balls of radius $\beta\epsilon$. ∎

We now apply this result to a covering of X to obtain the bound on the fractal dimension.

Proof. By definition we have

$$\omega_{d+1}(F, x) < 1$$

for all $x \in X$. In fact, since F is C^1, $\omega_{d+1}(F, x)$ is a continuous function of x, and so there exists a $\gamma < 1$ such that

$$\omega_{d+1}(F, x) \le \gamma \qquad \text{for all} \qquad x \in X.$$

It follows that for any $r \ge 1$,

$$\omega_{d+1}(F^r, x) \le \omega_{d+1}(F, x)\omega_{d+1}(F, F(x)) \ldots \omega_{d+1}(F, F^{r-1}(x)) \le \gamma^r.$$

Now, note that

$$\omega_{d+1}(F^r, x) \ge \alpha_{d+1}(F^r, x)^{d+1},$$

and hence

$$\alpha_{d+1}(F^r, x) \le \gamma^{r/(d+1)}.$$

Therefore, choosing r large enough that

$$\rho \equiv 2\gamma^{r/(d+1)} < \tfrac{1}{2}$$

we can apply Lemma B.3 to $G = F^r$. Condition (i) is clearly satisfied, and we just have to check (ii). However, $\omega_i(G, x)$ is continuous on X and so must take on its maximum value somewhere in X; we choose σ small enough that (ii) is satisfied.

We now start our iteration, observing that $\beta(x)$ is continuous and so uniformly continuous on X. Begin with a cover C_0, using N_0 balls of radius ϵ, where ϵ is such that

$$\tfrac{1}{2}\beta(y) \le \beta(x) \le 2\beta(y) \qquad \text{for all} \qquad x, y \in X \qquad \text{with} \qquad |x - y| \le \epsilon \qquad \text{(B.5)}$$

and is also less than the ϵ_0 required in Lemma B.3.

We will find a bound on the number of balls of radius $(2\rho)^m \epsilon$ required to cover X, and we will deduce a bound on the fractal dimension.

From our initial cover we now form a family of covers C_m by repeatedly applying Lemma B.3, that is, in the first step we apply Lemma B.3 to each of the original N_0 balls. The image of the ball $B_\epsilon(x_0)$ can be covered by at most $c\beta(x_0)^{-d}$ balls of radius $\beta(x_0)\epsilon$, with centres in $G(B_\epsilon(x_0))$. We then apply Lemma B.3 again to these balls – the ball centred at x_1 and of radius $\beta(x_0)\epsilon$ can be covered by $c\beta(x_1)^{-d}$ balls of radius $\beta(x_0)\beta(x_1)\epsilon$. Then C_m consists of all such balls, with centres x_n and radii $\beta(x_0)\beta(x_1)\ldots\beta(x_{n-1})\epsilon$, where the x_j are the centres of the balls used in each stage of the construction and $1 \le n \le m$.

We now show that the subcollection C'_m of balls in C_m whose radii lie in the range $[(\rho/2)^m \sigma \epsilon, (2\rho)^m \epsilon]$ still covers X. To see this, let y be a point in X and consider the points

$$y_0, y_1, y_2, \ldots, y_m \qquad \text{with} \qquad y_m = y \qquad \text{and} \qquad y_{j+1} = G(y_j).$$

Since $\beta(x) \in [\sigma, \rho]$ for all $x \in X$, we can choose an n with $0 \le n \le m - 1$ such that

$$\beta(y_n)\beta(y_{n+1}) \ldots \beta(y_{m-1}) \in [\sigma\rho^m, \rho^m].$$

Now take a ball, $B(x_0; \epsilon)$, from the original covering C_0 that contains y_n, choose x_1 to be the centre of a ball from C_1 that contains $y_{n+1} = G(y_n)$, and so on. Then the ball centred at x_{m-n} and of radius $\beta(x_{m-n-1}) \ldots \beta(x_0)\epsilon$ belongs to C_m and contains y. Not only that, but since $|x_l - y_{n+l}| \le \epsilon$ for each $0 \le l \le m - n$, we have from (B.5) that

$$\tfrac{1}{2}\beta(y_{n+l}) \le \beta(x_l) \le 2\beta(y_{n+l}),$$

and so

$$\beta(x_{m-n-1}) \ldots \beta(x_0)\epsilon$$
$$\in [2^{n-m}\beta(y_n)\ldots\beta(y_{m-1})\epsilon, 2^{m-n}\beta(y_n)\ldots\beta(y_{m-1})\epsilon]$$
$$\subset [(\rho/2)^m \sigma\epsilon, (2\rho)^m \epsilon].$$

Thus we have found a ball in C'_m containing y.

Finally we bound the number of balls in C'_m by using a weighting system. We assign a weight of 1 to the centres x_0 of the N_0 balls in the original cover. For every x_0, we divide the weight evenly among all the centres x_1 that result from an application of Lemma B.3. We continue in this way for each stage of the construction. Since we are dividing the weight equally, the sum of the weights for each collection of centres $x_n (1 \le n \le m)$ is at most N_0, and the sum of the weights of all the centres in C_m is at most $m N_0$.

Since the weights are divided equally, an upper bound on the number of x_{n+1} coming from a ball centred at x_n implies a lower bound on the weights of the centres in the new covering. The weight assigned to a centre at the nth stage is therefore at least

$$\frac{1}{c\beta(x_0)^{-d}c\beta(x_1)^{-d}\ldots c\beta(x_{n-1})^{-d}} = \frac{[\beta(x_0)\beta(x_1)\ldots\beta(x_{n-1})]^d}{c^n}.$$

Now, if a ball centred at x_n lies in C'_m then its weight must be at least

$$\frac{((\rho/2)^m\sigma)^d}{c^n} \ge \frac{((\rho/2)^m\sigma)^d}{c^m},$$

and since the total weight of all the balls in C'_m cannot exceed $m N_0$, the number of balls in C'_m cannot exceed

$$N_m = \frac{m N_0 c^m}{[(\rho/2)^m\sigma]^d}.$$

It follows that

$$d_f(X) \le \limsup_{m\to\infty} \frac{\log N_m}{-\log((2\rho)^m\epsilon)} = \frac{\log c - d\log(\rho/2)}{-\log(2\rho)}$$
$$= d - \frac{\log c + d\log 4}{\log 2\rho}.$$

Since ρ can be made arbitrarily small by choosing r large enough, we have that

$$d_f(X) \le d$$

as required. $\blacksquare$

Our previous Theorem 13.16 is a straightforward corollary of the above result.

Corollary B.4. *If* $TR_n(\mathcal{A}) < 0$ *then* $d_f(\mathcal{A}) \leq n$.

Proof. We will write $TR_n(\mathcal{A})$ in another form as

$$\limsup_{t \to \infty} \frac{1}{t} \log \Omega_n(t),$$

where

$$\Omega_n(t) = \exp \left[\sup_{x_0 \in \mathcal{A}} \sup_{P^{(n)}(0)} \int_0^t \text{Tr}\Big(L(s; x_0) P^{(n)}(s)\Big) \, ds \right]$$

is a bound on the volume of an n-dimensional volume element at time t.

Now, suppose that n is such that

$$TR_n(\mathcal{A}) < 0;$$

then for large enough T we must have

$$\Omega_n(T) < 1.$$

It follows that

$$\omega_n(S(T), x) \leq \Omega_n(T) < 1 \qquad \text{for all} \qquad x \in X,$$

and therefore the result follows from Theorem B.2. ∎

References

R. A. Adams (1975) *Sobolev Spaces* (Academic Press, New York).

S. Agmon (1965) *Lectures on Elliptic Boundary Value Problems*, Mathematical Studies (Van Nostrand, New York).

S. Agmon, A. Douglis, & L. Nirenberg (1959) Estimates near the boundary for solutions of elliptic partial differential equations satisfying general boundary conditions I, *Commun. Pure Appl. Math.* **12**, 623–727.

S. Agmon, A. Douglis, & L. Nirenberg (1964) Estimates near the boundary for solutions of elliptic partial differential equations satisfying general boundary conditions II, *Commun. Pure Appl. Math.* **17**, 35–92.

J. M. Arrieta & A. N. Carvalho (1999) Abstract parabolic problems with critical nonlinearities and applications to Navier-Stokes and heat equations, *Trans. Am. Math. Soc.* **325**, 285–310.

A. V. Babin & M. I. Vishik (1983) Attractors of partial differential equations and estimates of their dimension, *Russ. Math. Surv.* **38**, 151–213.

A. V. Babin & M. I. Vishik (1992) *Attractors of Evolution Equations* (North-Holland, Amsterdam).

A. Ben-Artzi, A. Eden, C. Foias, & B. Nicolaenko (1993) Hölder continuity for the inverse of Mañé's projection, *J. Math. Anal. Appl.* **178**, 22–29.

N. P. Bhatia & G. P. Szegö (1967) *Dynamical Systems: Stability Theory and Applications*, Springer Lecture Notes in Mathematics Volume 35 (Springer-Verlag, Berlin).

J. E. Billotti & J. P. LaSalle (1971) Dissipative periodic processes, *Bull. Am. Math. Soc.* **77**, 1082–1088.

M. A. Blinchevskaya & Y. S. Ilyashenko (1999) Estimate for the entropy dimension of the maximal attractor for k-contracting systems in an infinite-dimensional space, Preprint, Moscow State University.

A. Bloch & E. S. Titi (1990) On the dynamics of rotating elastic beams, in G. Conte, A. M. Perdon, & B. Wyman (eds), *Proceedings of New Trends in Systems Theory* (Birkhäuser Verlag, Boston).

B. Bollobás (1990) *Linear Analysis* (Cambridge University Press, Cambridge, England).

K. Borsuk (1969) Fundamental retracts and fundamental sequences, *Fund. Math.* **64**, 55–85.

K. Borsuk (1970) A note on the theory of shape of compacta, *Fund. Math.* **67**, 265–278.

T. Cazenave & A. Haraux (1998) *An Introduction to Semilinear Evolution Equations*,

Oxford Lecture Series in Mathematics and Its Applications 13 (Oxford University Press, Oxford, England).

N. Chaffee & E. F. Infante (1974) A bifurcation problem for a nonlinear parabolic equation, *J. Appl. Anal.* **4**, 17–37.

S.-N. Chow, K. Lu, & G. R. Sell (1992) Smoothness of inertial manifolds, *J. Math. Anal. Appl.* **169**, 283–312.

B. Cockburn, D. A. Jones, & E. S. Titi (1997) Estimating the number of asymptotic degrees of freedom for nonlinear dissipative systems, *Math. Comput.* **66**, 1073–1087.

E. A. Coddington & N. Levinson (1955) *Theory of Ordinary Differential Equations* (McGraw-Hill, New York).

P. Collet, J.-P. Eckmann, H. Epstein, & J. Stubbe (1993) A global attracting set for the Kuramoto–Sivashinsky equation, *Commun. Math. Phys.* **152**, 203–214.

P. Constantin (1989) A construction of inertial manifolds, *Contemp. Math.* **99**, 27–62.

P. Constantin & C. Foias (1985) Global Lyapunov exponents, Kaplan–Yorke formulas and the dimension of the attractors for 2D Navier–Stokes equations, *Commun. Pure Appl. Math.* **38**, 1–27.

P. Constantin & C. Foias (1988) *Navier–Stokes Equations* (University of Chicago Press, Chicago).

P. Constantin, C. Foias, & R. Temam (1985) Attractors representing turbulent flows, *Mem. Am. Math. Soc.* **53**.

P. Constantin, C. Foias, & R. Temam (1988) On the dimension of the attractors in two-dimensional turbulence, *Physica D* **30**, 284–296.

P. Constantin, C. Foias, B. Nicolaenko, & R. Temam (1989) *Integral Manifolds and Inertial Manifolds for Dissipative Partial Differential Equations*, Applied Mathematics Sciences Volume 70 (Springer-Verlag, Berlin).

H. Crauel & F. Flandoli (1994) Attractors for random dynamical systems, *Probability Theory and Related Fields* **100**, 365–393.

H. Crauel, A. Debussche, & F. Flandoli (1995) Random attractors, *J. Dyn. Diff. Eqns.* **9**, 307–341.

E. B. Davies (1995) *Spectral Theory and Differential Operators*, Cambridge Studies in Advanced Mathematics Volume 42 (Cambridge University Press, Cambridge, England.)

A. Debussche (1990) Inertial manifolds and Sacker's equation, *Diff. Integral Eqns.* **3**, 467–486.

A. Debussche (1998) Hausdorff dimension of a random invariant set, *J. Math. Pures Appl.* **77**, 967–988.

A. Debussche & R. Temam (1994) Convergent families of approximate inertial manifolds, *J. Math. Pures Appl.* **73**, 485–552.

C. R. Doering & J. D. Gibbon (1995) *Applied Analysis of the Navier–Stokes Equations*, Cambridge Texts in Applied Mathematics (Cambridge University Press, Cambridge, England).

A. Douglis & L. Nirenberg (1955) Interior estimates for elliptic systems of partial differential equations, *Commun. Pure Appl. Math.* **8**, 503–538.

A. Eden, C. Foias, B. Nicolaenko, & R. Temam (1990) Ensembles inertiels pour des équations d'évolution dissipatives, *C. R. Acad. Sci. Paris I* **310**, 559–562.

A. Eden, C. Foias, B. Nicolaenko, & Z. She (1993) Exponential attractors and their relevance to fluid dynamics systems, *Physica D* **63**, 350–360.

A. Eden, C. Foias, & B. Nicolaenko (1994a) Exponential attractors of optimal Lyapunov dimension for Navier–Stokes equations, *J. Dyn. Diff. Eqns.* **6**, 301–323.

A. Eden, C. Foias, B. Nicolaenko, & R. Temam (1994b) *Exponential Attractors for*

References   447

Dissipative Evolution Equations, Research in Applied Mathematics Series (Wiley, New York).

L. C. Evans (1998) *Partial Differential Equations*, Graduate Studies in Mathematics Volume 19 (American Mathematical Society, Providence, RI).

E. Fabes, M. Luskin, & G. R. Sell (1991) Construction of inertial manifolds by elliptic regularisation, *J. Diff. Eqns.* **89**, 355–387.

K. J. Falconer (1985) *The Geometry of Fractal Sets*, Cambridge Tracts in Mathematics 85 (Cambridge University Press, Cambridge, England).

K. J. Falconer (1990) *Fractal Geometry* (Wiley, Chichester, England).

H. Federer (1991) *Geometric Measure Theory*, Classics in Mathematics (reprint of 1969 edition) (Springer-Verlag, Berlin).

N. Fenichel (1971) Persistence and smoothness of invariant manifolds for flows, *Indiana Univ. Math. J.* **21**, 355–387.

B. Fiedler & C. Rocha (1996) Heteroclinic orbits of semilinear parabolic equations, *J. Diff. Eqns.* **125**, 239–281.

F. Flandoli & J. A. Langa (1998) Determining modes for dissipative random dynamical systems *Stoch. Stoch. Rep.* **66**, 1–25.

C. Foias & E. J. Olson (1996) Finite fractal dimension and Hölder–Lipschitz parametrization, *Indiana Univ. Math. J.* **45**, 603–616.

C. Foias & G. Prodi (1967) Sur le comportement global des solutions non stationnaires des équations de Navier–Stokes en dimension 2, *Rend. Sem. Mat. Univ. Padova* **39**, 1–34.

C. Foias & R. Temam (1979) Some analytic and geometric properties of the solutions of the Navier–Stokes equations, *J. Math. Pures Appl.* **58**, 339–368.

C. Foias & R. Temam (1984) Determination of the solutions of the Navier–Stokes equations by a set of nodal values, *Math. Comput.* **43**, 117–133.

C. Foias & R. Temam (1987) The connection between the Navier–Stokes equations, dynamical systems, and turbulence theory, in *Directions in Partial Differential Equations* (Academic Press, New York), 55–73.

C. Foias, O. Manley, R. Temam, & Y. Treve (1983) Asymptotic analysis of the Navier–Stokes equations, *Physica D* **9**, 157–188.

C. Foias, G. R. Sell, & R. Temam (1985) Variétés inertielles des équations différentielles dissipatives, *C. R. Acad. Sci. Paris I* **301**, 139–141.

C. Foias, G. R. Sell, & R. Temam (1988a) Inertial manifolds for nonlinear evolution equations, *J. Diff. Eqns.* **73**, 309–353.

C. Foias, B. Nicolaenko, G. R. Sell, & R. Temam (1988b) Inertial manifolds for the Kuramoto–Sivashinsky equation and an estimate of their lowest dimension, *J. Math. Pures Appl.* **67**, 197–226.

C. Foias, O. Manley, & R. Temam (1988c) Modelling of the interaction of small and large eddies in two dimensional turbulent flows, *Math. Modell. Numer. Anal.* **22**, 93–114.

C. Foias, G. R. Sell, & E. S. Titi (1989) Exponential tracking and approximation of inertial manifolds for dissipative nonlinear equations, *J. Dyn. Diff. Eqns.* **1**, 199–244.

F. G. Friedlander & M. Joshi (1999) *Introduction to the Theory of Distributions* (Cambridge University Press, Cambridge, England).

P. K. Friz & J. C. Robinson (1999) Smooth attractors have zero "thickness," *J. Math. Anal. Appl.* **240**, 37–46.

P. K. Friz & J. C. Robinson (2000) Parametrising the attractor of the two-dimensional Navier–Stokes equations with a finite number of nodal values (*Physica D*, to appear).

P. K. Friz, I. Kukavica, & J. C. Robinson (2000) Nodal parametrisation of attractors for Dirichlet boundary conditions, *Disc. Cont. Dyn. Sys.*, submitted.

G. Fusco & C. Rocha (1991) A permutation related to the dynamics of a scalar parabolic PDE, *J. Diff. Eqns.* **91**, 75–94.

G. P. Galdi (1994a) *An Introduction to the Mathematical Theory of Navier–Stokes Equations 1: Linearized Steady Problems* (Springer-Verlag, Berlin).

G. P. Galdi (1994b) *An Introduction to the Mathematical Theory of Navier–Stokes Equations 2: Nonlinear Steady Problems* (Springer-Verlag, Berlin).

B. M. Garay (1991) Strong cellularity and global asymptotic stability, *Fund. Math.* **138**, 147–154.

B. García-Archilla, J. Novo, & E.S. Titi (1998) Postprocessing the Galerkin method: A novel approach to approximate inertial manifolds, *SIAM J. Numer. Anal.* **35**, 941–972.

B. García-Archilla, J. Novo, & E. S. Titi (1999) An approximate inertial manifolds approach to postprocessing the Galerkin method for the Navier–Stokes equations, *Math. Comput.* **68**, 893–911.

R. Geoghegan & R. R. Summerhill (1973) Concerning the shapes of finite-dimensional compacta, *Trans. Am. Math. Soc.* **179**, 281–292.

J. D. Gibbon & E. S. Titi (1997). Attractor dimension and small length scale estimates for the three-dimensional Navier–Stokes equations, *Nonlinearity* **10**, 109–119.

D. Gilbarg & N. S. Trudinger (1983) *Elliptic Partial Differential Equations of Second Order*, Grundlehren der mathematischen Wissenschaften Volume 224 (Springer-Verlag, Berlin).

P. A. Glendinning (1994) *Stability, Instability and Chaos*, Cambridge Texts in Applied Mathematics (Cambridge University Press, Cambridge, England).

J. Goodman (1994) Stability of the Kuramoto–Sivashinsky equation and related systems, *Commun. Pure Appl. Math.* **47**, 293–306.

C. Guillopé (1982) Comportement à l'infini des solutions des équations de Navier–Stokes et propriété des ensembles fonctionnels invariantes (ou attracteurs), *Ann. Inst. Fourier (Grenoble)* **32**, 1–37.

B. Günther (1995) Construction of differentiable flows with prescribed attractor, *Topol. Appl.* **62**, 87–91.

B. Günther & J. Segal (1993) Every attractor of a flow on a manifold has the shape of a finite polyhedron, *Proc. Am. Math. Soc.* **119**, 321–329.

H. Haken (1978) *Synergetics* (Springer-Verlag, New York).

J. K. Hale (1969) *Ordinary Differential Equations* (Wiley, Baltimore).

J. K. Hale (1988) *Asymptotic Behavior of Dissipative Systems*, Mathematical Surveys and Monographs Number 25 (American Mathematical Society, Providence, RI).

J. K. Hale & G. Raugel (1989) Lower semicontinuity of attractors for approximations of semigroups and partial differential equations, *Math. Comput.* **50**, 89–123.

J. K. Hale, L. T. Magalhães, & W. M. Oliva (1984) *An Introduction to Infinite-Dimensional Dynamical Systems - Geometric Theory*, Springer Applied Mathematical Sciences Vol 47 (Springer-Verlag, Berlin).

J. K. Hale, X.-B. Lin, & G. Raugel (1988) Upper semicontinuity of attractors for approximation of semigroups and partial differential equations, *J. Math. Comput.* **50**, 89–123.

A. Haraux (1988) Attractors of asymptotically compact processes and applications to nonlinear partial differential equations, *Commun. PDEs* **13**, 1383–1414.

G. H. Hardy & E. M. Wright (1960) *An Introduction to the Theory of Numbers* (Oxford University Press, Oxford, England).

P. Hartman (1973) *Ordinary Differential Equations* (Wiley, Baltimore).

D. Henry (1984) *Geometric Theory of Semilinear Parabolic Equations*, Springer Lecture Notes in Mathematics Volume 840 (Springer-Verlag, Berlin).

J. G. Heywood & R. Rannacher (1982) Finite approximation of the nonstationary Navier–Stokes problem. Part I: Regularity of solutions and second-order error estimates for spatial discretization, *SIAM J. Numer. Anal.* **19**, 275–311.

J. G. Heywood & R. Rannacher (1993) On the question of turbulence modeling by approximate inertial manifolds and the nonlinear Galerkin method, *SIAM J. Numer. Anal.* **30**, 1603–1621.

M. Hirsch, C. Pugh, & M. Shub (1977) *Invariant Manifolds*, Lecture Notes in Mathematics Volume 583 (Springer-Verlag, Berlin).

E. Hopf (1951) Über die Aufgangswertaufgave für die hydrodynamischen Grundliechungen, *Math. Nachr.* **4**, 213–231.

B. Hunt (1996) Maximal local Lyapunov dimension bounds the box dimension of chaotic attractors, *Nonlinearity* **9**, 845–852.

B. Hunt & V. Y. Kaloshin (1999) Regularity of embeddings of infinite-dimensional fractal sets into finite-dimensional spaces, *Nonlinearity* **12**, 1263–1275.

M. Hurley (1983) Bifurcation and chain recurrence, *Ergod. Th. Dyn. Sys.* **13**, 123–129.

J. M. Hyman & B. Nicolaenko (1986) The Kuramoto–Sivashinsky equation: A bridge between PDEs and dynamical systems, *Physica D* **18**, 113–126.

J. M. Hyman, B. Nicolaenko, & S. Zaleski (1986) Order and complexity in the Kuramoto–Sivashinsky model of weakly turbulence interfaces, *Physica D* **23**, 265–292.

J. S. Il'yashenko (1992) Global analysis of the phase portrait for the Kuramoto–Sivashinsky equation, *J. Dyn. Diff. Eqns.* **4**, 585–615.

F. Jauberteau, C. Rosier, & R. Temam (1990) The nonlinear Galerkin method in computational fluid dynamics, *Appl. Numer. Anal.* **6**, 361–370.

M. S. Jolly, I. G. Kevrekidis, & E. S. Titi (1990) Approximate inertial manifolds for the Kuramoto–Sivashinsky equation: Analysis and computations, *Physica D* **44**, 38–60.

D. A. Jones & E. S. Titi (1993) Upper bounds on the number of determining modes, nodes, and volume elements for the Navier–Stokes equations, *Indiana Univ. Math. J.* **42**, 1–12.

D. A. Jones, L. G. Margolin, & E. S. Titi (1995) On the effectiveness of the approximate inertial manifold – a computational study, *Theor. Comput. Fluid Dyn.* **7**, 243–260.

D. A. Kamaev (1984) Hopf's conjecture for a class of chemical kinetics equations, *J. Sov. Math.* **25**, 836–849.

I. G. Kevrekidis, B. Nicolaenko, & J. C. Scovel (1990) Back in the saddle again: A computer assisted study of the Kuramoto-Sivashinsky equation, *SIAM J. Appl. Math.* **50**, 760–790.

R. H. Kraichnan (1967) Inertial ranges in two-dimensional turbulence, *Phys. Fluids* **10**, 1417–1423.

E. Kreyszig (1978) *Introductory Functional Analysis with Applications* (Wiley, New York).

S. Kuksin & A. Shirikyan (2000) Stochastic dissipative PDEs and Gibbs measures, *Commun. Math. Phys.*, to appear.

Y. Kuramoto & T. Tsuzuki (1976) Persistent propagation of concentration waves in dissipative media far from equilibrium, *Prog. Theor. Phys.* **55**, 365–369.

O. A. Ladyzhenskaya (1963) *The Mathematical Theory of Viscous Incompressible Flow* (Gordon & Breach, New York).

O. A. Ladyzhenskaya (1975) On the dynamical system generated by the Navier–Stokes equations, *J. Sov. Math.* **3**, 4 [English translation of a 1972 Russian paper in *Zap. Nauchnich Seminarovs LOMI. Leningrad* **27**, 91–114].

O. A. Ladyzhenskaya (1982) On finite dimensionality of bounded invariant sets for the Navier–Stokes equations and some other dissipative systems, *Zap. Nauchnich Seminarovs LOMI* **115**, 137–155.

O. A. Ladyzhenskaya (1991) *Attractors for Semigroups and Evolution Equations* (Cambridge University Press, Cambridge, England).

J. A. Langa & J. C. Robinson (1999) Determining asymptotic behavior from the dynamics on attracting sets, *J. Dyn. Diff. Eqns.* **11**, 319–331.

J. Leray (1933) Etude de diverses équations intégrales non linéaires et de quelques problèmes que pose l'hydrodynamique, *J. Math. Pures Appl.* **12**, 1–82.

J. Leray (1934a) Essai sur les mouvements plans d'un liquide visqueux que limitent des parois, *J. Math. Pures Appl.* **13**, 331–418.

J. Leray (1934b) Essai sur le mouvement d'un liquide visqueux emplissant l'espace, *Acta Math.* **63**, 193–248.

J. L. Lions (1969) *Quelques Méthodes de Résolution des Problèmes aux Limites non Linéaires* (Dunod Gauthier-Villars, Paris).

P. L. Lions (1994) *Mathematical Topics in Fluid Mechanics, Volume 1: Incompressible Models* (Oxford University Press, Oxford, England).

P. L. Lions (1996) *Mathematical Topics in Fluid Mechanics, Volume 2: Compressible Models* (Oxford University Press, Oxford, England).

E. Lorenz (1963) Deterministic nonperiodic flow, *J. Atmos. Sci.* **20**, 448–464.

A. Mahalov, E. S. Titi, & S. Leibovitch (1990) Invariant helical subspaces for the Navier–Stokes equations, *Arch. Ration. Mech. Anal.* **112**, 193–222.

J. Mallet-Paret & G. R. Sell (1988) Inertial manifolds for reaction–diffusion equations in higher space dimensions, *J. Am. Math. Soc.* **1**, 805–866.

J. Mallet-Paret, G. R. Sell, & Z. Shao (1993) Obstructions to the existence of normally hyperbolic inertial manifolds, *Indiana Univ. Math. J.* **42**, 1027–1055.

R. Mañé (1977) Reduction of semilinear parabolic equations to finite dimensional C^1 flows, in *Geometry and Topology*, Lecture Notes in Mathematics Volume 1248 (Springer-Verlag, New York).

R. Mañé (1981) On the dimension of the compact invariant sets of certain nonlinear maps, *Springer Lecture Notes in Math.* **898**, 230–242.

M. Marion (1987) Attractors for reaction–diffusion equations: Existence and estimate of their dimension, *Appl. Anal.* **25**, 101–147.

M. Marion (1989) Approximate inertial manifolds for reaction–diffusion equations in high space dimension, *J. Dyn. Diff. Eqns.* **1**, 245–267.

M. Marion & R. Temam (1989) Nonlinear Galerkin methods, *SIAM J. Numer. Anal.* **26**, 1139–1157.

R. A. McCoy (1973) Cells and cellularity in infinite-dimensional normed linear spaces, *Trans. Am. Math. Soc.* **176**, 401–410.

E. J. McShane (1934) Extension of the range of functions, *Bull. Am. Math. Soc.* **40**, 837–842.

R. Meise & D. Vogt (1997) *Introduction to Functional Analysis* (Oxford University Press, Oxford, England).

X. Mora (1983) Finite dimensional attracting manifolds in reaction diffusion equations, *Contemp. Math.* **16**, 353–360.

B. Nicolaenko, B. Scheurer, & R. Temam (1985) Some global properties of the Kuramoto Sivashinsky equation: Nonlinear stability and attractors, *Physica D* **16**, 155–183.

H. Ninomiya (1992) Some remarks on inertial manifolds, *J. Math. Kyoto Univ.* **32**, 667–688.

L. C. Piccinini, G. Stampacchia, & G. Vidossich (1984) *Ordinary Differential Equations in* $\mathbb{R}^n$, Springer Applied Mathematical Sciences Vol 39 (Springer-Verlag, Berlin).

H. A. Priestley (1997) *Introduction to Integration* (Oxford University Press, Oxford, England).

M. Renardy & R. C. Rogers (1992) *An Introduction to Partial Differential Equations*, Texts in Applied Mathematics Volume 13 (Springer-Verlag, New York).

I. Richards (1982) On the gaps between numbers which are the sum of two squares, *Adv. Math.* **46**, 1–2.

C. Robinson (1995) *Dynamical Systems: Stability, Symbolic Dynamics, and Chaos* (CRC Press, London).

J. C. Robinson (1993) Inertial manifolds and the cone condition, *Dyn. Sys. Appl.* **2**, 311–330.

J. C. Robinson (1995) A concise proof of the "geometric" construction of inertial manifolds, *Phys. Lett. A* **200**, 415–417.

J. C. Robinson (1996) The asymptotic completeness of inertial manifolds, *Nonlinearity* **9**, 1325–1340.

J. C. Robinson (1999a) Solutions of continuous ODEs obtained as the limit of solutions of Lipschitz ODEs, *Nonlinearity* **12**, 555–561.

J. C. Robinson (1999b) Global attractors: Topology and finite dimensional dynamics, *J. Dyn. Diff. Eqns.* **11**, 557–581.

A. Rodriguez Bernal (1990) Inertial manifolds for dissipative semiflows in Banach spaces, *Appl. Anal.* **37**, 95–141.

C. A. Rogers (1970) *Hausdorff Measures* (Cambridge University Press, Cambridge, England).

W. Rudin (1974) *Real and Complex Analysis* (McGraw-Hill, New York).

W. Rudin (1976) *Principles of Mathematical Analysis* (McGraw-Hill, Tokyo).

W. Rudin (1991) *Functional Analysis* (McGraw-Hill, New York).

T. Sauer, J. A. Yorke, & M. Casdagli (1993) Embedology, *J. Stat. Phys.* **71**, 529–547.

K. Schmalfuß (1992) Backward cocycles and attractors and attractors of stochastic differential equations, in V. Reitmann, T. Redrich, and N. Kosch (eds.), *International Seminar on Applied Mathematics – Nonlinear Dynamics: Attractor Approximation and Global Behaviour*, 185–192 (Teubner, Leipzig).

G. R. Sell (1967) Non-autonomous differential equations and topological dynamics I, II, *Trans. Am. Math. Soc.* **127**, 241–262 and 263–283.

G. R. Sell (1996) Global attractors for the three-dimensional Navier–Stokes equations, *J. Dyn. Diff. Eqns.* **8**, 1–33.

G. I. Sivashinsky (1977) Nonlinear analysis of hydrodynamics instability in laminar flames – I. Derivation of basic equations, *Acta Astron.* **4**, 1177–1206.

S. Smale (1967) Differentiable dynamical systems, *Bull. Am. Math. Soc.* **73**, 747–817.

J. Smoller (1983) *Shock Waves and Reaction-Diffusion Equations* (Springer-Verlag, New York).

C. T. Sparrow (1982) *The Lorenz Equations: Bifurcations, Chaos, and Strange Attractors* (Springer-Verlag, New York).

E. M. Stein (1970). *Singular Integrals and Differentiability Properties of Functions* (Princeton University Press, Princeton, NJ).

A. M. Stuart & A. R. Humphries (1996) *Dynamical Systems and Numerical Analysis* (Cambridge University Press, Cambridge, England).

W. A. Sutherland (1975) *Introduction to Metric and Topological Spaces* (Oxford University Press, Oxford, England).

K. Taira (1995) *Analytic Semigroups and Semilinear Initial Boundary Value Problems*, London Mathematical Society Lecture Note Series Number 223 (Cambridge University Press, Cambridge, England).

References

M. E. Taylor (1996) *Partial Differential Equations: Basic Theory*, Texts in Applied Mathematics Volume 23 (Springer-Verlag, New York).

R. Temam (1984) *Navier–Stokes Equations* (North-Holland, Amsterdam).

R. Temam (1985) *Navier–Stokes Equations and Nonlinear Functional Analysis* (Society for Industrial and Applied Mathematics, Philadelphia).

R. Temam (1988) *Infinite-Dimensional Dynamical Systems in Mechanics and Physics*, Springer Applied Mathematical Sciences Volume 68 (Springer-Verlag, Berlin).

R. Temam, ed. (1995) *Inertial manifolds and their application to the simulations of turbulence*, special issue of *Theoretical and Computational Fluid Dynamics* **7**(3).

R. Temam (1997) second edition of Temam (1988).

E. S. Titi (1990) On approximate inertial manifolds to the Navier–Stokes equations, *J. Math. Anal. Appl.* **149**, 540–557.

M. Ukhovskii & V. I. Yudovitch (1995). Axially symmetric flows of ideal and viscous fluids filling the whole space, *J. Appl. Math. Mech.* **32**, 52–62.

M. I. Vishik (1992) *Asymptotic Behaviour of Solutions of Evolutionary Equations* (Cambridge University Press, Cambridge, England).

J. H. Wells & L. R. Williams (1975) *Embeddings and Extensions in Analysis* (Springer-Verlag, New York).

S. Wiggins (1994) *Normally Hyperbolic Invariant Manifolds in Dynamical Systems*, Applied Mathematical Sciences Vol 105 (Springer-Verlag, New York).

K. Yosida (1980) *Functional Analysis*, Springer Classics in Mathematics (Springer-Verlag, Berlin).

N. Young (1988) *Hilbert Spaces* (Cambridge University Press, Cambridge, England).

Index

a.e., 21, 22
absorbing set, 264, 281
  for Kuramoto–Sivashinsky equation, 430
  for Lorenz equations, 271–2
  for Navier–Stokes equations
    in two dimensions
      in $\mathbb{H}^1$, 311
      in $\mathbb{H}^2$, 313
      in $\mathbb{L}^2$, 310
    in three dimensions, if well-posed
      in $\mathbb{H}^1$, 318
      in $\mathbb{H}^2$, 322, 324
  for reaction–diffusion equation
    in H_0^1, 289
    in L^2, 286
adjoint, 88
Agmon's inequality, 317, 323
Alaoglu compactness theorem, 105, 203, 208,
    223, 228, 246, 253, 319
almost
  every, 21
  everywhere, 22
"appropriate preparation", 391, 392, 403
  of reaction–diffusion equation, 396–8
approximation
  of continuous ODE by Lipschitz ODEs, 49
Arzelà–Ascoli theorem, 49, 52, 57, 58, 59,
    69, 144
asymptotic
  completeness, 404, 405
  dynamics, 6, 266–7
    on the attractor, 276–7
    finite-dimensional, 387, 403, 406.
      See also Utopian theorem
attractor
  exponential, 374–6, 384
  global. *See* global attractor

$\underline{B}(X, \epsilon)$, 17
$\overline{B}(X, \epsilon)$, 17

backwards uniqueness. *See* injectivity
  of semigroup
Baire category theorem, 66–7, 408, 410
Banach space, 11
  -valued
    function spaces, 188–9
    integration, 189, 210, 211
    Sobolev spaces, 189
      and continuous functions, 190
    weak derivatives, 190
      and regularity in time, 191, 193
basis
  in Banach space, 85
  of eigenvectors of A
    for H, 76
    for $R(A)$, 75
  in Hilbert space, 37, 107
    separable, 38
  for $L^2(\Omega \times \Omega)$, 85
  for $L_w^2(a, b)$, 77
bi-Laplacian, 427
bilinear form
  from bi-Laplacian, 427
  from Laplacian, 161, 164
  from Stokes problem, 236
boundary
  conditions, 145
    Dirichlet, 159
    Neumann, 186
    periodic, 149, 159
  straightening, 129, 147, 180, 186
  values
    for $u \in H^1(\Omega)$, 146
    for $u \in H_0^1(\Omega)$, 147

c^0, 97
$C^0(\Omega)$, 14
$C^0(\overline{\Omega})$, 14
$C_b^0(\Omega)$, 15
$C^0(I, X)$, 188

453

Printed in the United States
By Bookmasters